Sposito, V. A.

Linear and nonlinear programming

LINEAR AND NONLINEAR PROGRAMMING

LINEAR AND NONLINEAR PROGRAMMING

V. A. Sposito

THE IOWA STATE UNIVERSITY PRESS / AMES

1975

V. A. Sposito is Associate Professor of Statistics and a member of the staff of the Computation Center at Iowa State University where his fields of specialization include mathematical programming, statistical computations, numerical analysis, and operations research. He has served as special consultant for various firms and institutes, some of which include the Center for Agricultural and Rural Development, Cyphernetics Corporation, Ford Foundation, ERS, and U.S. Department of Agriculture. Dr. Sposito is currently Supervisor of Applications Software Development at the Statistical Laboratory at Iowa State University.

Dr. Sposito received the B.A. degree with honors in mathematics from California State University of Sacramento in 1965, and the M.S. degree in statistics at Iowa State University in 1968. He later received the Ph.D. degree with honors in statistics at Iowa State in 1970. He is a member of the American Statistical Association, Mathematical Programming Society, SIGMAP, Mu Sigma Rho, Pi Mu Epsilon, and Sigma Xi. He has authored numerous software documents as well as technical papers in various scientific journals, and is a frequent participant in professional meetings.

Composed by Science Press
Printed by The Iowa State University Press

First edition, 1975

Library of Congress Cataloging in Publication Data

Sposito, V A 1936–
Linear and nonlinear programming.

Includes index.
1. Linear programming. 2. Nonlinear programming. I. Title.
T57.74.S68 519.7 75-15668

ISBN 0-8138-1015-9

To: **S. F. BRUNK, M.D.**
UNIVERSITY HOSPITALS, THE UNIVERSITY OF IOWA

A N D

MARLENE, LORI, and PHILIP

CONTENTS

P R E F A C E

This book has been purposefully designed as a needed introductory text in linear and nonlinear programming with applications in statistics as well as in other disciplines. In preparing the text for easy use by teachers in several fields, the author has grouped sections involving special topics.

This book is based, in part, on notes for a three-quarter sequence in mathematical programming for advanced undergraduate and first-year graduate students in economics, statistics, operations research, mathematics, and business which the author conducted at Iowa State University.

Emphasis has been placed on the importance or relationship of certain theory to computational techniques. In the first three chapters, particular emphasis is placed on the basic concepts and properties of linear programming through a hypothetical production example and a scheduling problem. The graphic method is used to provide a conceptual understanding of certain salient properties of linear programming. Thus this book begins at a relatively elementary level so students will grasp easily the main ideas of linear programming which form the core of mathematical programming. This approach permits topics such as sensitivity analysis, parametric programming, and other computational techniques to be covered in a manner easily understood by students from any discipline.

The theory of linear duality is then presented followed by the Kuhn-Tucker theory and quadratic duality. Subsequently, algorithms in nonlinear programming, such as quadratic and geometric programming, are discussed. Other topics in nonlinear programming that emphasize principles of statistics are covered also. This approach differs from the approach of some existing texts which emphasize topics in transportation problems and/or integer programming. No attempt has been made to discuss such topics since existing literature is saturated with such material.

The appendix includes use of the IBM Mathematical Programming System, MPSX, and/or the Management Science System, MPS-III. The author has found that most students, especially at the graduate level, will seek some sort of computer package to solve problems pertaining to their own personal research. Since most colleges and universities have IBM computers, the author has elected to discuss some of the computer packages associated with such equipment. The procedures related to the earlier chapters, such as sensitivity analysis and parametric programming, are included.

If teachers must select material for a one-year course at the undergraduate level, then probably all of Chapters 1, 2, 3, Appendix A, and various parts of the remaining chapters should be covered. For instance, in Chapter 4 linear duality can be studied without the proof of Lemma 4.6 or Appendix B, which discusses relevant concepts over convex cones. In the first half of the text, emphasis is placed on linear programming models and computational techniques; also covered is the theoretical derivation of the simplex method and certain such related linear programming topics as sensitivity analysis, parametric programming, alternative criteria in curve fitting, and chance constraint programming.

Students should have some knowledge of Chapters 1–3 or at least a good background in linear programming and matrix theory if the text is to be used at the graduate level. A one-quarter graduate level course would be based on Chapters 4, 6, 7, and 8. Chapter 5 should be included if the students have completed one or two courses in statistics.

The author is deeply grateful to S. F. Brunk, M.D., to whom this text is dedicated, for his devotion to his profession and his help to many persons, myself for one. I can only say, "Thank you very much for all your time, consideration, and help."

The author is indebted also to his family for helping him pursue all his personal objectives including the writing of this text.

A personal thanks to Carol Dunn for clarifying certain unclear passages in the manuscript; all remaining unresolved clarity rests with the author.

My other indebtedness extends to many, not least of all to Iowa State University, the Statistical Laboratory, and the Computation Center, all of which provided facilities, intellectual stimulation, and encouragement without which this text might never have been written. A special thanks to Oscar Kempthorne and H. T. David for their constant encouragement as well as for valuable comments made after critically reading parts of this manuscript. I am indebted also to H. T. David and D. J. Soults who introduced me to this field of study and who taught me more than I know how to acknowledge.

The author is indebted also to R. J. Klemm, B. Bowerman, Sister Nona Allard, and Dan Dvoskin for reading parts of this manuscript and for their useful suggestions; and finally, to Kristie Whitaker for her superb technical typing and Carol Sanderson for her valuable suggestions given in the final editing phase of this manuscript.

LINEAR AND NONLINEAR PROGRAMMING

C H A P T E R   O N E

INTRODUCTION

1.1 Historical Perspective of Mathematical Programming

Determining the minimum (or maximum) numerical value of some mathematical expression in some restricted space was attempted frequently by many of our early scientists.

These models could take many forms depending on the problem at hand, but the general mathematical programming problem can be expressed as

$$\begin{aligned} &\text{maximize} \quad (\text{or minimize}) \quad f(\mathbf{x}) \\ &\text{subject to} \quad g_i(\mathbf{x}) \le b_i \qquad i = 1, 2, \ldots, m \end{aligned} \tag{1.1}$$

Here $f(\mathbf{x})$ and $\{g_i(\mathbf{x})\}$ are real-valued functions of $\mathbf{x}$ and the vector $\mathbf{b}$ is known. Special cases of this problem are linear programming problems where, in (1.1), $f(\mathbf{x})$ and $\{g_i(\mathbf{x})\}$ are linear functions of $\mathbf{x}$; in particular

$$\begin{aligned} \text{maximize} \quad & \textstyle\sum_{i=1}^{n} c_i x_i \\ \text{subject to} \quad & a_{11}x_1 + a_{12}x_2 + \cdots + a_{1n}x_n \le b_1 \\ & a_{21}x_1 + a_{22}x_2 + \cdots + a_{2n}x_n \le b_2 \\ & \qquad \vdots \\ & a_{m1}x_1 + a_{m2}x_2 + \cdots + a_{mn}x_n \le b_m \\ & \qquad x_1, x_2, \ldots, x_n \ge 0 \end{aligned}$$

which we state in matrix form as the problem

$$\begin{aligned} \text{maximize} \quad & \mathbf{c}'\mathbf{x} \\ \text{subject to} \quad & A\mathbf{x} \le \mathbf{b} \\ & \mathbf{x} \ge 0 \end{aligned}$$

This was one of the earliest models to be considered; Fourier, a French mathematician, had formulated these types of models for use in mechanics and probability theory around 1826. In 1939 the Russian mathematician Kantorovich formulated production problems as linear programming problems and suggested a possible way of solving such models.

The general area of research in linear programming type problems came into its own during World War II when large-scale military operations required careful planning of logistic support. This led to mathematical techniques known as operations research techniques. Linear programming was one of these techniques. The primary contributor in solving linear programming models was George Dantzig. His general algorithm is known as the simplex algorithm, developed in 1947. Subsequently, interest in this area grew and, in 1949, T. C. Koopmans organized in Chicago the Cowles Commission Conference on Linear Programming. The papers presented in this conference were collected by Koopmans in 1951 in the book titled *Activity Analysis of Production and Allocation.*

A. Charnes, W. W. Cooper, and A. Henderson wrote the first book on linear programming. Since then the number of texts in this area has grown tremendously with applications in fields of economics, engineering, statistics, mathematics, and business.

H. W. Kuhn and A. W. Tucker in 1950 addressed themselves to the nonlinear mathematical programming problem. Their work has proven to be a classic. Algorithms for the solutions of quadratic programming problems have been based on their theory, i.e., models with linear constraints and objective functions of the form $\mathbf{c}'\mathbf{x} + \mathbf{x}'D\mathbf{x}$. Moreover, the Kuhn-Tucker theory has led to the area of duality through the saddle value or Lagrangian approach since duality is essentially an analytic concept. It is shown that associated with a mathematical programming problem is another problem which, when solved, yields the optimal solution to the original problem, and conversely. The saddle value problem is formulated with Lagrangian functions for inequality restrictions.

Since the late forties and early fifties the area of mathematical programming has branched to many types of models. For one, geometric programming in the late 1960s has come into its own. These problems have proven quite popular in engineering type models. The early pioneers in this area were R. J. Duffin, E. L. Peterson, and C. Zener.

1.2 Examples of Programming Problems

This section displays certain problems which can be formulated as linear or nonlinear programming problems.

Consider the problem of dividing a positive quantity C into two parts in such a way that the product of the two parts is a maximum. This problem can be stated as a nonlinear programming problem: determine

x_1 and x_2 which will

$$\begin{aligned} \text{maximize} \quad & x_1 x_2 \\ \text{subject to} \quad & x_1 + x_2 = C \\ & x_1, \quad x_2 \geq 0 \end{aligned}$$

Let us now consider the problem of determining regression estimates for the linear regression model

$$y_h = \beta_0 + \beta_1 x_h + u_h$$

where y_h is the hth observation on a dependent variable, x_h the hth observation on an independent variable, and u_h is a random disturbance. Then the classical least squares theory leads to estimates of β_0 and β_1 which will minimize

$$Q(\beta_0, \beta_1) = \sum_h (y_h - \beta_0 - \beta_1 x_h)^2 = (Y - X\beta)'(Y - X\beta)$$

or minimize the sum of squares of errors. This is commonly expressed in matrix notation as $Q(\beta_0, \beta_1) = Q(\beta) = (\mathbf{Y} - X\beta)'(\mathbf{Y} - X\beta)$ where X is an $n \times 2$ matrix; here n denotes the total number of observations. $Q(\beta_0, \beta_1)$ is a quadratic function of β_0 and β_1, i.e., $Q(\beta_0, \beta_1) = \sum_h y_h^2 + \mathbf{c}'\beta + \beta' D \beta$ where $\mathbf{c}' = -2\mathbf{Y}'X$ and $D = X'X$.

In this case β is unrestricted so one can find estimates which minimize $Q(\beta_0, \beta_1)$ by solving the well-known normal equations, i.e.

$$X'X\beta = X'\mathbf{Y}$$

In certain cases, certain limits or constraints are imposed on β_0, β_1, or both. For example, say the researcher desires a positive intercept and a value of the slope between ℓ and u. Then we must

$$\begin{aligned} \text{minimize} \quad & Q(\beta_0, \beta_1) \\ \text{subject to} \quad & \ell \leq \beta_1 \leq u \\ & \beta_0 \geq 0 \end{aligned}$$

which is now a quadratic programming problem.

Another researcher with the same set of measurements might also want to find estimates of β_0 and β_1 for the above linear model, but believes it is more reasonable to determine values of β_0 and β_1 which will minimize the sum of absolute deviations, i.e.

$$\text{minimize} \quad \textstyle\sum_h | y_h - \beta_0 - \beta_1 x_h |$$

Whether or not we have restrictions imposed on our model, the problem of finding the values of β_0 and β_1 which minimize the sum of absolute deviations can be reformulated as a linear programming problem. To see this, let ϵ_{1h} denote the positive deviation (and ϵ_{2h} denote the negative deviation) above (and below) the fitted line for the hth observation. Then for any h

$$y_h = \beta_0 + \beta_1 x_h + \epsilon_{1h} - \epsilon_{2h}$$

where at most one ϵ_{1h} or ϵ_{2h} can be nonzero by the nature of the problem. Also, $\epsilon_{1h} + \epsilon_{2h}$ is the absolute deviation between the fitted equation and y_h. We therefore have the following linear programming problem

$$\begin{aligned} \text{minimize} \quad & \textstyle\sum_h (\epsilon_{1h} + \epsilon_{2h}) \\ \text{subject to} \quad & X\beta + I\epsilon_1 - I\epsilon_2 = \mathbf{y} \\ & \epsilon_1, \quad \epsilon_2 \geq 0 \end{aligned}$$

which will yield the best estimates of β_0 and β_1 under the criterion of minimizing the sum of absolute deviations.

Another example of a programming problem is an investor trying to determine a security portfolio from a set of n securities which the investor hopes will provide at least a certain return at the least possible risk. Here, we assume the variance based on past performances of a security is a measure of the risk that the realized rate of return deviates from the expected rate of return μ_i. The covariance σ_{ij} of their returns provides a measure of the correlation between the rates of return on securities i and j. When σ_{ij} is positive the returns on the securities tend to go up and down together. Therefore, a rational investor would not invest too heavily in a set of securities that seem to move together. In this case, we would do well to diversify our portfolio; sacrificing return for risk reduction and measuring covariance between securities will enable us to achieve this required diversification.

To minimize risk arising from variability within a security and risk from an undiversified portfolio, we minimize the function $\mathbf{x}'S\mathbf{x}$ where S is the covariance matrix determined from past performances of the securities. Terms of the type $x_i^2 s_{ii}$ account for variability within a security and terms of the type $x_i x_j s_{ij}$ account for the covariance between securities. If the investor wants at least a rate of return of p percent based on d dollars, then we have the following quadratic programming problem

$$\begin{aligned}
\text{minimize} \quad & \mathbf{x}'S\mathbf{x} \\
\text{subject to} \quad & \textstyle\sum \mu_i x_i \geq p \\
& \textstyle\sum x_i = 1 \\
& x_i \geq 0 \qquad i = 1, 2, \ldots, n
\end{aligned}$$

where x_i denotes the fractional amount of d dollars to be invested in security i.

The above examples clearly show that many problems can be formulated as mathematical programming problems that are either linear or nonlinear in structure. The plan of this book is to develop the theory of linear programming and procedures or algorithms to obtain the optimal solutions for such problems in the first five chapters. Subsequent chapters discuss algorithms and theory pertinent to certain nonlinear programming problems such as quadratic and geometric programming problems.

CHAPTER TWO

LINEAR PROGRAMMING

2.1 Introduction

Linear programming is by far the most widely used optimization technique. Linear programming deals with the problem of determining feasible plans which are optimal with respect to a certain agreed-upon linear objective function; in particular, it determines a plan which maximizes or minimizes some linear function over all possible feasible plans. The *feasible* plans are those that satisfy certain restrictions which are usually in the form of a system of linear inequalities. Hence linear programming is defined in terms of a mathematical model composed of linear functions, in which "programming" is used as a synonym for planning; linear programming involves choosing activities (plans, schedules, allocations) in such a way as to obtain an optimal program. An *optimal* program is defined as a feasible plan which maximizes (or minimizes) some linear function, or objective function, from among all possible feasible plans.

The great variety of problems to which linear programming can be applied is indeed remarkable. It is used in curve fitting problems under such different criteria as minimizing the sum of squares, minimizing the maximum deviation, or minimizing the sum of absolute deviations. It is used in portfolio or investment problems, shipping problems, allocation problems, production scheduling, game problems, and so on, almost ad infinitum. In particular, consider the problem of finding $x_1, x_2, \ldots, x_n$ which

maximizes (or minimizes) the linear function $z = c_1x_1 + c_2x_2 + \cdots + c_nx_n$

$$\text{subject to} \quad \begin{aligned} a_{11}x_1 + a_{12}x_2 + \cdots + a_{1n}x_n &\leq b_1 \\ a_{21}x_1 + a_{22}x_2 + \cdots + a_{2n}x_n &\leq b_2 \\ &\vdots \\ a_{m1}x_1 + a_{m2}x_2 + \cdots + a_{mn}x_n &\leq b_m \\ x_1, x_2, \ldots, x_n &\geq 0 \end{aligned}$$

where a_{ij}, b_i, and c_j are known constants. The unknown variables

x_j $(j = 1, 2, \ldots, n)$ can be obtained by linear programming techniques.

Expressed in matrix notation, the problem is to find a vector $\mathbf{x}$ in E^n which

$$\begin{aligned} &\text{maximizes (minimizes)} \quad \mathbf{c}'\mathbf{x} \\ &\text{subject to} \quad A\mathbf{x} \leq \mathbf{b} \\ &\qquad\qquad\qquad \mathbf{x} \geq 0 \end{aligned} \tag{2.1}$$

Here, A is an $m \times n$ matrix.

The next section illustrates some problems that can be expressed in terms of a linear programming model. Section 2.3 considers some geometric interpretations of linear programming that give valuable insights to certain fundamental properties and concepts. Section 2.4 discusses a procedure based on the Gauss-Jordan technique for obtaining a consistent solution of a system of linear equations; and moreover, how one can determine a vector which maximizes the specified objective function. The theoretical derivation of the simplex procedure is deferred until later in the chapter.

2.2 Linear Programming Models

EXAMPLE 2.1. Activity-analysis problem (Karlin 1959, p. 174)

A manufacturer has the option of using one or more of four types of production processes. The first and second processes yield items A and the third and fourth yield items B. The inputs for each of these processes are labor measured in man-weeks, pounds of raw material X, and boxes of raw material Y. Since each process varies as to input requirements, the profits of the processes differ, even for processes producing the same item. Now suppose further that the manufacturer has fixed amounts of these three resources, i.e., labor, pounds of raw material X, and boxes of raw material Y. Hence the manufacturer, in deciding on a week's production schedule, is limited in the range of possibilities by the available amounts of the three resources. The manufacturer wants to determine how much of the two products should be manufactured and which processes should be used to maximize his profits.

Table 2.1, below, gives a breakdown of how much of each resource must be used to produce an item by each process, the profit associated with that item, and the limitation on each resource.

This problem has four decision variables and three restrictions. One is tempted to use the process (or processes) that produces items with the highest unit profit in order to maximize profits. However, the interdependence encountered in order to allocate the resources in the best manner makes an intuitive approach invalid as will be shown later in the chapter. Presently we

TABLE 2.1. **Activity-Analysis Problem (Ex. 2.1)**

	One Item of A		One Item of B		Limitation on
	Process 1	Process 2	Process 3	Process 4	Resources
Man-Weeks	1	1	1	1	15
Pounds of Material X	7	5	3	2	90
Boxes of Material Y	3	4	10	8	100
Profit Units	6	5.1	9	6	
Production Levels	x_1	x_2	x_3	x_4	

are interested only in the proper formulation of the linear programming model.

Let x_1, x_2, x_3, and x_4 represent the production levels of processes 1, 2, 3, and 4, respectively. In this problem we want to maximize profits specified in Table 2.1. Therefore, the objective function z we want to maximize is

$$z = 6x_1 + 5.1x_2 + 9x_3 + 6x_4 \tag{2.2}$$

Each production plan must be such that it satisfies the available resources. In particular, let us first consider the labor constraint. There are only 15 man-weeks available and, for each unit of either product A or B made, it takes 1 man-week (one unit of resource one, labor). Hence, the first restriction imposed on the production plan states that the total allowable labor must be less than or equal to 15 man-weeks, or

$$x_1 + x_2 + x_3 + x_4 \leq 15 \tag{2.3}$$

It should be noted that the relationship (2.3) is stated in terms of an inequality, because it may be more profitable to determine an optimal production schedule which does not make use of all of resource one.

Now, with respect to the restriction imposed by the availability of the pounds of raw material X, we see that every item of product A produced by process 1 makes use of 7 units of resource two (material X). Alternatively, when an item of product A is produced by process 2, only 5 units of resource two are used. Likewise, every unit of product B produced by process 3 uses 3 units of resource two; and when produced by process 4, it uses only 2 units of resource two. Hence, since there are only 90 units of resource two available, it follows that any feasible production schedule must satisfy the following relationship

$$7x_1 + 5x_2 + 3x_3 + 2x_4 \leq 90 \tag{2.4}$$

Similarly, the other restriction related to the availability of boxes of raw material Y is expressed as

$$3x_1 + 4x_2 + 10x_3 + 8x_4 \leq 100 \tag{2.5}$$

One further set of restrictions must be made on the decision variables. Since we do not allow negative production, then each unknown production level must be either zero or positive, i.e.

$$x_1 \geq 0 \quad x_2 \geq 0 \quad x_3 \geq 0 \quad x_4 \geq 0 \tag{2.6}$$

Now, considering (2.2) through (2.6), this production scheduling problem can be formulated as the following linear programming problem

$$\begin{array}{ll} \text{maximize} & z = 6x_1 + 5.1x_2 + 9x_3 + 6x_4 \\ \text{subject to} & x_1 + x_2 + x_3 + x_4 \leq 15 \\ & 7x_1 + 5x_2 + 3x_3 + 2x_4 \leq 90 \\ & 3x_1 + 4x_2 + 10x_3 + 8x_4 \leq 100 \\ & x_1, \quad x_2, \quad x_3, \quad x_4 \geq 0 \end{array} \tag{2.7}$$

The linear programming model (2.7) demonstrates two properties characteristic of any linear programming model, i.e., additivity and proportionality. These two properties are intuitively obvious. In particular, the amount of profit we obtain from selling 1 unit of item A made with process 1 is 6 profit units, and with process 2 we obtain 5.1 profit units. Hence, if 4 and 10 items are produced with processes 1 and 2, then the total profit units are 75. This illustrates the *additivity* property of linear relationships: levels of activities (or returns) are additive in their effects. The *proportionality* property states that there exists a multiplicative relationship between the units of resource required and the number of units produced. For instance, every unit of item A produced by process 1 requires 7 units of resource two, or in general, x_1 units of item A produced by process 1 require $7x_1$ units of resource two.

Before considering another example, we should state that equations (2.3) through (2.6) define the set of all possible production plans. Moreover, these constraints define the feasible region for this particular model.

EXAMPLE 2.2 Transportation problem

Transportation problems are problems of determination of optimal shipping patterns between origins (or sources) and destinations. They arise quite frequently in practice and have nearly the same basic structure. Nonclassical transportation problems are those in which the direct shipping path of source i to destination j is deleted from the problem.

The general form of the transportation problem is as follows:

Suppose a manufacturer wants to ship a number of items from several sources to a number of destinations. Each destination requires a specific number of units of the item while each source can supply up to a certain number. In particular let

m = number of sources indexed by $i = 1, 2, \ldots, m$
n = number of destinations indexed by $j = 1, 2, \ldots, n$
a_i = total number of items available for shipment at source i
b_j = number of items requested by destination j
x_{ij} = number of items shipped from source i to destination j

The x_{ij} are the unknown shipping amounts to be determined. The manufacturer wants to determine how many units should be sent from each source to each destination so that the total shipping cost is a minimum.

If the manufacturer knows the cost c_{ij} of shipping one unit of the item from source i to destination j and if the shipping costs are proportional to the amount shipped, then the linear cost function or objective function z to be minimized is

$$z = \sum_{i=1}^{m} \sum_{j=1}^{n} c_{ij}x_{ij} \tag{2.8}$$

To determine the restrictions, consider first the relation that defines the number of units shipped from any source i. We see that the total amount shipped from source i must be less than or equal to a_i, and this relationship can be expressed as

$$\sum_{j=1}^{n} x_{ij} \leq a_i \qquad i = 1, 2, \ldots, m \tag{2.9}$$

These m restrictions state that the sum of the amounts shipped from any origin to each of the destinations must not exceed the amount available at the source (origin).

Consider the constraints that define the amount requested by each destination. If each demand must be met, the sum of the amounts shipped from every origin to each destination must equal the amount requested by the destinations, b_j. Hence

$$\sum_{i=1}^{m} x_{ij} = b_j \qquad j = 1, 2, \ldots, n \tag{2.10}$$

Also, since a negative x_{ij} would imply a shipment from destination j to source i, we must also impose the following positivity restriction on all the variables

$$x_{ij} \geq 0 \tag{2.11}$$

Combining (2.8) through (2.11), we have the linear programming model

$$\begin{aligned}
\text{minimize} \quad & z = \sum_{i=1}^{m} \sum_{j=1}^{n} c_{ij}x_{ij} \\
\text{subject to} \quad & \sum_{j=1}^{n} x_{ij} \leq a_i \qquad i = 1, 2, \ldots, m \\
& \sum_{i=1}^{m} x_{ij} = b_j \qquad j = 1, 2, \ldots, n \\
& x_{ij} \geq 0 \quad \text{for all } i \text{ and } j
\end{aligned}$$

It is left as an exercise to show that if the total amount available is equal to the total amount requested, that is, if

$$\sum_{i=1}^{m} a_i = \sum_{j=1}^{n} b_j$$

then the above linear programming problem has a feasible solution.

EXAMPLE 2.3. Investment problem (Hiller and Lieberman 1967, p. 168, Ex. 3)

Suppose an investor has \$10,000 and wants to determine an investment plan that will maximize the amount of money he can accumulate by the end of a five-year period.

Money-making activities A and B are available at the beginning of each of the next five years. In these activities, each dollar invested in A at the beginning of any year returns a profit of 40 cents two years later that can be immediately reinvested. Similarly, each dollar invested in B at the beginning of any year returns a profit of 70 cents three years later that also can be reinvested immediately.

The investor must also consider money-making activities C and D which will be available at a time in the future. Each dollar invested in C at the beginning of the second year returns a profit of \$1 four years later. And each dollar invested in D at the beginning of the fifth year returns a profit of 30 cents one year later. Let

x_{ij} = dollars invested in activity i $(i = A, B, C, D)$ in period j $(j = 1, 2, 3, 4, 5)$

r_j = dollars not invested in period j $(j = 1, 2, 3, 4, 5)$

It is, perhaps, not apparent at this point how one can define the objective function to be maximized, so we shall defer its definition for a moment and first define the feasible region. This region is defined by a system of linear equations relating the amount invested in years 1, 2, 3, 4, and 5.

In particular, consider the first year (period 1) in which the investor must define how to allocate the original investment of \$10,000. There are several immediate options available: invest a certain amount in activity A or in activity B or both or not to invest all funds in this period if this happens to yield the optimal policy. Hence, the basic relationship defining activities in period 1 is

$$x_{A1} + x_{B1} + r_1 = 10{,}000 \tag{2.12}$$

Now, consider the beginning of the second year (period 2). We see again that the investor can invest in either activity A or B or both. In this period there is also the opportunity to invest in activity C, and again, the option of not investing all the funds available. By available funds, we mean that money which was not invested in period 1, but can be invested, if desired, in period 2. Hence, the linear relationship for period 2 is defined as

$$-r_1 + x_{A2} + x_{B2} + x_{C2} + r_2 = 0 \tag{2.13}$$

Here the activity r_1 is a *transfer activity* which makes unused funds in period 1 available for investment purposes in period 2. The same type of activity must be incorporated in the restrictions for periods 3, 4, and 5.

Therefore, part of the algebraic expressions for periods 3, 4, and 5 will take the form

$$-r_{j-1} + x_{Aj} + x_{Bj} + r_j \qquad j = 3, 4, 5 \tag{2.14}$$

However, each period (3, 4, and 5) has additional funds made available by money invested in an earlier period, and these must be incorporated in (2.14). In particular, for each dollar invested in A in periods 1, 2, and 3, then \$1.40 becomes available in periods 3, 4, and 5. Likewise, each dollar invested in B in periods 1 and 2 will result in \$1.70 being available for investment in periods 4 and 5. Thus, the investment opportunities in periods 3 and 4 can be expressed as

$$-1.4x_{A1} \qquad\qquad - r_2 + x_{A3} + x_{B3} + r_3 = 0 \tag{2.15}$$

$$-1.7x_{B1} - 1.4x_{A2} \qquad - r_3 + x_{A4} + x_{B4} + r_4 = 0 \tag{2.16}$$

The only additional variable that needs to be introduced in period 5 is activity D. Reasoning similar to that used above results in expressing the restriction for period 5 as

$$-1.7x_{B2} - 1.4x_{A3} - r_4 + x_{A5} + x_{B5} + x_{D5} + r_5 = 0 \tag{2.17}$$

Combining (2.12), (2.13), (2.15), (2.16), (2.17), and a nonnegativity restriction on all x_{ij} and r_j we have

$$
\begin{aligned}
x_{A1} + x_{B1} + r_1 &= 10{,}000 \\
-r_1 + x_{A2} + x_{B2} + x_{C2} + r_2 &= 0 \\
-1.4x_{A1} \qquad -r_2 + x_{A3} + x_{B3} + r_3 &= 0 \\
-1.7x_{B1} \quad -1.4x_{A2} \qquad -r_3 + x_{A4} + x_{B4} + r_4 &= 0 \\
-1.7x_{B2} \quad -1.4x_{A3} \qquad -r_4 + x_{A5} + x_{B5} + x_{D5} + r_5 &= 0 \\
x_{Aj}, x_{Bj}, x_{Cj}, x_{Dj} \geq 0 \qquad j &= 1, 2, \ldots, 5 \\
r_j \geq 0 \qquad j &= 1, 2, \ldots, 5
\end{aligned}
$$

The objective function to be maximized is the total number of dollars accumulated at the end of the fifth year. Its formulation should now be apparent considering each variable one at a time. The profit coefficients c_{ij} for $x_{C2}, x_{B3}, x_{A4}, x_{D5}$, and r_5 are 2, 1.7, 1.4, 1.3, and 1, respectively. The other variables have profit coefficients of zero. For example, the funds accumulated from activities x_{Aj} ($j = 1, 2, 3$) have been transferred to the following periods for reinvestment purposes and need not be considered in the objective function. If money is invested in activity A in period 4, we have a sum of \$1.40 per dollar invested at the end of period 5; therefore $c_{A4} = 1.4$. Also, $c_{A5} = 0$, since money invested in activity A in period 5 gives no noticeable return until the end of period 6. The process of verifying the other coefficients is left to the reader.

Example 2.4. Scheduling

A local firm decides to sponsor a 3-hour television show featuring a group of comedians and a well-known band. The company insists on at least 18 minutes of commercials. The TV network requires that commercials must not exceed 36 minutes, and this time must not exceed the time allocated to the group of comedians. The local firm also insists that the time allocated to the comedians must not exceed 2 hours, so the band must be used to fill in the remaining time. It will cost \$50 per minute for the commercials, and \$150 and \$100 per minute for the group of comedians and band, respectively. It has been estimated, from past experience, that for every minute scheduled for the comedians 4,000 additional viewers will tune in; for every minute scheduled for the band 2,000 additional viewers will tune in; and for every minute the commercials are on 1,000 viewers will tune out.

How should the sponsor schedule the TV special to

(i) obtain the maximum number of viewers, or
(ii) produce the show at minimum cost, or
(iii) obtain a minimum-cost program with the largest number of viewers?

Let

$$x_1 = \text{number of minutes allocated to the comedians}$$
$$x_2 = \text{number of minutes allocated to commercials}$$
$$180 - x_1 - x_2 = \text{number of minutes allocated to the band}$$

Then $x_1 \geq 0$, $x_2 \geq 0$, $180 - x_1 - x_2 \geq 0$ since the time allocated must not be negative.

If we consider the restrictions imposed by the sponsor and the TV network, respectively, we have

$$x_2 \geq 18 \qquad x_2 \leq 36$$
$$x_1 \leq 120 \qquad x_2 \leq x_1$$

To meet the criteria specified in (i) and (ii), we must determine appropriate objective functions which will represent the number of viewers watching the show, f, and the cost of the program, g. Both functions can be expressed in terms of x_1 and x_2, namely

$$f = 4{,}000x_1 + 2{,}000(180 - x_1 - x_2) - 1{,}000x_2 \qquad (2.18a)$$

$$g = 150x_1 + 100(180 - x_1 - x_2) + 50x_2 \qquad (2.18b)$$

Thus, (2.18a) will determine the number of viewers for any combination of x_1 and x_2, and (2.18b) will represent the cost of the program associated with any values of x_1 and x_2.

Hence, the problem in terms of criteria (i) and (ii) can be expressed in the form: determine $x_1 \geq 0$ and $x_2 \geq 0$ so that

$$\begin{aligned} x_1 + x_2 &\leq 180 \\ x_2 &\geq 18 \\ x_1 \qquad &\leq 120 \\ x_2 &\leq 36 \\ x_1 - x_2 &\geq 0 \end{aligned} \qquad (2.19)$$

and f is a maximum or g is a minimum.

If the sponsor is interested in scheduling a program at the smallest possible cost which will obtain the maximum number of viewers, the structure of the problem becomes slightly modified. A "trick" commonly used is to incorporate one of these objectives into the set of restrictions. Hence, consider the additional restriction, $z \leq f$, and the modified objective function, $g - z$. Then minimizing $-z$ is equivalent to making z as large as possible, i.e., forcing f to be as large as possible. Therefore, to find a minimal cost plan that will yield the maximum number of viewers one should solve the following linear programming problem

$$\begin{aligned} &\text{minimize} && g - z \\ &\text{subject to} && (2.19) \\ & && f - z \geq 0 \end{aligned}$$

It should be noted that in the above presentation we imposed the restriction that the number of minutes allocated to the band be represented by $180 - x_1 - x_2$. This requires that the following restriction be imposed on our problem

$$x_1 + x_2 \leq 180$$

An alternative way of handling this constraint would be

$$x_1 + x_2 + x_3 = 180$$

where x_3 = the number of minutes for the band. These formulations are equivalent, and it is true in general that establishing a linear programming model is not necessarily unique. The two-dimensional model is introduced so that a graphic solution can be derived; this is given in Section 2.3.

2.3 Geometric Interpretation

In the previous section we discuss how certain types of problems are formulated into a linear programming model. In subsequent sections we present the theory which leads to the computations of the solutions to linear programming problems. To give the reader a better understanding of certain fundamental terminology, we discuss these concepts in terms of some graphic representations.

Associated with every linear programming problem is a feasible region defined in terms of the unknown variables $x_1, x_2, \ldots, x_n$. To permit a graphic representation of the feasible regions let us confine our discussion to two-dimensional problems.

Consider the scheduling problem, Example 2.4, of the previous section. In this problem the feasible region is defined as

$$\begin{aligned} x_1 + x_2 &\leq 180 \\ x_2 &\geq 18 \\ x_1 \quad\quad &\leq 120 \\ x_2 &\leq 36 \\ x_1 - x_2 &\geq 0 \\ x_1, \quad x_2 &\geq 0 \end{aligned}$$

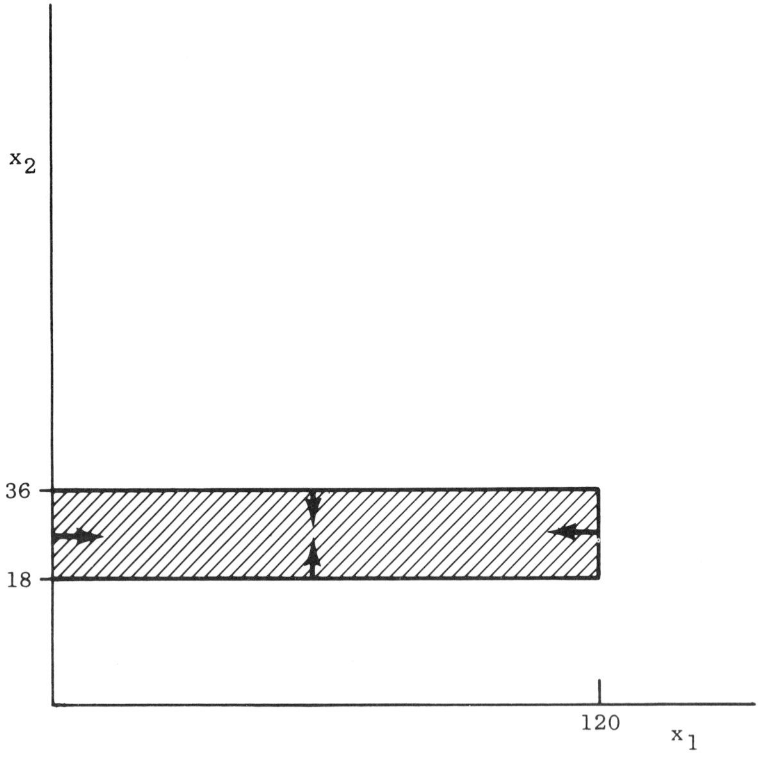

F'gure 2.1. Initial feasible region.

Since the variables must be nonnegative, we know that we are restricted to the nonnegative orthant or quadrant in E^2, two-dimensional space. Moreover, x_1 and x_2 are restricted to lie in the closed intervals [0, 120] and [18, 36], respectively. Hence our feasible region is immediately confined to the shaded area shown in Figure 2.1. Let us now consider the remaining two constraints, namely

$$x_1 - x_2 \geq 0$$
$$x_1 + x_2 \leq 180$$

If we first look at the boundary defined by $x_1 - x_2 \geq 0$, we see that the equation $x_1 - x_2 = 0$ intersects the feasible region at the points (18, 18) and (36, 36) as shown in Figure 2.2. Since this restriction is expressed as an inequality, we know that the set of feasible solutions has been reduced to either area I or area II, including the boundary, as shown in Figure 2.2. A

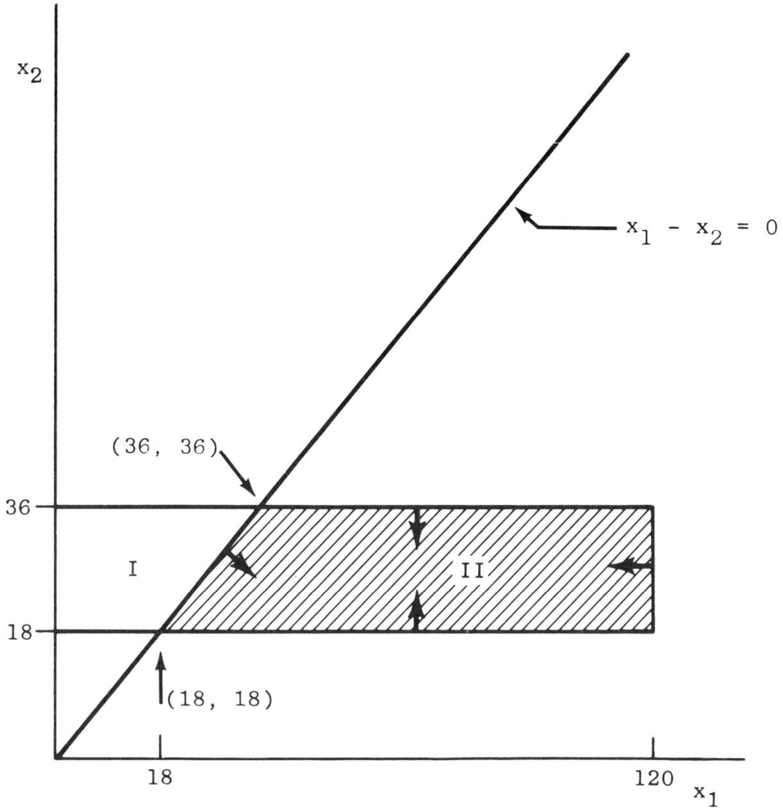

Figure 2.2. Reduced feasible region.

simple way to determine the appropriate region is to consider any point which does not lie on the boundary but lies in either I or II. Let us select a point in I, say $(x_1^*, x_2^*) = (0, 20)$; then

$$x_1^* - x_2^* = -20 \not\geq 0$$

Therefore, (x_1^*, x_2^*) violates the restriction $x_1 - x_2 \geq 0$, and it is clear that the new constraint has reduced the feasible region to region II.

Consider in the same manner the constraint

$$x_1 + x_2 \leq 180$$

Here, the equation defined by the boundary does not intersect region II. Two conclusions can be made immediately: either (i) the feasible region

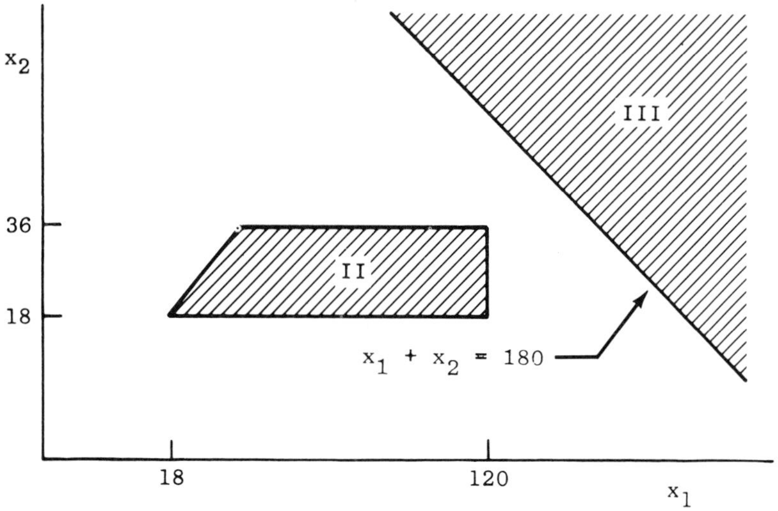

Figure 2.3. Final feasible region.

is null or (ii) the feasible region has not been changed by introducing this new constraint. Possibility (i) implies that there exist no points (x_1, x_2) which satisfy all the restrictions. The nonexistence of a solution which satisfies all the constraints indicates that the problem is *infeasible*. In such a situation the feasible region would look like Figure 2.3, where region III is defined as

$$\{x \in E^2 \mid x_1 + x_2 \geq 180\}$$

However, since the inequality is less than or equal to 180, the feasible region remains the same as shown in Figure 2.2. A feasible scheduling plan is represented by all points in region II.

How do we find an optimal solution? Let us consider the two agreed-upon objective functions

$$f = 1{,}000(2x_1 - 3x_2) + 360{,}000$$
$$g = 50(x_1 - x_2) + 18{,}000$$

Inasmuch as f and g are multiples of 1,000 and 50, respectively, we see that the variables which maximize f or minimize g will also maximize $2x_1 - 3x_2$ or minimize $x_1 - x_2$. If we treat $2x_1 - 3x_2$ and $x_1 - x_2$ as parameters, i.e., examine each objective function for some particular value of $z = f$(or g), we obtain two families of parallel lines as shown in Figure

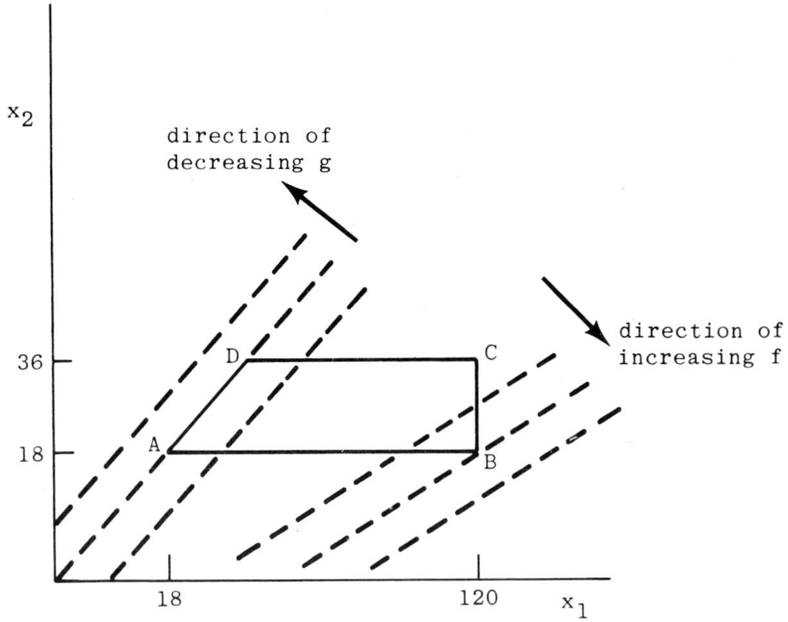

Figure 2.4. Direction of *z* with respect to *f* and *g*.

2.4. The direction of increasing (decreasing) z with respect to f (or g) is toward point B (or D). Hence, a unique maximum is attained at B = (120, 18) and this corresponds to 546,000 viewers at a cost of \$23,100. A minimum-cost program is attained at point D with z = \$18,000. It should be noted that the same cost is achieved at point A = (18, 18). Hence, there exist alternative solutions under a minimum-cost criterion and a unique solution if one is interested in the maximum number of viewers without regard to the cost of the program. It can easily be verified that any point on the line segment $\overline{AD}$ is a minimum-cost solution.

If the sponsor is interested, as noted in Example 2.4, in a solution under two objectives, i.e., a minimum-cost program with the largest obtainable audience, then he is essentially restricting his set of solutions to the line segment $\overline{AD}$. At points A and D one obtains 342,000 and 324,000 viewers, respectively, at the same cost of \$18,000. Hence, the optimal solution under these two objectives is reached at point A.

Using the following definition we see that the set of feasible solutions, commonly called the feasible region, is convex.

DEFINITION 2.1. *A set S is convex if for any two points or vectors* $\overline{\mathbf{x}}$ *and* $\hat{\mathbf{x}}$ *in S, then* $\hat{\hat{\mathbf{x}}} = \alpha\overline{\mathbf{x}} + (1 - \alpha)\hat{\mathbf{x}}$ *is in S for any* $\alpha \in [0, 1]$.

It is now possible to identify some of the important facts present in the above discussion which are also true in the n-dimensional case. These are

1. A programming problem may have no solution; when this happens we say the problem is infeasible.
2. The feasible region is generated by the intersection of a finite number of half planes.
3. The feasible region is a convex polyhedral set, i.e., it contains a finite number of sides and any convex combination of any two points in this region also lies in this region.
4. An optimal solution is a feasible solution which maximizes (or minimizes) the objective function.
5. An optimal solution lies at a corner or extreme point of the feasible region and not in its interior.
6. A linear programming problem may have more than one optimal solution; and, since the feasible region is convex, the set of optimal solutions also forms a convex set.

One more important situation should be discussed before presenting techniques to solve higher-dimensional problems. Besides having linear programming problems without feasible solutions or with optimal solutions, it is possible to have situations in which the feasible region is nonempty but has feasible points that will yield an objective function with no finite maximum. For instance, consider the problem

$$\begin{aligned} \text{maximize} \quad & x_1 \\ \text{subject to} \quad & x_1 - x_2 \geq 0 \\ & x_1, \quad x_2 \geq 0 \end{aligned}$$

This problem is shown graphically in Figure 2.5, and one can readily see that x_1 can be made arbitrarily large. When the objective function of the linear problem can be made arbitrarily large we say that the problem has an *unbounded* solution.

If one considers the objective function $-x_1 + x_2$, we see that one optimal solution is reached at the extreme point $(x_1^*, x_2^*) = (0, 0)$, and $-x_1^* + x_2^* = 0$. However, any point along the boundary is also optimal, i.e., $(1, 1), (2, 2), \ldots, (1{,}000, 1{,}000)$. Even though the maximum value of the objective function cannot exceed zero, the value of (x_1, x_2) can be made arbitrarily large. One should not classify such problems as being unbounded.

It is left as an exercise to construct a linear programming problem such that the program has an optimal solution with a variable that can obtain a value of infinity.

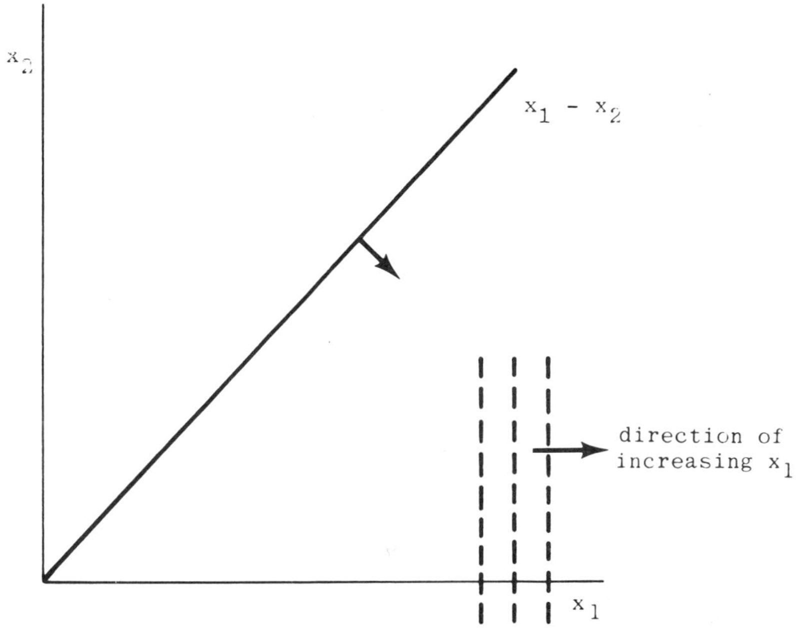

Figure 2.5. Unbounded solution.

2.4 An Illustration of the Simplex Method

In the previous sections the basic structure and some of the characteristic properties of linear programming were introduced. It was shown that the geometric approach could be used to obtain solutions to linear programming problems involving two variables. However, for larger problems the geometric approach would fail us completely. In this section we illustrate how larger-sized problems can be solved in a systematic and efficient manner. This procedure, known as the simplex method, is based on solving a system of linear equations with the Gauss-Jordan procedure. A good discussion of the Gauss-Jordan procedure is given by Gass (1969). The theoretical derivation of the simplex method is developed in Section 2.5.

To apply the Gauss-Jordan procedure the feasible region of the linear programming problem must be converted to a system of equalities. Hence, if a constraining relation is of the form

$$g(x_1, x_2, \ldots, x_n) \leq k \qquad k = \text{constant}$$

then the relation may be converted to an equality by the addition of a nonnegative *slack* variable x^s so that

$$g(x_1, x_2, \ldots, x_n) + x^s = k \qquad x^s \geq 0$$

If the constraining relation is of the form

$$g(x_1, x_2, \ldots, x_n) \geq k$$

then the relation can be converted to an equality by the subtraction of a nonnegative *surplus* variable

$$g(x_1, x_2, \ldots, x_n) - x^s = k \qquad x^s \geq 0$$

The cost coefficient of x^s in the objective function is then given a value of zero.

2.4.1 *Solution of Example 2.1*

Consider the feasible region defined in Example 2.1. We explicitly include unused materials or idle man-weeks as variables

$$\begin{aligned} x_5 &= \text{slack (unused) man-weeks} \\ x_6 &= \text{slack (unused) material } X \\ x_7 &= \text{slack (unused) material } Y \end{aligned} \tag{2.20}$$

where again

$$x_5 \geq 0 \qquad x_6 \geq 0 \qquad x_7 \geq 0 \tag{2.21}$$

Then, representing z by x_0, we can convert the relations in (2.7) to

$$\begin{array}{llllllllll} \text{Row 0} & x_0 - 6x_1 - & 5.1x_2 - & 9x_3 - & 6x_4 & & & & = & 0 \\ \text{Row 1} & x_1 + & x_2 + & x_3 + & x_4 & + x_5 & & & = & 15 \\ \text{Row 2} & 7x_1 + & 5x_2 + & 3x_3 + & 2x_4 & & + x_6 & & = & 90 \\ \text{Row 3} & 3x_1 + & 4x_2 + & 10x_3 + & 8x_4 & & & + x_7 & = & 100 \end{array}$$

$$x_1, x_2, x_3, x_4, x_5, x_6, x_7 \geq 0 \tag{2.22}$$

Our entire model then consists of (2.22) where we want to maximize profits x_0. Introducing x_0 into our constraint set allows us to determine total profits for various feasible solutions. As described in this section, introducing this profit equation allows us to derive a feasible production plan that is better than the present production plan.

Observe that (2.22) consists of four linear equations in eight unknowns. Before discussing an algebraic approach for finding one such set of values, it would be advisable to review a few simple rules for solving simultaneous linear equations.

Suppose we wish to solve the following two equations in three unknowns

$$\begin{aligned} x_1 \quad + 5x_3 &= 2 \\ x_2 + 2x_3 &= 4 \end{aligned} \tag{2.23}$$

Although there are actually an infinite number of possible solutions (given any two solutions, any weighted average of these two solutions is also a solution), the easiest method for obtaining one such solution is to let $x_3 = 0$; immediately we conclude that $x_1 = 2$ and $x_2 = 4$.

If we want to solve for x_1 and x_3, a technique (which will prove useful later) is to convert (2.23) into a form such that, when we let $x_2 = 0$, we can once again solve for the relevant unknowns without further computations. If we divide the second equation by 2, i.e.

$$(1/2)x_2 + x_3 = 2$$

then multiply the new second equation by -5, and add it to the first equation, we have

$$\begin{aligned} x_1 - 5(1/2)x_2 \qquad &= -8 \\ (1/2)x_2 + x_3 &= 2 \end{aligned}$$

Now, letting $x_2 = 0$, we know immediately that $x_1 = -8$ and $x_3 = 2$.

In general, if we have m linear equations in n unknowns where n is larger than m, set $n - m$ of the unknowns equal to zero and solve for the remaining m. Suppose we are solving for the first m unknowns; we multiply or divide appropriate equations by constants, and add or subtract one equation from another so as to put the equations into the form

$$\begin{array}{llllll} x_1 & & & + a_{1,m+1}x_{m+1} + \cdots + a_{1n}x_n & = b_1 \\ & x_2 & & + a_{2,m+1}x_{m+1} + \cdots + a_{2n}x_n & = b_2 \\ & & \ddots & \vdots \qquad\qquad\qquad \vdots & \ \ \vdots \\ & & x_m & + a_{m,m+1}x_{m+1} + \cdots + a_{mn}x_n & = b_m \end{array}$$

We let $x_{m+1} = 0, x_{m+2} = 0, \ldots, x_n = 0$, and consequently

$$x_1 = b_1, x_2 = b_2, \ldots, x_m = b_m$$

We are now ready to obtain the first solution to the sample problem. The easiest first solution to (2.22) is to let x_1, x_2, x_3, $x_4 = 0$, yielding the first basic solution: $x_0 = 0$, $x_5 = 15$, $x_6 = 90$, and $x_7 = 100$. The word basic refers to the fact that we shall consider only solutions including four unknown variables, four being the number of equations, and such that the values of these four are uniquely determined.

The interpretation of the solution is immediate: no production is called for, i.e., only slack or unused inputs appear in the tentative schedule. Hence, profits are zero! The coefficients in row 0 have special significance as they indicate a better solution:

The interpretation of coefficients in row 0: Each coefficient represents the increase (negative coefficients) or decrease (positive coefficients) in profits with a unit increase of the associated unknown x_i. For example, if it is possible to increase x_1 by 2 units, then profits will be increased by $2 \times \$6 = \$12$. If x_3 can be increased by 10 units, then profits will rise by \$90. Since the current solution consists of all slack variables, using any process will be profitable. Which process should we try?

Simplex Criterion I. If there are unknowns not in the present basic solution having a negative coefficient in row 0, select any unknown with the largest per unit potential gain (say x_j). If all unknowns not in the present basic solution have positive or zero coefficients in row 0, an optimal solution has been obtained.

Therefore, in the present case, we shall try x_3. How large can we make x_3 without violating our available supplies and manpower? In other words, how large can we make x_3 without violating our constraints? The problem is to determine x_3 such that x_0 is made as large as possible under the following constraints

$$\begin{array}{lrcrcrcrcrcrcrcl}
\text{Row 0} & x_0 & - & 6x_1 & - & 5.1x_2 & - & 9x_3 & - & 6x_4 & & & & & = & 0 \\
\text{Row 1} & & & x_1 & + & x_2 & + & \boxed{x_3} & + & x_4 & + & x_5 & & & = & 15 \\
\text{Row 2} & & & 7x_1 & + & 5x_2 & + & \boxed{3x_3} & + & 2x_4 & & & + & x_6 & = & 90 \\
\text{Row 3} & & & 3x_1 & + & 4x_2 & + & \boxed{10x_3} & + & 8x_4 & & & & + x_7 & = & 100 \\
\end{array}$$

$$x_1, x_2, x_3, x_4, x_5, x_6, x_7 \geq 0$$

We can make x_3 as large as we wish as long as rows 1, 2, and 3 are not violated and $x_1, x_2, x_3, x_4, x_5, x_6, x_7 \geq 0$. At present, $x_1, x_2, x_4 = 0$ and in the next solution they will also be zero, since x_3 (not $x_1, x_2,$ or x_4) is the variable to be entered into the solution. Thus, the values of the new variables in the next solution can be determined by considering

$$\begin{aligned}
&\text{Row 1} \quad & x_3 + x_5 \qquad\qquad &= 15\\
&\text{Row 2} \quad & 3x_3 \qquad + x_6 \qquad &= 90\\
&\text{Row 3} \quad & 10x_3 \qquad\qquad + x_7 &= 100\\
& & x_1, x_2, x_4 &= 0
\end{aligned} \tag{2.24}$$

If we enter x_3 into the solution it will have a positive value. We can enter as much of x_3 as we wish as long as x_5, x_6, and x_7 are nonnegative in the new solution. But from rows 1, 2, and 3 the new values of x_5, x_6, and x_7 are as follows, with the required constraints shown

$$\begin{aligned}
x_5 &= 15 - x_3 \geq 0\\
x_6 &= 90 - 3x_3 \geq 0\\
x_7 &= 100 - 10x_3 \geq 0
\end{aligned} \tag{2.25}$$

These conditions are equivalent to

$$\begin{aligned}
x_3 &\leq 15\\
3x_3 &\leq 90 \qquad x_3 \leq 90/3 = 30\\
10x_3 &\leq 100 \qquad x_3 \leq 10
\end{aligned} \tag{2.26}$$

The only way for x_5, x_6, and x_7 to remain nonnegative is if simultaneously $x_3 \leq 15$, $x_3 \leq 30$, and $x_3 \leq 10$, which will happen if and only if $x_3 \leq 10$. For example, if $x_3 = 15$, the new value of x_7 is $x_7 = 100 - 10(15) = 100 - 150 = -50$, a violation of the positivity constraint on x_7. Since x_3 must be ≤ 10 and since we want x_3 to be as large as possible, the new value of x_3 is 10. The new values of x_5, x_6, and x_7 thus become from equation (2.25)

$$\begin{aligned}
x_5 &= 15 - 10 = 5\\
x_6 &= 90 - 3(10) = 60\\
x_7 &= 100 - 10(10) = 0
\end{aligned}$$

We note that since the new value of x_7 is zero this variable leaves the solution, and becomes nonbasic.

Now, on comparing equations (2.24) with equations (2.26), we see that the ratios in equations (2.26), which determine the maximum amount of x_3 that can be entered into the solution, can be obtained methodologically by taking the ratios in equations (2.24). It is, then, the minimum ratio which equals the new value of x_3 and corresponds to the variable x_7 which will leave the solution.

Before we continue, let us pretend that equations (2.27) read as follows

			Ratio	
Row 1	$0x_3 + x_5$	$= 15$	$15/0$	
Row 2	$3x_3 \quad + x_6$	$= 90$	$90/3$	
Row 3	$-10x_3 \quad + x_7$	$= 100$	$-100/10$	(2.27)

The point of looking at this modification is that the reader should never blindly follow a set of rules, but rather should analyze the problem in terms of being sure that x_5, x_6, and x_7 remain positive in the new solution. From equations (2.27), we see that the new values of x_5, x_6, and x_7 are given by

$$\begin{aligned} x_5 &= 15 - 0x_3 \geq 0 \\ x_6 &= 90 - 3x_3 \geq 0 \\ x_7 &= 100 - (-10)x_3 = 100 + 10x_3 \geq 0 \end{aligned} \tag{2.28}$$

We see, then, that no matter how large x_3 is, x_5 and x_7 will always be nonnegative. Also, x_6 is the only variable that could be negative. Equation (2.28) is equivalent to $90 - 3x_3 \geq 0$ or $x_3 \leq 30$. Thus, 30 units of x_3 can enter the solution and the new value of x_6 is $90 - 3(30) = 0$. The point, methodologically, then, is that since zero and negative coefficients of x_3 in the constraint equations do not affect the positivity of the new value of the variable currently in the solution, these coefficients should be ignored in computing the ratios. Therefore, in summary we have

Simplex Criterion II. (a) Take the ratios of the present basic solution to the coefficients of the entering unknown x_j (ignore ratios with zero or negative numbers in the denominator). (b) Select the minimum ratio which corresponds to x_k (this ratio may not be unique, but for now we shall assume that in case of ties select any corresponding x_k). If the minimum ratio occurs for variable x_k in the present solution, we then set $x_k = 0$ in the new solution.

From the above discussion it should be noted that if the coefficients of the entering unknown x_j were nonpositive, then the problem would be unbounded as noted in (2.28). Table 2.2 illustrates the computations specified in Criterion II. Since the minimum ratio occurs for x_7, x_7 will not

TABLE 2.2 Initial Iteration (Ex. 2.1)

	Solution	Coeffs. of x_3	Ratio	Min	Next Solution
x_0	0	-9	–		
x_5	15	1	15		
x_6	90	3	30		
x_7	100	10	10	10	$x_3 = 10, x_7 = 0$

appear in the next solution, x_3 will take its place and will equal 10. The second tentative program will still contain the unknowns x_0, x_5, and x_6 and we set the remaining unknowns equal to zero.

To find the new solution, we perform several row operations on (2.22) analogous to our manipulations of the two equations in three unknowns (2.23). First we divide row 3 by 10. This causes the coefficient of x_3 to be 1.

$$\begin{array}{lrcl}
\text{Row 0} & x_0 - 6x_1 - 5.1x_2 - 9x_3 - 6x_4 & = & 0 \\
\text{Row 1} & x_1 + x_2 + x_3 + x_4 + x_5 & = & 15 \\
\text{Row 2} & 7x_1 + 5x_2 + 3x_3 + 2x_4 + x_6 & = & 90 \\
\text{Row 3} & .3x_1 + .4x_2 + x_3 + .8x_4 + .1x_7 & = & 10
\end{array}$$

We now create zero coefficients for x_3 in rows 0, 1, and 2 as follows:

row 0	multiply row 3 by 9 and add to row 0
row 1	multiply row 3 by -1 and add to row 1
row 2	multiply row 3 by -3 and add to row 2

The result is

$$\begin{array}{lrcl}
\text{Row 0} & x_0 - 3.3x_1 - 1.5x_2 + 1.2x_4 + .9x_7 & = & 90 \\
\text{Row 1} & .7x_1 + .6x_2 + .2x_4 + x_5 - .1x_7 & = & 5 \\
\text{Row 2} & 6.1x_1 + 3.8x_2 - .4x_4 + x_6 - .3x_7 & = & 60 \\
\text{Row 3} & .3x_1 + .4x_2 + x_3 + .8x_4 + .1x_7 & = & 10
\end{array} \qquad (2.29)$$

Therefore, by letting $x_1 = 0$, $x_2 = 0$, $x_4 = 0$, and $x_7 = 0$, we derive the second basic solution, $x_0 = 90$, $x_5 = 5$, $x_6 = 60$, and $x_3 = 10$. The new solution has profits equal to \$90, which follows from the fact that new profits = (profits from previous solution) + (number of units of new pro-

TABLE 2.3. **Second Iteration (Ex. 2.1)**

	Solution	Coeffs. of x_1	Ratios	Min	Next Solution
x_0	90	−3.3	–		
x_5	5	.7	7.143	7.143	$x_1 = 7.143, x_5 = 0$
x_6	30	6.1	4.918		
x_3	10	.3	33.333		

cess) × (per unit potential gain for this process [given by the profit coefficient 9 in row 0]).

Now, we might ask, have we found the most profitable program? Criterion I indicates that we can still maximize profits; therefore we must carry out another iteration. From (2.29) both x_1 and x_2 look profitable, with x_1 promising the greatest gain. Consequently, we select x_1 as the new process to try. We perform the operations specified by Criterion II in Table 2.3.

To solve for x_0, x_1, x_3, and x_6, we manipulate the equations in (2.29) by dividing row 1 by .7; this causes the coefficient of x_1 to be 1.

$$
\begin{aligned}
&\text{Row 0} \quad x_0 - 3.3x_1 - 1.5\ x_2 + 1.2\ x_4 \\
&\qquad\qquad + .9\ x_7 = 90 \\
&\text{Row 1} \quad x_1 + .857x_2 + .286x_4 + 1.429x_5 \\
&\qquad\qquad - .143x_7 = 7.143 \\
&\text{Row 2} \quad 6.1x_1 + 3.8\ x_2 - .2\ x_4 + x_6 \\
&\qquad\qquad - .3\ x_7 = 60 \\
&\text{Row 3} \quad .3x_1 + .4\ x_2 + x_3 + .8\ x_4 \\
&\qquad\qquad + .1\ x_7 = 10
\end{aligned}
$$

We next create zero coefficients for x_1 in rows 0, 2, and 3 as follows

row 0	multiply row 1 by 3.3 and add to row 0
row 2	multiply row 1 by −6.1 and add to row 2
row 3	multiply row 1 by −.3 and add to row 3

which yields

$$
\begin{aligned}
&\text{Row 0} \quad x_0 + 1.328x_2 + 2.144x_4 + 4.716x_5 \\
&\qquad\qquad + .428x_7 = 113.571 \\
&\text{Row 1} \quad x_1 + .857x_2 + .286x_4 + 1.429x_5 \\
&\qquad\qquad - .143x_7 = 7.143 \\
&\text{Row 2} \quad - 1.428x_2 - 2.143x_4 - 8.714x_5 + x_6 \\
&\qquad\qquad + .572x_7 = 16.428
\end{aligned}
$$

$$\text{Row 3} \qquad + .143x_2 + x_3 + .714x_4 - .429x_5 + .143x_7 = 7.857$$

The third basic solution is $x_0 = 113.571$, $x_1 = 7.143$, $x_6 = 16.428$, and $x_3 = 7.857$. Criterion I indicates that we have at last found an optimal solution. In this example, the solution is unique. If any coefficient in row 0 were zero besides those attached to the variables already in the solution, we would know that we could find an alternative optimal solution (i.e., equally as profitable), provided these associated variables could be made basic. All available man-weeks and material Y are used in this plan, and since the slack variable x_6^0 is positive then this represents the unused pounds of material X associated with this optimal plan.

It is always a good idea to check the final solution by substituting it into the original equations (2.22) to determine if any arithmetical errors have been committed

$$\begin{aligned}
113.571 - 6(7.143) - 9(7.857) &= 0\\
1(7.143) + 1(7.857) &= 15\\
7(7.143) + 3(7.857) + 1(16.428) &= 90\\
3(7.143) + 10(7.857) &= 100
\end{aligned}$$

We now give "economic" reasoning behind the so-called $z_j - c_j$ criterion to be derived in a later section where the $z_j - c_j$ are the coefficients in row 0. Consider equations (2.29) repeated below

$$\begin{aligned}
x_0 - 3.3x_1 - 1.5x_2 \qquad + 1.2x_4 \qquad\qquad + .9x_7 &= 90\\
.7x_1 + .6x_2 \qquad + .2x_4 + x_5 \qquad - .1x_7 &= 5\\
6.1x_1 + 3.8x_2 \qquad - .4x_4 \qquad + x_6 - .3x_7 &= 60\\
.3x_1 + .4x_2 + x_3 + 1.8x_4 \qquad\qquad + .1x_7 &= 10
\end{aligned}$$

and the original objective function

$$x_0 = 6x_1 + 5.1x_2 + 9x_3 + 6x_4 + 0x_5 + 0x_6 + 0x_7$$

We remember the coefficients of x_i in the first equation are obtained from the previous set of equations by the Gauss-Jordan elimination technique. There is another way to get these coefficients. They may be derived mathematically or reasoned out economically (this method helps in understanding the economic interpretation of linear programming).

Pretend we do not know the coefficient of x_1 in row 0 is -3.3, and ask if we can determine this coefficient. We reason this way: Equations

(2.29) can be rewritten (remembering that if we enter x_1, the new values of x_2, x_4, and x_7 will be zero) as follows

$$\begin{aligned} x_3 &= 10 - .3x_1 \\ x_5 &= 5 - .7x_1 \\ x_6 &= 90 - 6.1x_1 \end{aligned}$$

Remembering that x_3, x_5, and x_6 are worth 9, 0, and 0 per unit, we see that for each unit of x_1 we enter, it would take .3 units of x_3 away at a cost of $9(.3) = 2.7$, .7 units of x_5 away at a cost of $0(.7) = 0$, 6.1 units of x_6 away at a cost of $0(6.1) = 0$; the total cost of adding one unit of x_1 is 2.7.

But the total profit for adding each unit of x_1 is 6. The marginal profit for adding 1 unit of x_1 is $6 - 2.7 = 3.3$ and the marginal cost of adding 1 unit of x_1 is then -3.3; i.e., there is a profit if x_1 is entered into the solution. Now, defining z_1 = cost of adding 1 unit of x_1 = $9(.3) + 0(.7) + 0(6.1) = 2.7$, and c_1 = profit of adding 1 unit of $x_1 = 6$, we see $z_1 - c_1 = 2.7 - 6 = -3.3$, the marginal cost of adding 1 unit of x_1. Thus, we should add x_1 into our solution since we will make a marginal profit of 3.3 per unit added.

Note -3.3 is the coefficient of x_1. This is always the case. In fact, the reader can verify that

$$\begin{aligned} z_2 - c_2 &= -1.5 \\ z_3 - c_3 &= 0 \\ z_4 - c_4 &= 1.2 \\ z_5 - c_5 &= 0 \\ z_6 - c_6 &= 0 \\ z_7 - c_7 &= 0.9 \end{aligned}$$

are the respective coefficients of the variables in row 0 in equations (2.29).

2.5 Theoretical Derivation of the Simplex Method

Let us consider the set S

$$\{\mathbf{x} \mid A\mathbf{x} = \mathbf{b}, \mathbf{x} \geq 0\} \text{ in } E^{n+m}$$

where A is an $m \times (n + m)$ matrix generated from the original constraints after introducing appropriate slack and surplus variables. In this section we show that any solution to our original linear programming problem corresponds to an extreme point of the constraint set S.

DEFINITION 2.2. *$\mathbf{x} \in S$ is an extreme point of S if and only if there does not exist $\mathbf{y} \in S, \mathbf{z} \in S (\mathbf{y} \neq \mathbf{z})$, and $\eta \in (0, 1)$ such that*

$$\mathbf{x} = \eta \mathbf{y} + (1 - \eta)\mathbf{z}$$

DEFINITION 2.3. *A feasible solution is any vector $\mathbf{x}$ in E^{n+m} which satisfies S; i.e., $\mathbf{x} \in S$.*

DEFINITION 2.4. *A basic feasible solution of S is a feasible solution with no more than m positive x_j.*

DEFINITION 2.5 *A nondegenerate basic feasible solution is a basic feasible solution with exactly m positive x_j.*

Hence, in view of Definitions 2.4 and 2.5, a degenerate basic feasible solution has fewer than m positive x_j.

A more convenient way to write S is

$$S = \left\{ \lambda \,\middle|\, \sum_{i=1}^{n+m} \lambda_i \mathbf{P}_i = \mathbf{P}_0, \lambda_i \geq 0 \right\}$$

where $\mathbf{P}_i$ denotes the ith column of A and $\mathbf{P}_0 = \mathbf{b}$. In particular, if S is expressed as a linear combination of the basic variables, then an optimal solution of our original problem can be expressed as

$$\mathbf{x}^0 = \sum_{j=1}^{m} \lambda_j \mathbf{P}_j \tag{2.30}$$

where the $\mathbf{P}_j$ have been arranged so that the first m constitute an optimal basis for A.

We assume in this section that the m constraints in A are linearly independent. This is known as the *nondegeneracy assumption* and implies, in view of Definition 2.5, that each λ_j in (2.30) is positive for $j = 1, 2, \ldots, m$ and zero for $j = m + 1, \ldots, n + m$. After presenting the simplex algorithm under this assumption, it is shown how we can drop this assumption and still obtain an optimal solution. The problems encountered in the simplex algorithm with degenerate basic feasible solutions are addressed in Section 2.8.

LEMMA 2.1. *λ^0 is an extreme point of S if and only if the nonzero λ_i^0 are coefficients of linearly independent vectors.*

Proof. Let there be r nonzero components of λ^0 and let us rearrange the components so that the first r are nonzero. The columns of A are also rearranged in a corresponding fashion.

Now suppose that the first r vectors are dependent. Then there exist scalars $z_1, z_2, \ldots, z_r$ not all equal to zero, such that

$$\sum_{j=1}^{r} z_j \mathbf{P}_j = 0 \tag{2.31}$$

But with $\lambda^0 \in S$, (2.31) implies that

$$\sum_{j=1}^{r} (\lambda_j^0 \mathbf{P}_j + z_j \mathbf{P}_j) = \mathbf{b}$$

and it follows that for any $\epsilon > 0$

$$\sum_{j=1}^{r} (\lambda_j^0 \pm \epsilon z_j)\mathbf{P}_j = \mathbf{b}$$

Since $\lambda_j^0 > 0$ for $j = 1, 2, \ldots, r$, we can choose $\epsilon > 0$ so that $\lambda_j^0 - \epsilon z_j > 0$ and $\lambda_j^0 + \epsilon z_j > 0$. Let

$$\lambda^1 = \begin{bmatrix} \lambda_1^0 - \epsilon z_1 \\ \vdots \\ \lambda_r^0 - \epsilon z_r \end{bmatrix} \qquad \lambda^2 = \begin{bmatrix} \lambda_1^0 + \epsilon z_1 \\ \vdots \\ \lambda_r^0 + \epsilon z_r \end{bmatrix}$$

then

$$\lambda^0 = (1/2)(\lambda^1 + \lambda^2)$$

and λ^0 is not an extreme point of S. Hence it follows if the first r components of λ^0 are nonzero, then the first set of r vectors are linearly independent.

We show now that if the first r vectors of A are linearly independent, then λ^0 is an extreme point of S. Suppose that λ^0 is not an extreme point; then there exist $\lambda^1, \lambda^2 \in S(\lambda^1 \neq \lambda^2)$, and $\eta \in (0, 1)$ such that

$$\lambda^0 = \eta\lambda^1 + (1 - \eta)\lambda^2 \qquad 0 < \eta < 1$$

For $j = r + 1, r + 2, \ldots, n + m$

$$0 = \lambda_j^0 = \eta\lambda_j^1 + (1 - \eta)\lambda_j^2$$

Therefore $\lambda_j^1 = \lambda_j^2 = 0$ for $j = r + 1, \ldots, n + m$ since $\lambda_j^1 \geq 0$, $\lambda_j^2 \geq 0$. For $j = 1, 2, \ldots, r$

$$\sum_{j=1}^{r} \lambda_j^1 \mathbf{P}_j = \mathbf{P}_0 \tag{2.32}$$

$$\sum_{j=1}^{r} \lambda_j^2 \mathbf{P}_j = \mathbf{P}_0 \tag{2.33}$$

Subtracting (2.33) from (2.32) we have

$$0 = \sum_{j=1}^{r} (\lambda_j^1 - \lambda_j^2)\mathbf{P}_j$$

But $\mathbf{P}_1, \mathbf{P}_2, \ldots, \mathbf{P}_r$ are linearly independent; therefore $\lambda_j^1 = \lambda_j^2$ for all j. This contradicts the fact that $\boldsymbol{\lambda}^1 \neq \boldsymbol{\lambda}^2$, since the r vectors are linearly independent; hence $\boldsymbol{\lambda}^0$ is an extreme point of S.

COROLLARY 2.1.1. *If $\boldsymbol{\lambda}^0$ is an extreme point of S, then $\boldsymbol{\lambda}^0$ has at most m nonzero coefficients.*

COROLLARY 2.1.2. *The set S has a finite number of extreme points.*

Proof. This follows since there are only a finite number of bases of A, which is bounded by $\binom{n+m}{m}$ if the rank of A is m.

LEMMA 2.2. *The extreme point associated with a basis is unique.*

Proof. Suppose that there exist $\boldsymbol{\lambda}^1, \boldsymbol{\lambda}^2 \in S$ $(\boldsymbol{\lambda}^1 \neq \boldsymbol{\lambda}^2)$ such that

$$\sum_{j=1}^{m} \lambda_j^1 \mathbf{P}_j = \mathbf{P}_0$$

$$\sum_{j=1}^{m} \lambda_j^2 \mathbf{P}_j = \mathbf{P}_0$$

Then, this implies that

$$0 = \sum_{j=1}^{m} (\lambda_j^1 - \lambda_j^2)\mathbf{P}_j$$

or that

$$\lambda_j^1 = \lambda_j^2 \qquad j = 1, 2, \ldots, m$$

Therefore there exists only one extreme point for a given basis.

LEMMA 2.3. *The set of all feasible solutions of S is a convex set.*

Proof. To show that S is a convex set, we need to show that if $\bar{\mathbf{x}}$ and $\hat{\mathbf{x}}$ are any two points in S, then any convex combination of these two vectors is also in S.

First, we note that for $\alpha \in [0, 1]$

$$\hat{\bar{\mathbf{x}}} = \alpha\bar{\mathbf{x}} + (1 - \alpha)\hat{\mathbf{x}} \geq 0$$

Hence we now only need to show that $A\hat{\bar{\mathbf{x}}} = \mathbf{b}$. This is trivial since $A\bar{\mathbf{x}} = \mathbf{b}$, $A\hat{\mathbf{x}} = \mathbf{b}$, and $\alpha \in [0, 1]$ imply that

$$\begin{aligned} A\hat{\bar{\mathbf{x}}} &= \alpha A\bar{\mathbf{x}} + (1 - \alpha)A\hat{\mathbf{x}} \\ &= \alpha\mathbf{b} + (1 - \alpha)\mathbf{b} = \mathbf{b} \end{aligned}$$

Hence $\hat{\bar{\mathbf{x}}} \in S$.

LEMMA 2.4. *If the maximum* $\mathbf{c}'\mathbf{x}$ *for* $\mathbf{x} \in S$ *is equal to* $\mathbf{c}'\mathbf{x}^0$ *for some* $\mathbf{x}^0 \in S$, *then the maximum* $\mathbf{c}'\mathbf{x}$ *for* $\mathbf{x} \in S$ *is equal to* $\mathbf{c}'\mathbf{x}^*$ *where* $\mathbf{x}^*$ *is an extreme point of S.*

Proof. Let $\mathbf{x}^{(1)}, \mathbf{x}^{(2)}, \ldots, \mathbf{x}^{(k)}$ be the extremes of S; then any solution $\mathbf{x}^0$ can be written as $\sum_{i=1}^{k} \lambda_i \mathbf{x}^{(i)}$ with each $\lambda_i \geq 0$ and $\sum_{i=1}^{k} \lambda_i = 1$. Now

$$\begin{aligned} \mathbf{c}'\mathbf{x}^0 &= \mathbf{c}' \sum_{i=1}^{k} \lambda_i \mathbf{x}^{(i)} \\ &= \sum_{i=1}^{k} \lambda_i \mathbf{c}'\mathbf{x}^{(i)} \leq \sum_{i=1}^{k} \lambda_i (\max_k \mathbf{c}'\mathbf{x}^{(k)}) \\ &= \max_k \mathbf{c}'\mathbf{x}^{(k)} = \mathbf{c}'\mathbf{x}^* \end{aligned}$$

But in view of the fact that $\mathbf{x}^* \in S$, and $\mathbf{c}'\mathbf{x}^0 \geq \mathbf{c}'\mathbf{x}$ for all $\mathbf{x} \in S$, it follows that $\mathbf{c}'\mathbf{x}^0 \geq \mathbf{c}'\mathbf{x}^*$. Hence $\mathbf{c}'\mathbf{x}^0 = \mathbf{c}'\mathbf{x}^*$.

From the previous lemmas we can summarize the following pertinent results with respect to S.

1. Every basic feasible solution of S corresponds to an extreme point of a convex set of feasible solutions.
2. Every extreme point is associated with m linearly independent vectors from the set $\mathbf{P}_j$ ($j = 1, 2, \ldots, n + m$).
3. There is some extreme point at which the objective function $z = \mathbf{c}'\mathbf{x}^0$ takes on its maximum value.

From the foregoing we know that we only need to consider the set of extreme points to identify our optimal linear programming solution. However, even though this set is finite, it becomes computationally difficult to examine every possible extreme point when m and $n + m$ are large. We now discuss how to get a new extreme point from an old extreme point in a way that reduces the number of extreme points under consideration.

2.5.1 · Getting a Better Extreme Point from a Feasible Extreme Point

Assume that

$$\lambda^0 = \begin{bmatrix} \lambda_1^0 \\ \lambda_2^0 \\ \vdots \\ \lambda_m^0 \\ 0 \\ \vdots \\ 0 \end{bmatrix}$$

is an extreme point of S with $\lambda_i^0 > 0$ ($i = 1, 2, \ldots, m$). Let $\mathbf{P}_1, \mathbf{P}_2, \ldots, \mathbf{P}_m$ be the basic vectors associated with a particular extreme point of S. Therefore any vector, say $\mathbf{P}_k$, can be expressed as some combination of $\mathbf{P}_1, \mathbf{P}_2, \ldots, \mathbf{P}_m$. In particular, there exist x_{jk} such that

$$\mathbf{P}_k = \sum_{j=1}^{m} \mathbf{P}_j x_{jk} \qquad k = 1, 2, \ldots, n + m \tag{2.34}$$

In terms of matrix notation, there exists a matrix X such that

$$P = BX \tag{2.35}$$

where P is an $m \times (n + m)$ matrix composed of the vectors in (2.34) and B is a basis of A. It should be noted that in view of (2.35), $X = B^{-1}P$ or

$$\mathbf{x}_k = B^{-1}\mathbf{P}_k \tag{2.36}$$

for each vector of X.

Now consider $\lambda^0 \in S$. Then by definition of S

$$\sum_{j=1}^{m} \lambda_j^0 \mathbf{P}_j = \mathbf{P}_0 \tag{2.37}$$

From (2.34), (2.37), and for any $\theta > 0$ where $k > m$

$$\sum_{j=1}^{m} \lambda_j^0 \mathbf{P}_j + \theta \left(\mathbf{P}_k - \sum_{j=1}^{m} \mathbf{P}_j x_{jk} \right) = \mathbf{P}_0$$

or

$$\sum_{j=1}^{m} (\lambda_j^0 - \theta x_{jk}) \mathbf{P}_j + \theta \mathbf{P}_k = \mathbf{P}_0 \tag{2.38}$$

Let

$$\lambda^* = \begin{bmatrix} \lambda_1^0 - \theta x_{1k} \\ \lambda_2^0 - \theta x_{2k} \\ \vdots \\ \lambda_m^0 - \theta x_{mk} \\ 0 \\ \vdots \\ 0 \\ \theta \\ 0 \\ \vdots \end{bmatrix}$$

Then

$$\begin{aligned} \mathbf{c}'\boldsymbol{\lambda}^* &= \sum_{j=1}^{m} (\lambda_j^0 - \theta x_{jk})c_j + \theta c_k \\ &= \mathbf{c}'\boldsymbol{\lambda}^0 + \theta\left(c_k - \sum_{j=1}^{m} x_{jk}c_j\right) \\ &= \mathbf{c}'\boldsymbol{\lambda}^0 + \theta(c_k - z_k) \end{aligned} \tag{2.39}$$

where $z_k = \sum_{j=1}^{m} x_{jk}c_j$. Therefore, a new extreme point $\boldsymbol{\lambda}^*$ which increases the optimum can be obtained for $\theta > 0$ if and only if $z_k - c_k < 0$ for some k, i.e.

$$\mathbf{c}'\boldsymbol{\lambda}^* > \mathbf{c}'\boldsymbol{\lambda}^0$$

Let us consider the following cases.

Case 1. The coefficient $z_{k_0} - c_{k_0} < 0$ for some k_0 with $x_{jk_0} > 0$ for at least one j.

If $x_{jk_0} > 0$, then from (2.38), $\lambda_j^0 - \theta x_{jk_0} \geq 0$ if and only if $\theta \leq \lambda_j^0/x_{jk_0}$ for all $j = 1, 2, \ldots, m$ such that $x_{jk_0} > 0$. Hence

$$\max \theta = \min_j \{\lambda_j^0/x_{jk_0} \mid x_{jk_0} > 0\} = \lambda_r^0/x_{rk_0}$$

When $\theta = \lambda_r^0/x_{rk_0}$, then (2.38) can be written as

$$\sum_{\substack{j=1 \\ j \neq r}}^{m} (\lambda_j^0 - \theta x_{jk_0})\mathbf{P}_j + [\lambda_r^0 - (\lambda_r^0/x_{rk_0})x_{rk_0}]\mathbf{P}_r + \theta \mathbf{P}_{k_0} = \mathbf{P}_0$$

and our new basis is

$$(\mathbf{P}_1, \mathbf{P}_2, \ldots, \mathbf{P}_{r-1}, \mathbf{P}_{r+1}, \ldots, \mathbf{P}_m, \mathbf{P}_{k_0})$$

Hence we can obtain a new basis, which improves the value of the objective function. Moreover, when θ is selected to be

$$\min_j \{\lambda_j^0/x_{jk_0} \mid x_{jk_0} > 0\}$$

then we are assured that the new basic solution will also be a feasible solution; i.e., all components of the basic solution will be nonnegative.

Case 2. The coefficient $z_{k_0} - c_{k_0} < 0$ for some k_0 and $x_{jk_0} \leq 0$ for $j = 1, 2, \ldots, m$ and less than zero for at least one j.

In this situation, for any $\theta > 0$

$$\lambda_j^0 - \theta x_{jk_0} \geq 0 \qquad j = 1, 2, \ldots, m$$

This implies that as θ tends to infinity, $\mathbf{c}'\boldsymbol{\lambda}^*$ will increase without limit; hence the problem is unbounded.

Case 3. The coefficient $z_k - c_k \geq 0$ for all k. We show this implies that $\boldsymbol{\lambda}^0$ is the optimal solution of our linear programming problem.

For any $\boldsymbol{\lambda} \in S$

$$\begin{aligned}
\mathbf{c}'\boldsymbol{\lambda} &= \sum_{k=1}^{n+m} c_k \lambda_k \leq \sum_{k=1}^{n+m} z_k \lambda_k \\
&= \sum_{k=1}^{n+m} (\mathbf{c}_B' \mathbf{x}_k) \lambda_k \\
&= \mathbf{c}_B' \sum_{k=1}^{n+m} \mathbf{x}_k \lambda_k \\
&= \mathbf{c}_B' \sum_{k=1}^{n+m} B^{-1} \mathbf{P}_k \lambda_k \quad \text{from (2.36)} \\
&= \mathbf{c}_B' B^{-1} \mathbf{b} \quad \text{from (2.37)} \\
&= \mathbf{c}'\boldsymbol{\lambda}^0
\end{aligned}$$

where $\mathbf{c}_B$ denotes the coefficients corresponding to the basic variables. Hence, $\boldsymbol{\lambda}^0$ is an optimal solution of our maximization problem.

Before considering an example, let us reconsider (2.39). We have shown that one can determine a vector to bring into the basis which will increase the value of the optimum by choosing any $z_j - c_j$ which is negative. Usually, one will prefer in practice to select the $z_j - c_j$ which is the most negative. An alternate rule, if one desires to bring in a vector that will indeed give us the largest possible increase in z, is to select the vector that will make

$$\theta(c_k - z_k) = (\lambda_j^0 / x_{jk_0})(c_{k_0} - z_{k_0})$$

as small as possible for $\theta > 0$. This will indeed give us the greatest increase in z, but at the expense of increasing the number of computations since it requires the value of λ_j^0/x_{jk_0} for every nonbasic vector with negative $z_j - c_j$. Hence, in practice usually one considers only the $z_j - c_j$ coefficients of the nonbasic variables.

EXAMPLE 2.5

Consider the following system without regard to its objective function

$\mathbf{P}_1$	$\mathbf{P}_2$	$\mathbf{P}_3$	$\mathbf{P}_4$	$\mathbf{P}_5$	$\mathbf{P}_6$		$\mathbf{P}_0$
$3x_1$	$-\ x_2$	$+\ 2x_3$	$+\ x_4$			$=$	7
$2x_1$	$-\ 4x_2$			$+\ x_5$		$=$	12
$-\ 4x_1$	$-\ 3x_2$	$+\ 8x_3$			$+\ x_6$	$=$	10

where all $x_i \geq 0$.

The initial extreme point solution is $\lambda^0 = (0, 0, 0, 7, 12, 10)$ with corresponding basic solution of $x_4^0 = 7$, $x_5^0 = 12$, and $x_6^0 = 10$.

In vector notation

$$7\,\mathbf{P}_4 + 12\,\mathbf{P}_5 + 10\,\mathbf{P}_6 = \mathbf{P}_0 \qquad (2.40)$$

Suppose we wanted to introduce $\mathbf{P}_1$ to obtain a new extreme point solution. We have

$$3\,\mathbf{P}_4 + 2\,\mathbf{P}_5 + (-4)\,\mathbf{P}_6 = \mathbf{P}_1 \qquad (2.41)$$

Multiplying (2.41) by θ and subtracting from (2.40), we have

$$(7 - 3\theta)\mathbf{P}_4 + (12 - 2\theta)\mathbf{P}_5 + (10 + 4\theta)\mathbf{P}_6 + \theta\mathbf{P}_1 = \mathbf{P}_0$$

Define

$$\theta = \min\{b_i^0/x_{i1} \mid x_{i1} > 0\} = \min\{7/3, 12/2\} = 7/3$$

Then

$$(22/3)\mathbf{P}_5 + (58/3)\mathbf{P}_6 + (7/3)\mathbf{P}_1 = \mathbf{P}_0$$

and our second extreme point solution is $(7/3, 0, 0, 0, 22/3, 58/3)$ with corresponding basic vectors of $\mathbf{P}_1$, $\mathbf{P}_5$, and $\mathbf{P}_6$.

It should be noted that if θ was chosen to be $10/-4 = -5/2$, then we would have a basic solution with negative components.

A more efficient approach of determining which vectors will leave and enter the basis is to set up the problem in tableau form. In particular

Basis	$\mathbf{P}_1$	$\mathbf{P}_2$	$\mathbf{P}_3$	$\mathbf{P}_4$	$\mathbf{P}_5$	$\mathbf{P}_6$	$\mathbf{P}_0$	θ
$\mathbf{P}_4$	[3]	−1	2	1	0	0	7	7/3
$\mathbf{P}_5$	2	−4	0	0	1	0	12	6
$\mathbf{P}_6$	−4	−3	8	0	0	1	10	

If $\mathbf{P}_1$ is chosen to enter the system, then the pivot element $a_{11} = 3$ is determined by taking the ratio of the appropriate right-hand components with the positive entries in $\mathbf{P}_1$. Since this occurs in row 1, then $\mathbf{P}_1$ enters the basis and it will replace $\mathbf{P}_4$.

Updating this tableau with respect to the new basis we have

Basis	$\mathbf{P}_1$	$\mathbf{P}_2$	$\mathbf{P}_3$	$\mathbf{P}_4$	$\mathbf{P}_5$	$\mathbf{P}_6$	$\mathbf{P}_0$
$\mathbf{P}_1$	1	−1/3	2/3	1/3	0	0	7/3
$\mathbf{P}_5$	0	−10/3	−4/3	−2/3	1	0	22/3
$\mathbf{P}_6$	0	−13/3	32/3	4/3	0	1	58/3

Thus $\lambda_1^0 = x_1^0 = 7/3$, $\lambda_5^0 = x_5^0 = 22/3$, and $\lambda_6^0 = x_6^0 = 58/3$, corresponding to the nonzero solutions in the second extreme point solution.

2.5.2 Simplex Transformation

Let us consider the following tableau, in which λ^0 denotes the present basic solution associated with the basis $\mathbf{P}_1, \mathbf{P}_2, \ldots, \mathbf{P}_m$.

Basis	$\mathbf{P}_1$	$\mathbf{P}_2$	$\cdots$	$\mathbf{P}_k$	$\cdots$	$\mathbf{P}_{n^*}$	$\mathbf{P}_0$
$\mathbf{P}_1$	x_{11}	x_{12}	$\cdots$	x_{1k}	$\cdots$	x_{1n^*}	λ_1^0
$\mathbf{P}_2$	x_{21}	x_{22}	$\cdots$	x_{2k}	$\cdots$	x_{2n^*}	λ_2^0
$\vdots$	$\vdots$	$\vdots$	$\cdots$	$\vdots$	$\cdots$	$\vdots$	$\vdots$
$\vdots$	$\vdots$	$\vdots$	$\cdots$	$\vdots$	$\cdots$	$\vdots$	$\vdots$
$\mathbf{P}_m$	x_{m1}	x_{m2}	$\cdots$	x_{mk}	$\cdots$	x_{mn^*}	λ_m^0
$\mathbf{z}$	z_1	z_2	$\cdots$	z_k	$\cdots$	z_{n^*}	
$\mathbf{z} - \mathbf{c}$	$z_1 - c_1$	$z_2 - c_2$	$\cdots$	$z_k - c_k$	$\cdots$	$z_{n^*} - c_{n^*}$	

where $n^* = n + m$.

Any column $\mathbf{P}_i$ can be expressed in terms of the present basis I, and can be written as

$$\mathbf{P}_i^I = \sum_{j=1}^{m} \mathbf{P}_j x_{ji} \tag{2.42}$$

where the $\mathbf{P}_j$ form a basis in the present tableau. Consider a new basis in which $\mathbf{P}_k$ has replaced $\mathbf{P}_r$, i.e., $(\mathbf{P}_1, \mathbf{P}_2, \ldots, \mathbf{P}_{r-1}, \mathbf{P}_{r+1}, \ldots, \mathbf{P}_m, \mathbf{P}_k)$. Then the ith vector associated with the new basis can be expressed as

$$\mathbf{P}_i^{II} = \sum_{\substack{j=1 \\ j \neq r}}^{m} \mathbf{P}_j y_{ji} + y_{ri} \sum_{j=1}^{m} \mathbf{P}_j x_{jk} \quad \text{from (2.42)}$$

$$= \sum_{\substack{j=1 \\ j \neq r}}^{m} (y_{ji} + y_{ri} x_{jk}) \mathbf{P}_j + y_{ri} \mathbf{P}_r x_{rk} \tag{2.43}$$

Now

$$0 = \mathbf{P}_i^{II} - \mathbf{P}_i^I$$

$$= \sum_{\substack{j=1 \\ j \neq r}}^{m} (y_{ji} + y_{ri} x_{jk} - x_{ji}) \mathbf{P}_j + (y_{ri} x_{rk} - x_{ri}) \mathbf{P}_r$$

In view of Lemma 2.1, the coefficients are coefficients of a linearly independent set of vectors. Therefore

$$y_{ji} + y_{ri} x_{jk} - x_{ji} = 0 \qquad j = 1, \ldots, m (j \neq r) \tag{2.44}$$

$$y_{ri} x_{rk} - x_{ri} = 0 \tag{2.45}$$

This implies that in the updated tableau the new coefficients are calculated in the following manner, using the pivot element x_{rk}.

$$y_{ri} = x_{ri} / x_{rk} \quad \text{for } j = r$$

$$y_{ji} = x_{ji} - x_{jk}(x_{ri}/x_{rk}) \quad \text{for } j \neq r$$

The right-hand vector $\mathbf{P}_0$ is transformed in the same manner, letting $i = 0$. Recall in our scheduling problem solved in Section 2.4 that the objective function was incorporated as a row in our linear system. Therefore, denoting this row by 0, we have that the updated coefficients in the updated tableau are

$$y_{0i} = x_{0i} - x_{0k}(x_{ri}/x_{rk}) = -c_i + c_k(y_{ri}) = -c_i + z_i$$

Before giving an example, let us consider how one can easily find an initial extreme point.

2.5.3 *Obtaining an Initial Extreme Point*

Let us first consider the constraints associated with the following linear programming problem

$$A\boldsymbol{\lambda} \le \mathbf{b}$$
$$\boldsymbol{\lambda} \ge 0$$

where the vector **b** has all nonnegative components. Then adding slack variables the expanded matrix is

$$\begin{bmatrix} a_{11} & \cdots & a_{1n} & 1 & 0 & \cdots & 0 \\ a_{21} & \cdots & a_{2n} & 0 & 1 & 0 & 0 \\ \vdots & \vdots & \vdots & \vdots & & \ddots & \vdots \\ a_{m1} & \cdots & a_{mn} & \cdots & \cdots & \cdots & 1 \end{bmatrix} = [A, I]$$

In this case an initial basis is the set of vectors $(\mathbf{P}_{n+1}, \mathbf{P}_{n+2}, \ldots, \mathbf{P}_{n+m})$ with initial extreme point $(0, \ldots; 0, b_1, b_2, \ldots, b_m)$.

Let us now consider the case where there exists at least one constraint which is not an inequality restriction. Again we are assuming the necessary condition that all components of **b** are nonnegative.

Assume there are r greater-than inequality constraints which are formulated as equalities by the addition of r surplus variables; then the expanded matrix can be expressed as $[A, I_{m-r}, P_r]$ where

$$P_r = \begin{bmatrix} -1 & 0 & \cdots & 0 \\ 0 & -1 & \cdots & 0 \\ \vdots & & \ddots & \vdots \\ & & & -1 \\ 0 & & \cdots & 0 \\ \vdots & & & \vdots \\ 0 & & & 0 \end{bmatrix}_{m \times r}$$

Note at this point that if any of these vectors is considered to be in the basis, then necessarily one of the initial extreme point components would be negative. In this situation, to obtain an initial extreme point we introduce "artificial variables" into the problem to artificially give us an initial extreme point. This device of adding artificial variables is always applicable if no other initial extreme point is available.

The procedure of incorporating the artificial variables is as follows:

After adding all slack and surplus variables to the problem, add a dummy vector with all components zero except for a 1 in a row of a corresponding greater-than or equality constraint; do this for all such rows. Choose the associated profit coefficients to be some negative value M so that the simplex algorithm will replace these vectors by more profitable, reasonable vectors.

If in the final tableau, an artificial variable is in the final solution at a nonzero level, then the problem is infeasible.

EXAMPLE 2.6

Consider the problem

$$\begin{aligned} \text{minimize} \quad & 2x_1 + x_2 \\ \text{subject to} \quad & 3x_1 + x_2 \le 4 \\ & 4x_1 + 3x_2 \ge 6 \\ & x_1 + 2x_2 = 3 \\ & x_1, \quad x_2 \ge 0 \end{aligned}$$

Since the solution of linear programming problems presented so far was for maximization problems, we must maximize $-(2x_1 + x_2)$ or $(-2x_1 - x_2)$, and then multiply the optimum by -1. After also adding the slack and surplus variables we have

$$\begin{aligned} \text{maximize} \quad & -2x_1 - x_2 \\ \text{subject to} \quad & \end{aligned}$$

$$\begin{bmatrix} 3 & 1 & 1 & 0 \\ 4 & 3 & 0 & -1 \\ 1 & 2 & 0 & 0 \end{bmatrix} \begin{bmatrix} x_1 \\ x_2 \\ s_1 \\ s_2 \end{bmatrix} = \begin{bmatrix} 4 \\ 6 \\ 3 \end{bmatrix}$$

with $x_1, x_2, s_1, s_2 \ge 0$.

Introducing artificial vectors in order to obtain a full identity in the expanded matrix, we have the following initial tableau where the profit coefficient M associated with each artificial variable is chosen to be 1,000.

TABLEAU 0

c	Basis	P_0	P_1	P_2	P_3	P_4	P_5	P_6	θ
			−2	−1	0	0	−1,000	−1,000	
0	P_3	4	3	1	1	0	0	0	4
−1,000	P_5	6	4	3	0	−1	1	0	2
−1,000	P_6	3	1	[2]	0	0	0	1	3/2
	z	−9,000	−5,000	−5,000	0	1,000	−1,000	−1,000	
	z − c		−4,998	−4,999	0	1,000	0	0	

Since the smallest $z_j - c_j$ associated with $\mathbf{P}_2$ is −4,999, then $\mathbf{P}_2$ enters the basis; and since this vector yields a minimum positive value of θ in row 3, $\mathbf{P}_6$ leaves the solution. Applying the simplex transformation yields the next tableau.

TABLEAU 1

c	Basis	P_0	P_1	P_2	P_3	P_4	P_5	P_6	θ
			−2	−1	0	0	−1,000	−1,000	
0	P_3	5/2	5/2	0	1	0	0	−1/2	1
−1,000	P_5	3/2	[5/2]	0	0	−1	1	−3/2	3/5
−1	P_2	3/2	1/2	1	0	0	0	1/2	3
	z	−3,003/2	−5,001/2	−1	0	1,000	−1,000	2,999/2	
	z − c		−4,997/2	0	0	1,000	0	4,999/2	

The only negative $z_j - c_j$ now occurs with $\mathbf{P}_1$ and yields a minimum value of θ in row 2. Therefore $\mathbf{P}_1$ enters the basis and $\mathbf{P}_5$ leaves the basis. So we have

TABLEAU 2

c	Basis	P_0	P_1	P_2	P_3	P_4	P_5	P_6
			−2	−1	0	0	−1,000	−1,000
0	P_3	1	0	0	1	1	−1	1
−2	P_1	3/5	1	0	0	−2/5	2/5	−3/5
−1	P_2	6/5	0	1	0	1/5	−1/5	4/5
	z	−12/5	−2	−1	0	3/5	−3/5	2/5
	z − c		0	0	0	3/5	4,997/5	5,002/5

In this tableau, all $z_j - c_j$ are nonnegative, and this final tableau does not contain any nonzero artificial variables, namely, those associated with

$\mathbf{P}_5$ and $\mathbf{P}_6$. Therefore the optimal solution is $x_1^0 = 3/5$, $x_2^0 = 6/5$, $s_1^0 = 1$, and $s_2^0 = 0$; also, the value of the optimum is $-(-12/5) = 12/5$, since we originally changed the objective function from a minimization problem to a maximization problem.

In Tableau 0

$$B = \begin{bmatrix} 1 & 0 & 0 \\ 0 & 1 & 0 \\ 0 & 0 & 1 \end{bmatrix} = [\mathbf{P}_3, \mathbf{P}_5, \mathbf{P}_6]$$

The basis associated with Tableau 1 is

$$B = \begin{bmatrix} 1 & 0 & 1 \\ 0 & 1 & 3 \\ 0 & 0 & 2 \end{bmatrix} = [\mathbf{P}_3, \mathbf{P}_5, \mathbf{P}_2]$$

since $\mathbf{P}_2$ replaced $\mathbf{P}_6$.

It should be noted that now

$$B^{-1} = \begin{bmatrix} 1 & 0 & -1/2 \\ 0 & 1 & -3/2 \\ 0 & 0 & 1/2 \end{bmatrix}$$

With knowledge of B^{-1}, Tableau 1 could be easily computed as follows (see 2.36)

$$\mathbf{x}_1 = B^{-1}\mathbf{P}_1 = \begin{bmatrix} 1 & 0 & -1/2 \\ 0 & 1 & -3/2 \\ 0 & 0 & 1/2 \end{bmatrix} \begin{bmatrix} 3 \\ 4 \\ 1 \end{bmatrix} = \begin{bmatrix} 5/2 \\ 5/2 \\ 1/2 \end{bmatrix}$$

$$\mathbf{x}_0 = B^{-1}\mathbf{P}_0 = B^{-1} \begin{bmatrix} 4 \\ 6 \\ 3 \end{bmatrix} = \begin{bmatrix} 5/2 \\ 3/2 \\ 3/2 \end{bmatrix}$$

In the final tableau

$$B = \begin{bmatrix} 1 & 3 & 1 \\ 0 & 4 & 3 \\ 0 & 1 & 2 \end{bmatrix} = [\mathbf{P}_3, \mathbf{P}_1, \mathbf{P}_2] \qquad B^{-1} = \begin{bmatrix} 1 & -1 & 1 \\ 0 & 2/5 & -3/5 \\ 0 & -1/5 & 4/5 \end{bmatrix}$$

Hence

$$\mathbf{x}_0 = B^{-1}\mathbf{P}_0 = B^{-1}\begin{bmatrix} 4 \\ 6 \\ 3 \end{bmatrix} = \begin{bmatrix} 1 \\ 3/5 \\ 6/5 \end{bmatrix}$$

Likewise all components or vectors in the final tableau could be determined from the original tableau if one had knowledge of B^{-1}.

An alternative procedure for handling artificial vectors is described in Section 2.6.

2.6 Pseudo-Objective Function

This procedure defines a new objective function, called a pseudo-objective function, in which artificial vectors are used to establish a starting basis. The original objective function is still defined in our original tableau and is updated as usual as we proceed from one tableau to the next. This allows us to break up the simplex algorithm into two stages, Phase I and Phase II. In Phase I, via the simplex algorithm we determine or try to determine a feasible solution by only considering the pseudo-objective function. Once a logical feasible solution has been established, then the pseudo-objective function is no longer needed, and the original updated objective function is used in computing an optimal solution.

To define the pseudo-objective function, let us first recall that when artificial variables are used in our model, the objective function is modified with appropriate cost coefficients for these variables. Then with the simplex algorithm we proceed to remove these variables from the basic solution, and when this is done the solution is logically feasible. Another way to accomplish the same end is to consider the sum of the infeasibilities (the sum of artificial values) and to maximize this expression; this expression is called the pseudo-objective function.

To see clearly how this objective function is created, let us consider the following problem

$$\begin{array}{lrrrl}
\text{maximize} & x_1 & - \; x_2 & & \\
\text{subject to} & x_1 \; + & x_2 \; + & x_3 & \geq 4 \\
 & & 2x_2 \; - & x_3 & \leq 1{,}000 \\
 & 5x_1 & & + \; 2x_3 & \geq 10 \\
 & x_1, & x_2, & x_3 & \geq 0
\end{array}$$

Choosing $M \gg 0$ for each artificial variable that is introduced we have the following tableau

Row 0	$z_0 - x_1 + x_2 + Mx_7^a + Mx_8^a = 0$
Row 1	$x_1 + x_2 + x_3 - x_4^s + x_7^a = 4$
Row 2	$2x_2 - x_3 + x_5^s = 1{,}000$
Row 3	$5x_1 + 2x_3 - x_6^s + x_8^a = 10$

Now consider the sum of artificial variables which is calculated by taking row 0 and subtracting from it M times row 1 and M times row 3, i.e., row 0 $-$ M(row 1) $-$ M(row 3). This gives us, letting $M = 1$

$$z_0 - 7x_1 - 3x_3 + x_4 + x_6 = -14$$

This expression is used as our pseudo-objective function.

In Phase I, the pseudo-objective function is used as an objective function row; once it has all components greater than or equal to zero, we switch to the updated objective function and continue with the simplex algorithm.

This procedure is useful when trying to solve two problems in which the second problem is different in structure only with respect to the resource vector. In particular, if upon using B^{-1} in the second problem we encounter an infeasible solution, i.e.

$$B^{-1}\mathbf{b}_2 = \mathbf{x}^0 \ngeq 0$$

then we can use the following procedure to get things back into standard form

(1) multiply each row whose basic variable is negative by -1
(2) introduce an artificial variable for each of these rows
(3) form a pseudo-objective function and do a Phase I operation

2.7 Associated Dual Problem

Considering the inverse of the matrix B associated with the final tableau of the simplex procedure, one can derive an optimal solution to another linear programming problem. Chapter 4 shows in detail that this associated linear programming problem is in fact the *dual* of the original problem. By this we mean that if one problem has an optimal solution, then so does the other, and their respective objective function values, or optima, are equal. This is demonstrated by considering the following two problems.

Problem I

$$\begin{aligned} &\text{maximize} \quad \mathbf{c}'\boldsymbol{\lambda} \\ &\text{subject to} \quad A\boldsymbol{\lambda} = \mathbf{b} \\ &\qquad\qquad\quad \boldsymbol{\lambda} \geq 0 \end{aligned}$$

with an optimal solution $\boldsymbol{\lambda}^0$ and

Problem II

$$\begin{aligned} &\text{minimize} \quad \mathbf{b}'\mathbf{y} \\ &\text{subject to} \quad A'\mathbf{y} \geq \mathbf{c} \end{aligned}$$

In Chapter 4, these two problems are discussed in greater detail, but for now let us prove the following lemma.

LEMMA 2.5. *Let B be the solution basis for Problem I associated with $\boldsymbol{\lambda}^0$, then $\mathbf{y}^{*\prime} = \mathbf{c}_B' B^{-1}$ is a feasible solution of Problem II, and $\mathbf{c}'\boldsymbol{\lambda}^0 = \mathbf{b}'\mathbf{y}^*$.*

Proof. To show first that $\mathbf{y}^*$ is a feasible solution, i.e., $A'\mathbf{y}^* \geq \mathbf{c}$, let B be the first m vectors in A. Then

$$\begin{aligned} \mathbf{y}^{*\prime} A &= \mathbf{y}^{*\prime}[B, \bar{B}] \\ &= [\mathbf{y}^{*\prime} B, \mathbf{y}^{*\prime} \bar{B}] \\ &= [\mathbf{c}_B' B^{-1} B, \mathbf{c}_B' B^{-1} \bar{B}] \\ &= \mathbf{c}_B'[I, X] \\ &= \mathbf{z}' \\ &\geq \mathbf{c}' \end{aligned}$$

since in the final tableau associated with Problem I, $z_j - c_j \geq 0$ for all j. Thus $\mathbf{y}^*$ is a feasible solution of Problem II.

To now show that the optima are equal

$$\begin{aligned} \mathbf{y}^{*\prime}\mathbf{b} &= \mathbf{c}_B' B^{-1}\mathbf{b} \\ &= \mathbf{c}_B' \boldsymbol{\lambda}_B^0 \\ &= \mathbf{c}'\boldsymbol{\lambda}^0 \end{aligned}$$

since each λ_i^0 associated with the nonbasic variables is zero.

This lemma shows that when solving a linear programming problem via the simplex algorithm one is also determining a vector for Problem II which yields the same value or optimum. Moreover, it is shown in Chapter 4 that $\mathbf{y}^*$ is the optimal solution of Problem II. This implies that when one solves Problem I, a by-product of the simplex algorithm is the solution to Problem II, and vice versa. This fact is very important computationally since the problem with the least number of constraints is usually the easiest problem to solve, and one can solve either Problem I or Problem II and obtain the solution to the other problem.

Consider again Example 2.6. In the final tableau the basic matrix is composed of $\mathbf{P}_3$, $\mathbf{P}_1$, and $\mathbf{P}_2$; then $\mathbf{y}^{*\prime} = (0, -3/5, 2/5)$ solves Problem II, as can easily be verified by the reader. The dual solution appears in the final tableau upon computing z_j for $j = 3, 5, 6$.

Another important result is established from Lemma 2.5. In particular, if any $z_j - c_j = 0$ in the final tableau, what can one say about Problem II, the dual problem? In view of Lemma 2.5 this question can be answered by the following two corollaries.

COROLLARY 2.5.1. *If all $z_j - c_j = 0$ in the final tableau, then all the constraints in the dual problem (Problem II) are binding.*

COROLLARY 2.5.2. *For any $z_j - c_j = 0$ in the final tableau, then the corresponding constraint in the dual problem is binding when evaluated at $\mathbf{y}^0$.*

2.8 Degeneracy and Cycling

The development presented so far in this chapter is under the assumption that no degenerate solutions exist; that is, all basic variables are positive. This assumption is necessary in order to show that the value of the objective function increases as one proceeds from one iteration to the next. If there exist degenerate solutions, and the leaving variable has a value of zero, then the updated value of the objective function will not increase, but only remain the same. In such situations, it could possibly happen that after a number of iterations with the objective function nonincreasing that a previous known extreme point is rediscovered. In such situations it is said that the simplex algorithm has "cycled"; and this implies that the algorithm might not converge in problems when degeneracy occurs.

An example is given below, from Beale (1955), which illustrates that cycling can occur and then a slight modification of the simplex method, so that the simplex algorithm converges to an optimal solution when one exists, is discussed.

Assume Tableau 0 is as follows where for computational convenience, the value of the objective function is associated with the vector $\mathbf{V}_0$.

Basis	$\mathbf{P}_0$	$\mathbf{P}_1$	$\mathbf{P}_2$	$\mathbf{P}_3$	$\mathbf{P}_4$	$\mathbf{P}_5$	$\mathbf{P}_6$	$\mathbf{P}_7$	$\mathbf{V}_0$
$\mathbf{V}_0$	0	$-3/4$	20	$-1/2$	6	0	0	0	1
$\mathbf{P}_5$	0	[1/4]	-8	-1	9	1	0	0	0
$\mathbf{P}_6$	0	[1/2]	-12	$-1/2$	3	0	1	0	0
$\mathbf{P}_7$	1	0	0	1	0	0	0	1	0

Then since $z_1 - c_1 = -3/4$, the vector $\mathbf{P}_1$ is selected to enter the basis. However, when comparing the quantities $0/(1/4)$ and $0/(1/2)$ that are associated with $\mathbf{P}_5$ and $\mathbf{P}_6$, there is no unique choice as to which vector should leave the basis. Selecting $\mathbf{P}_5$, we have

TABLEAU 1

Basis	$\mathbf{P}_0$	$\mathbf{P}_1$	$\mathbf{P}_2$	$\mathbf{P}_3$	$\mathbf{P}_4$	$\mathbf{P}_5$	$\mathbf{P}_6$	$\mathbf{P}_7$	$\mathbf{V}_0$
$\mathbf{V}_0$	0	0	-4	$-7/2$	33	3	0	0	1
$\mathbf{P}_1$	0	1	-32	-4	36	4	0	0	0
$\mathbf{P}_6$	0	0	[4]	$3/2$	-15	-2	1	0	0
$\mathbf{P}_7$	1	0	0	1	0	0	0	1	0

Note that in this tableau, the value of the objective function has remained at zero. In Tableau 1, $\mathbf{P}_2$ enters the basis, and the leaving vector $\mathbf{P}_6$ is unique since $\theta = 0/4$ with respect to row 2. Applying the simplex transformation we have

TABLEAU 2

Basis	$\mathbf{P}_0$	$\mathbf{P}_1$	$\mathbf{P}_2$	$\mathbf{P}_3$	$\mathbf{P}_4$	$\mathbf{P}_5$	$\mathbf{P}_6$	$\mathbf{P}_7$	$\mathbf{V}_0$
$\mathbf{V}_0$	0	0	0	-2	18	1	1	0	1
$\mathbf{P}_1$	0	1	0	[8]	-84	-12	8	0	0
$\mathbf{P}_2$	0	0	1	$3/8$	$-15/4$	$-1/2$	$1/4$	0	0
$\mathbf{P}_7$	1	0	0	1	0	0	0	1	0

Selecting $\mathbf{P}_3$ to enter the basis, we see that the minimum ratio θ is again not unique since $\theta = 0$ for rows 1 and 2. Selecting the first of these gives us

TABLEAU 3

Basis	$\mathbf{P}_0$	$\mathbf{P}_1$	$\mathbf{P}_2$	$\mathbf{P}_3$	$\mathbf{P}_4$	$\mathbf{P}_5$	$\mathbf{P}_6$	$\mathbf{P}_7$	$\mathbf{V}_0$
$\mathbf{V}_0$	0	1/4	0	0	−3	−2	3	0	1
$\mathbf{P}_3$	0	1/8	0	1	−21/2	−3/2	1	0	0
$\mathbf{P}_2$	0	−3/64	1	0	[3/16]	1/16	−1/8	0	0
$\mathbf{P}_7$	1	−1/8	0	0	21/2	3/2	−1	1	0

In this tableau, there are now two negative $z_j - c_j$ components in row 0. If $\mathbf{P}_4$ is selected to enter the basis, then $\mathbf{P}_2$ leaves the basis, giving us

TABLEAU 4

Basis	$\mathbf{P}_0$	$\mathbf{P}_1$	$\mathbf{P}_2$	$\mathbf{P}_3$	$\mathbf{P}_4$	$\mathbf{P}_5$	$\mathbf{P}_6$	$\mathbf{P}_7$	$\mathbf{V}_0$
$\mathbf{V}_0$	0	−1/2	16	0	0	−1	1	0	1
$\mathbf{P}_3$	0	−5/2	56	1	0	[2]	−6	0	0
$\mathbf{P}_4$	0	−1/4	16/3	0	1	1/3	−2/3	0	0
$\mathbf{P}_7$	1	5/2	−56	0	0	−2	6	1	0

Now $\mathbf{P}_5$ is selected to enter the basis, but again there is no unique non-negative quotient. We shall again choose the first of these $\mathbf{P}_3$ giving us

TABLEAU 5

Basis	$\mathbf{P}_0$	$\mathbf{P}_1$	$\mathbf{P}_2$	$\mathbf{P}_3$	$\mathbf{P}_4$	$\mathbf{P}_5$	$\mathbf{P}_6$	$\mathbf{P}_7$	$\mathbf{V}_0$
$\mathbf{V}_0$	0	−7/4	44	1/2	0	0	−2	0	1
$\mathbf{P}_5$	0	−5/4	28	1/2	0	1	−3	0	0
$\mathbf{P}_4$	0	1/6	−4	−1/6	1	0	[1/3]	0	0
$\mathbf{P}_7$	1	0	0	1	0	0	0	1	0

In Tableau 5, $\mathbf{P}_6$ enters the basis and $\mathbf{P}_4$ leaves. Applying the simplex transformation to this tableau yields our original tableau, Tableau 0.

Charnes (1952) has shown how cycling can be avoided by what is called a *perturbation* procedure. In the initial tableau the resource vector **b** is modified by adding a small number ϵ. In particular, the modified right-hand side is

$$b_1 + \epsilon$$
$$b_2 + \epsilon^2$$
$$b_3 + \epsilon^3$$
$$\vdots$$
$$b_m + \epsilon^m$$

This ensures that when the next vector leaves the basis the value of the objective function will increase at each iteration. After an optimal solution is reached using the modified right-hand side, then to update to the correct optimal solution one only has to multiply B^{-1} by the original right-hand vector, i.e.

$$\mathbf{x}^0 = B^{-1}\mathbf{P}_0$$

Applying this procedure to Tableau 0 we have that now the following ratios are compared: $\epsilon/(1/4)$ and $\epsilon^2/(1/2)$ for rows 1 and 2, respectively. Hence, $\mathbf{P}_6$ leaves the basis and we have

TABLEAU 1A

Basis	$\mathbf{P}_0$	$\mathbf{P}_1$	$\mathbf{P}_2$	$\mathbf{P}_3$	$\mathbf{P}_4$	$\mathbf{P}_5$	$\mathbf{P}_6$	$\mathbf{P}_7$	$\mathbf{V}_0$
$\mathbf{V}_0$	$3\epsilon^2/2$	0	2	$-5/4$	$21/2$	0	$3/2$	0	1
$\mathbf{P}_5$	$\epsilon - \epsilon^2/2$	0	-2	$-3/4$	$15/2$	1	$-1/2$	0	0
$\mathbf{P}_1$	$2\epsilon^2$	1	-24	-1	6	0	2	0	0
$\mathbf{P}_7$	$1 + \epsilon^3$	0	0	[1]	0	0	0	1	0

Here, $\mathbf{P}_3$ enters the basis and replaces $\mathbf{P}_7$ giving us

TABLEAU 2a

Basis	$\mathbf{P}_0$	$\mathbf{P}_1$	$\mathbf{P}_2$	$\mathbf{P}_3$	$\mathbf{P}_4$	$\mathbf{P}_5$	$\mathbf{P}_6$	$\mathbf{P}_7$	$\mathbf{V}_0$
$\mathbf{V}_0$	$5/4 + 3\epsilon^2/2 + 5\epsilon^3/4$	0	2	0	$21/2$	0	$3/2$	$5/4$	1
$\mathbf{P}_5$	$3/4 + \epsilon - \epsilon^2/2 + 3\epsilon^3/4$	0	-2	0	$15/2$	1	$-1/2$	$3/4$	0
$\mathbf{P}_1$	$1 + \epsilon^3 + 2\epsilon^2$	1	-24	0	6	0	2	1	0
$\mathbf{P}_3$	$1 + \epsilon^3$	0	0	1	0	0	0	1	0

Therefore, the optimal value of the objective function x_0^0 and optimal solution $\mathbf{x}^0$ is

$$\begin{bmatrix} x_0^0 \\ \mathbf{x}^0 \end{bmatrix} = \begin{bmatrix} 1 & 0 & 3/2 & 5/4 \\ 0 & 1 & -1/2 & 3/4 \\ 0 & 0 & 2 & 1 \\ 0 & 0 & 0 & 1 \end{bmatrix} \begin{bmatrix} 0 \\ 0 \\ 0 \\ 1 \end{bmatrix} = \begin{bmatrix} 5/4 \\ 3/4 \\ 1 \\ 1 \end{bmatrix}$$

In particular, the same solution is obtained by letting $\epsilon = 0$ in Tableau 2a.

The above discussion shows that if a linear program has an optimal solution and cycling does not occur, then there exists a finite sequence of pivot steps in the simplex algorithm. We have assumed in this presentation that at each iteration the minimum value of the quotients is positive. If, however, at some iteration the minimum positive quotient is not unique, then the next basic feasible solution will be degenerate. In practice, this situation has never been a problem. A number of alternative techniques exist in the literature which address the problem of determining a unique positive quotient (Dantzig 1951; Dantzig, Orden, and Wolfe 1954; and Gass 1969). We demonstrate below, using a modification of the above perturbation technique, that ties cannot occur when determining the minimum quotient.

Let us perturbate all the components of any extreme point solution by some positive ϵ. Then, as can be verified by the reader

$$\mathbf{x}^*(\epsilon) = B^{-1}\mathbf{P}_0 + \epsilon B^{-1}\mathbf{P}_1 + \epsilon^2 B^{-1}\mathbf{P}_2 + \cdots + \epsilon^{n+m} B^{-1}\mathbf{P}_{n+m} \tag{2.46}$$

Since we can always rearrange the vectors in our problem so that the first m vectors are basic, it follows that (2.46) can be expressed equivalently as

$$x_i^*(\epsilon) = x_i^* + \epsilon^i + \sum_{j=m+1}^{n} \epsilon^j x_{ij} \tag{2.47}$$

where $\mathbf{x}_i = B^{-1}\mathbf{P}_i$ for each i.

Clearly, for some small positive value of ϵ, it follows that $x_i^*(\epsilon)$ can be made positive for each i. Therefore, the minimum value of the positive quotients is determined for some vector, say ℓ, where

$$\frac{x_\ell^*(\epsilon)}{x_{\ell k}} = \min_i \frac{x_i(\epsilon)}{x_{\ell k}} = \frac{x_i^* + \epsilon^i + \sum_{j=m+1}^{n} \epsilon^j x_{ij}}{x_{\ell k}} \qquad \text{for } x_{\ell k} > 0$$

In view of the fact that $x_{\ell}^{*}(\epsilon)$ is the only variable involving ℓ, it follows from (2.47) that the minimum quotient is unique utilizing this procedure.

Exercises

1. Set up the model in Example 2.4 so that appropriate slack and surplus variables are added to the model under the assumption that the sponsor was interested in producing the show at minimum cost.
2. *a*) Does the tableau presented in Exercise 1 have an initial basic solution?
   b) How does this affect the procedure outlined in Section 2.4?
   c) A possible solution to the problems encountered above would be to add artificial variables to all the greater-than or equal-to constraints and give this variable a very large finite cost coefficient. In particular, if we are minimizing (maximizing), the cost coefficients would be M $(-M)$ where $M \gg 0$. (Note: What else is needed to make them basic?)
   d) Discuss the implications if, upon using the procedure in Section 2.4, an artificial variable appeared as a basic nonzero variable.
   e) What if a surplus or slack variable appeared in the final solution?
3. Solve Example 2.4 using the procedure outlined in Section 2.4 in view of Exercises 1 and 2 above.
4. With respect to the final tableau associated with Example 2.4, how could you identify and find another optimal solution?
5. Solve the following 2 problems graphically

$$\begin{array}{ll} \text{maximize} & x_1 + x_2 \quad \text{or minimize } -x_1 - x_2 \\ \text{subject to} & x_1 + 2x_2 \leq 3 \\ & 2x_1 + x_2 \leq 5 \\ & x_1, \quad x_2 \geq 0 \end{array}$$

   a) How do the two optimal solutions compare?
   b) How do the two optima compare?
6. Consider the general transportation problem described in Example 2.4 where the demand equals the supply. Show that in this situation the problem always has a feasible solution.
7. Would all optimal solutions for the class of problems in Exercise 6 have integer solutions? (Hint: Set up the initial tableau.)
8. Prove that if a vector is removed from the basis at some specified iteration, then it cannot reenter the basis at the next iteration.
9. Solve the following problem

$$\begin{array}{ll} \text{maximize} & 3x_1 + 5x_2 + 2x_3 \\ \text{subject to} & 2x_1 + 3x_2 + x_3 = 21 \\ & 3x_1 + x_2 + 4x_3 = 38 \\ & x_1, \quad x_2, \quad x_3 \geq 0 \end{array}$$

   Formulate a corresponding dual problem and identify its optimal solution.

10. Consider the following problems

a) maximize $x_1 + x_2$
subject to
$$\begin{aligned} 2x_1 + 3x_2 &\le 12 \\ |1x_2 - 1| &\le 3 \\ x_1, \quad x_2 &\ge 0 \end{aligned}$$

b) minimize $3x_1 - x_2$
subject to
$$\begin{aligned} 1 \le x_1 &\le 3 \\ 2 \le x_2 &\le 4 \\ 0 \le x_1 + x_2 &\le 6 \\ x_1, \quad x_2 &\ge 0 \end{aligned}$$

c) maximize $2x_1 + 3x_2$
subject to
$$\begin{aligned} x_1 - x_2 &= 0 \\ x_1 + x_2 &= 100 \\ x_1, \quad x_2 &\ge 0 \end{aligned}$$

Solve each of the above problems graphically and by the simplex method.

11. Use the simplex method to invert the following matrix

$$1/3 \begin{bmatrix} 1 & 1 & -1 \\ 1 & -2 & 2 \\ -1 & 2 & 1 \end{bmatrix}$$

12. Given is the following partial tableau at some stage of the simplex method. Assume the original constraints were all greater than or equal to and we are trying to find the optimal solution of a minimization problem.

c_B	Basis	V_0	P_1	P_2	P_3	P_4	P_5	P_6	λ^0
	V_0	1	0	8/3	−11	0	4/3	0	−8
−1	P_1	0	1	2/3	0	0	4/3	0	4
−3	P_4	0	0	−7/3	3	1	−2/3	0	2
1	P_6	0	0	−2/3	−2	0	2/3	1	2

If the inverse of the current basis is the matrix given in Exercise 11, then find the original tableau.

13. A farmer has 1,000 acres of land on which he can grow wheat or graze cattle. Assume that for a typical year, the farmer and his hired hands work 6,000 hours per year, and the farmer has $30,000 in working capital. Assume it takes a working capital of $10 per acre of wheat and $200 per head of cattle. One head of cattle requires 5 acres of grazing land. The average annual number of hours worked is 3 per acre of wheat planted and 12 per head of cattle. Assume the average net revenue of wheat is $15 per acre and $100 per head of cattle.

a) What is the farmer's optimal production plan?

b) Is the optimal solution determined in *a*) changed if the farmer also required that at least 100 acres of land be used to grow wheat?

c) Assume the net revenue per head of cattle increased to $150; would the solution determined in *a*) change? If so, determine the new solution.

14. Let $\Omega_I = \{\mathbf{x} \mid A\mathbf{x} = \mathbf{b}, \mathbf{x} \geq 0\}$ and $\Omega_{II} = \{\mathbf{y} \mid A'\mathbf{y} \geq \mathbf{c}\}$. Suppose that at some stage of the simplex procedure that $\mathbf{x}^* \in \Omega_I$ and $\mathbf{y}^{*\prime} = \mathbf{c}'_B B_*^{-1} \in \Omega_{II}$ where B_* is the present basic feasible matrix associated with $\mathbf{x}^*$.
   a) Show that $z_j - c_j \geq 0$ for all j.
   b) What does this imply about $\mathbf{x}^*$ and $\mathbf{y}^*$?

References

Arrow, K. J.; Hurwitz, L.; and Uzawa H. 1958. Studies in linear and nonlinear programming. *Stanford Math. Stud. Soc. Sci. II.* Stanford, Calif.

Beale, E. M. L. 1955. Cycling in the dual simplex algorithm. *Nav. Res. Logist. Q.* 2:269–76.

Charnes, A. 1952. Optimality and degeneracy in linear programming. *Econometrica* 20:160–70.

Charnes, A., and Cooper, W. W. 1961. *Management Models and Industrial Applications of Linear Programming.* Wiley, New York.

Charnes, A.; Cooper, W. W.; and Henderson, A. 1953. *An Introduction to Linear Programming.* Wiley, New York.

Cooper, L., and Steinberg, D. 1970. *Introduction to Methods of Optimization.* Saunders, Philadelphia.

Dantzig, G. B. 1951. Maximization of a linear function of variables subject to linear inequalities. In *Cowles Commission Monograph 13.* Ed. J. C. Koopmans. Wiley, New York.

———. 1963. *Linear Programming and Extensions.* Princeton Univ. Press, Princeton, N.J.

Dantzig, G. B.; Orden, A.; and Wolfe, P. 1954. Generalized simplex method for minimizing a linear form under linear inequality restraints. Rand Corp. Rept. RM-1264. Santa Monica, Calif.

Gass, S. L. 1969. *Linear Programming: Methods and Applications.* 3rd ed. McGraw-Hill, New York.

Hadley, G. 1962. *Linear Programming.* Reading, Mass.

Heady, E., and Chandler, W. 1964. *Linear Programming Methods.* Iowa State Univ. Press, Ames.

Hiller, F. S., and Lieberman, G. J. 1967. *Introduction to Operations Research.* Holden-Day, San Francisco.

Karlin, Samuel. 1959. *Mathematical Methods and Theory in Games, Programming, and Economics.* Reading, Mass.

Koopmans, T. C., ed. 1951. Activity analysis of production and allocation. In *Cowles Commission Monograph 13.* Wiley, New York.

Kwak, N. K. 1973. *Mathematical Programming with Business Applications.* McGraw-Hill, New York.

Van de Panne, C. 1971. *Linear Programming and Related Techniques.* American Elsevier, New York.

Wagner, H. M. 1970. *Principles of Management Science.* Prentice-Hall, Englewood Cliffs, N.J.

C H A P T E R   T H R E E

FURTHER COMPUTATIONAL ALGORITHMS AND TOPICS IN LINEAR PROGRAMMING

Some useful modifications of the simplex algorithm that reduce the number of computations and minimize the amount of computer memory needed to solve a typical linear programming problem are known as the revised simplex algorithm and the simple upper bounding method (see Lasdon 1970, Orchard-Hays 1968, and Van de Panne 1971). These techniques have proven useful in today's so-called linear programming computer packages.

A further feature of these procedures is that a considerable number of iterations are saved when solving a series of linear programming models similar in structure. Similar models are those linear programming problems with only small differences in the objective function and/or resource vector. Hence after one model is solved its optimal basis can be saved and used later to start the simplex procedure of another model near its optimal solution. This feature allows a researcher to analyze the model under various alternatives rather quickly and inexpensively.

In many applied problems, the unit cost or profit of an activity or available resource cannot be determined exactly; the model builder may not have perfect information on the coefficients in the model. For example, the price of a specific commodity might be a certain price today but a different price later. Under these conditions, one can analyze the model with the techniques covered in this chapter, commonly referred to as range or sensitivity analysis.

The final topic of this chapter is parametric programming, which deals with analyzing solutions to linear programming models when there exist certain relationships among unknown parameters. For example, if the profit of a certain commodity increases n cents per unit, then the profit of a related commodity will increase or decrease by some proportional amount. Parametric programming via the simplex algorithm can be used to analyze the solution of such problems.

3.1 Revised Simplex Algorithm

In the past two decades several algorithms have been developed to find optimal solutions of linear programming problems. The first of these, the simplex algorithm described earlier, was derived by Dantzig in 1947. Since then considerations have been made which reduce the number of computations necessary to obtain optimal solutions and minimize the storage needed to solve problems on a computer. The revised simplex is one such technique. Other techniques are discussed by Cooper and Steinberg (1970), Dantzig and Van Slyke (1967), Hirshfeld (1972), Lasdon (1970), and Orchard-Hays (1968). For convenience the statement of the problem is changed to

$$\begin{aligned} \text{maximize} \quad & z \\ \text{subject to} \quad & z - \mathbf{c}'\mathbf{x} = 0 \\ & A\mathbf{x} = \mathbf{b} \\ & \mathbf{x} \geq 0 \end{aligned} \tag{3.1}$$

after adding appropriate slack and surplus variables.

There are now $n + 1$ variables, $m + 1$ restrictions, and the basis is of rank $m + 1$. We can rewrite (3.1) equivalently as

$$\begin{aligned} \text{maximize} \quad & z \\ \text{subject to} \quad & \left[\begin{array}{c|c} 1 & -\mathbf{c}' \\ \hline 0 & A \end{array}\right] \begin{bmatrix} z \\ \mathbf{x} \end{bmatrix} = \begin{bmatrix} 0 \\ \mathbf{b} \end{bmatrix} \\ & \mathbf{x} \geq 0 \end{aligned} \tag{3.2}$$

If we arrange the columns of A so that the first m columns form a basis of rank m, then a basis of rank $m + 1$ is

$$B = \left[\begin{array}{c|c} 1 & -\mathbf{c}_B' \\ \hline 0 & B_m \end{array}\right] \tag{3.3}$$

where $\mathbf{c}_B$ is an m-vector in which the jth component is the original profit coefficient of the basic variable associated with the jth row.

The inverse of B can be easily verified to be

$$B^{-1} = \left[\begin{array}{c|c} 1 & \mathbf{c}_B' B_m^{-1} \\ \hline 0 & B_m^{-1} \end{array}\right] \tag{3.4}$$

Knowing B^{-1} we can easily determine the next tableau. We have

$$B^{-1}\left[\begin{array}{c|c|c} 1 & -\mathbf{c}' & 0 \\ \hline 0 & A & \mathbf{b} \end{array}\right] = \left[\begin{array}{c|cccc|c} 1 & z_1 - c_1 & z_2 - c_2 & \cdots & & \mathbf{c}'_B\mathbf{x}^0 \\ \hline 0 & \mathbf{x}_1 & \mathbf{x}_2 & \cdots & & \mathbf{x}^0 \end{array}\right] \tag{3.5}$$

where

$$B_m^{-1}\overline{\mathbf{P}}_j = \mathbf{x}_j \tag{3.6}$$

$$z_j = \mathbf{c}'_B\mathbf{x}_j = \mathbf{c}'_B B_m^{-1}\overline{\mathbf{P}}_j \tag{3.7}$$

Here, $\overline{\mathbf{P}}_j$ denotes the jth column of A and $\mathbf{P}_j$ denotes the jth column of (3.2).

Therefore, the jth column of (3.2) can be transformed by B^{-1} to

$$B^{-1}\mathbf{P}_j = B^{-1}\begin{bmatrix} -c_j \\ \overline{\mathbf{P}}_j \end{bmatrix}$$

$$= \begin{bmatrix} 1 & \mathbf{c}'_B B_m^{-1} \\ 0 & B_m^{-1} \end{bmatrix}\begin{bmatrix} -c_j \\ \overline{\mathbf{P}}_j \end{bmatrix}$$

$$= \begin{bmatrix} -c_j + \mathbf{c}'_B B_m^{-1}\overline{\mathbf{P}}_j \\ B_m^{-1}\overline{\mathbf{P}}_j \end{bmatrix} = \begin{bmatrix} z_j - c_j \\ \mathbf{x}_j \end{bmatrix} \tag{3.8}$$

and

$$B^{-1}\begin{bmatrix} 0 \\ \mathbf{b} \end{bmatrix} = \begin{bmatrix} \mathbf{c}'_B\mathbf{x}^0 \\ \mathbf{x}^0 \end{bmatrix} = \begin{bmatrix} z^0 \\ \mathbf{x}^0 \end{bmatrix} \tag{3.9}$$

This implies that if we know the original components of A, $\mathbf{b}$, $\mathbf{c}$, and any B^{-1} we have all the important information needed to determine which vectors enter and leave the basis. Moreover, the remaining values of this tableau do not need to be calculated.

From (3.8), the most negative $z_j - c_j$ is determined by multiplying the first row of B^{-1} by each $\begin{bmatrix} -c_j \\ \overline{\mathbf{P}}_j \end{bmatrix}$ defined in (3.2), namely

$$(B^{-1})_0 \begin{bmatrix} -c_j \\ \overline{\mathbf{P}}_j \end{bmatrix} = z_j - c_j \tag{3.10}$$

The question of how to determine or calculate B^{-1} remains. Recall that in the system $[B \mid I]$ the inverse of B can be computed using Gauss-elimination; that is, by elementary row operations the elements of B can be transformed to the identity matrix

$$[B \mid I] \rightarrow [I \mid D]$$

In doing this the by-product matrix D is B^{-1}, i.e.

$$B^{-1}[B \mid I] = [I \mid B^{-1}]$$

Let us consider any tableau and suppose we pivot on x_{rk}. Let y_{ij} be the (ij)th element of B^{-1}, the present inverse. After the rth row is divided by the pivot element, the kth vector of A and the jth vector of B^{-1} will be transformed accordingly

$$\begin{bmatrix} \vdots \\ x_{ik} \\ \vdots \\ x_{rk} \\ \vdots \end{bmatrix} \longrightarrow \begin{bmatrix} \vdots \\ x_{ik} \\ \vdots \\ 1 \\ \vdots \end{bmatrix} \tag{3.11}$$

$$\begin{bmatrix} \vdots \\ y_{ij} \\ \vdots \\ y_{rj} \\ \vdots \end{bmatrix} \longrightarrow \begin{bmatrix} \vdots \\ y_{ij} \\ \vdots \\ y_{rj}/x_{rk} \\ \vdots \end{bmatrix} \tag{3.12}$$

To complete the simplex transformation the components x_{ik} for $i = 1, 2, \ldots, m$ $(i \neq r)$ must be eliminated. In so doing the corresponding components of (3.12) for $i \neq r$ are changed to

$$Y_{ij} = -(y_{rj}/x_{rk})\,x_{ik} + y_{ij} \tag{3.13}$$

Therefore in the new inverse the jth column is updated as follows

$$Y_{ij} = y_{ij} - (x_{ik}/x_{rk})\,y_{rj} \qquad i \neq r \tag{3.13a}$$

$$Y_{rj} = (1/x_{rk})\,y_{rj} \tag{3.13b}$$

This implies that there exists some matrix E such that

$$EB_y^{-1} = B_Y^{-1} \tag{3.14}$$

Letting the matrix E be of the following form

$$\begin{bmatrix} 1 & 0 & \cdots & 0 & & 0 & \cdots & 0 \\ 0 & 1 & & \vdots & & & \cdots & 0 \\ & & \ddots & 0 & \eta_t & 0 & \cdots & \vdots \\ \vdots & & \cdots & 1 & & & & \\ & & & 0 & & 1 & \cdots & \\ & & & \vdots & & & \ddots & \\ 0 & 0 & \cdots & & & 0 & & 1 \end{bmatrix} \tag{3.15}$$

then in view of (3.15), the jth column of B_y^{-1} can be transformed to (3.13a) and (3.13b) if and only if

$$\eta_t = \frac{1}{x_{rk}} \begin{bmatrix} -x_{0k} \\ -x_{1k} \\ \vdots \\ 1 \\ -x_{r+1,k} \\ \vdots \end{bmatrix} = \frac{1}{x_{rk}} \begin{bmatrix} -(z_k - c_k) \\ -x_{1k} \\ \vdots \\ 1 \\ -x_{r+1,k} \\ \vdots \end{bmatrix} \tag{3.16}$$

If B is the initial identity matrix in tableau 0, then $B_0 = I = B_0^{-1}$. Furthermore,

$$\begin{aligned} B_t^{-1} &= E_t(B_{t-1}^{-1}) \\ &= E_t E_{t-1}(B_{t-2}^{-1}) \\ &\ \vdots \\ &= E_t E_{t-1} E_{t-2} \cdots E_1 I \end{aligned} \tag{3.17}$$

Therefore a history of E_t matrices is sufficient to calculate B_t^{-1} for any t. Notice that

$$E_t = [\mathbf{e}_0, \mathbf{e}_1, \ldots, \mathbf{e}_{r-1}, \eta_t, \mathbf{e}_{r+1}, \ldots, \mathbf{e}_m]$$

where $\mathbf{e}_i$ is a unit vector with 1 in the $(i + 1)$ component. Hence E_t is easily reconstructed from an explicit knowledge of η_t and r, and we need know only $(\eta_1, \eta_2, \ldots, \eta_t)$ and $(r_1, r_2, \ldots, r_t)$ to calculate B_t^{-1}.

EXAMPLE 3.1

Converting the following problem

$$\begin{array}{ll} \text{maximize} & 3x_1 + 5x_2 \\ \text{subject to} & x_1 \qquad\quad \le 4 \\ & 3x_1 + 2x_2 \le 18 \\ & x_1, \quad x_2 \ge 0 \end{array}$$

to tableau form, we have, after introducing two slack variables x_3^s and x_4^s

$$\begin{array}{llllllll} \text{Row 0} & z_0 & - 3x_1 & - 5x_2 & & & = & 0 \\ \text{Row 1} & & x_1 & & + x_3^s & & = & 4 \\ \text{Row 2} & & 3x_1 & + 2x_2 & & + x_4^s & = & 18 \end{array}$$

Since $z_2 - c_2 = -5$, the vector associated with x_2 enters the basis and the vector associated with x_4^s leaves the basis, giving us the final tableau

$$\begin{array}{llllllll} \text{Row 0} & z_0 & + (9/2)x_1 & & & + (5/2)x_4^s & = & 45 \\ \text{Row 1} & & x_1 & & + x_3^s & & = & 4 \\ \text{Row 2} & & (3/2)x_1 & + x_2 & & (1/2)x_4^s & = & 9 \end{array}$$

In this example

$$\eta_1 = \frac{1}{2}\begin{bmatrix} -(-5) \\ -0 \\ 1 \end{bmatrix} = \begin{bmatrix} 5/2 \\ 0 \\ 1/2 \end{bmatrix}$$

since the pivot element is 2 and η_1 is constructed from the vector corresponding to $\mathbf{P}_2$. Hence

$$E = \begin{bmatrix} 1 & 0 & 5/2 \\ 0 & 1 & 0 \\ 0 & 0 & 1/2 \end{bmatrix}$$

where η_1 is positioned in the rth column of E and r is the basic row of the leaving variable. In this case $E = B_1^{-1}$, which corresponds to vectors $\mathbf{V}_0$, $\mathbf{P}_4$, and $\mathbf{P}_2$ in the final tableau.

EXAMPLE 3.2

Assume we have the following tableau.

	$\mathbf{V}_0$	$\mathbf{P}_1$	$\mathbf{P}_2$	$\mathbf{P}_3$	$\mathbf{P}_4$	$\mathbf{P}_5$	$\mathbf{b}$
Row 0	1	1	−3	2	0	0	0
Row 1	0	3	−1	2	1	0	7
Row 2	0	−2	4	0	0	1	12

FIRST ITERATION

Step I. Initially $B_0 = I_3$.

$$z_j - c_j = (B_0^{-1})_0 \begin{bmatrix} -c_j \\ \overline{\mathbf{P}}_j \end{bmatrix} \quad \text{for } j = 1, 2, 3$$

$$= [1, 0, 0] \begin{bmatrix} 1 & -3 & 2 \\ 3 & -1 & 2 \\ -2 & 4 & 0 \end{bmatrix} = (1, -3, 2)$$

Hence the vector associated with x_2 enters the basis since the only negative $z_j - c_j$ component is $z_2 - c_2$.

Step II. Next we determine θ_r, the minimum nonnegative ratio of the right-hand coefficients to the positive components of the updated vector $\mathbf{P}_2$. Thus

$$B_0^{-1} \begin{bmatrix} -c_2 \\ \overline{\mathbf{P}}_2 \end{bmatrix} = \begin{bmatrix} -3 \\ -1 \\ 4 \end{bmatrix} \qquad B_0^{-1} \begin{bmatrix} 0 \\ 7 \\ 12 \end{bmatrix} = \begin{bmatrix} 0 \\ 7 \\ 12 \end{bmatrix}$$

Therefore $\theta_r = \min(12/4)$ with $r = 2$, and the pivot element is 4.

Step III.

$$\eta_1 = \frac{1}{4}\begin{bmatrix} -(-3) \\ -(-1) \\ 1 \end{bmatrix} = \begin{bmatrix} 3/4 \\ 1/4 \\ 1/4 \end{bmatrix}$$

so that

$$E_1 = \begin{bmatrix} 1 & 0 & 3/4 \\ 0 & 1 & 1/4 \\ 0 & 0 & 1/4 \end{bmatrix}$$

and $B_1^{-1} = E_1 B_0^{-1} = E_1$. This completes one iteration.

SECOND ITERATION

Step I. Now x_1, x_3, and x_5 are nonbasic, and we have the following consequences

$$(B_1^{-1})_0 \begin{bmatrix} -c_1 \\ \overline{\mathbf{P}}_1 \end{bmatrix} = [1, 0, 3/4] \begin{bmatrix} 1 \\ 3 \\ -2 \end{bmatrix} = -1/2 \quad \text{for } j = 1$$

$$(B_1^{-1})_0 \begin{bmatrix} -c_3 \\ \overline{\mathbf{P}}_3 \end{bmatrix} = 2 \quad \text{for } j = 3$$

$$(B_1^{-1})_0 \begin{bmatrix} -c_5 \\ \overline{\mathbf{P}}_5 \end{bmatrix} = 3/4 \quad \text{for } j = 5$$

implying that $\mathbf{P}_1$ enters the basis.

Step II.

$$B_1^{-1} \begin{bmatrix} -c_1 \\ \overline{\mathbf{P}}_1 \end{bmatrix} = \begin{bmatrix} 1 & 0 & 3/4 \\ 0 & 1 & 1/4 \\ 0 & 0 & 1/4 \end{bmatrix} \begin{bmatrix} 1 \\ 3 \\ -2 \end{bmatrix} = \begin{bmatrix} -1/2 \\ 5/2 \\ -1/2 \end{bmatrix}$$

and the updated right-hand side is

$$B_1^{-1}\mathbf{P}_0 = \begin{bmatrix} 9 \\ 10 \\ 3 \end{bmatrix}$$

giving $\theta_r = \theta_1 = 4$.

Step III.

$$\eta_2 = \frac{2}{5}\begin{bmatrix} -(-1/2) \\ 1 \\ -(-1/2) \end{bmatrix} = \begin{bmatrix} 1/5 \\ 2/5 \\ 1/5 \end{bmatrix} \qquad E_2 = \begin{bmatrix} 1 & 1/5 & 0 \\ 0 & 2/5 & 0 \\ 0 & 1/5 & 1 \end{bmatrix}$$

Therefore

$$B_2^{-1} = E_2 B_1^{-1}$$

$$= \begin{bmatrix} 1 & 1/5 & 0 \\ 0 & 2/5 & 0 \\ 0 & 1/5 & 1 \end{bmatrix}\begin{bmatrix} 1 & 0 & 3/4 \\ 0 & 1 & 1/4 \\ 0 & 0 & 1/4 \end{bmatrix} = \begin{bmatrix} 1 & 1/5 & 4/5 \\ 0 & 2/5 & 1/10 \\ 0 & 1/5 & 3/10 \end{bmatrix}$$

This completes the second iteration.

Third Iteration

Step I. The nonbasic variables are x_3, x_4, and x_5. Hence

$$(B_2^{-1})_0 \begin{bmatrix} -c_3 \\ \bar{\mathbf{P}}_3 \end{bmatrix} = [1, 1/5, 4/5] \begin{bmatrix} 2 \\ 2 \\ 0 \end{bmatrix} = 12/5 \quad \text{for } j = 3$$

$$(B_2^{-1})_0 \begin{bmatrix} -c_4 \\ \bar{\mathbf{P}}_4 \end{bmatrix} = 1/5 \quad \text{for } j = 4$$

$$(B_2^{-1})_0 \begin{bmatrix} -c_5 \\ \bar{\mathbf{P}}_5 \end{bmatrix} = 4/5 \quad \text{for } j = 5$$

Since all $z_j - c_j$ are nonnegative, we have an optimal solution given by

$$\begin{bmatrix} z^0 \\ \mathbf{x}^0 \end{bmatrix} = B_2^{-1}\begin{bmatrix} 0 \\ \mathbf{b} \end{bmatrix} = \begin{bmatrix} 1 & 1/5 & 4/5 \\ 0 & 2/5 & 1/10 \\ 0 & 1/5 & 3/10 \end{bmatrix}\begin{bmatrix} 0 \\ 7 \\ 12 \end{bmatrix} = \begin{bmatrix} 11 \\ 4 \\ 5 \end{bmatrix}$$

3.2 Solutions of Similar Problems

In many linear programming problems the model builder does not know the exact values of the coefficients in the objective function or in the resource vector. The researcher may want to consider solutions with different objective functions and right-hand vectors without completely resolving each model. In this section we discuss how one can solve similar problems without returning to the first step of the simplex algorithm.

Consider the following similar problems

$$\begin{aligned} \text{maximize} \quad & \mathbf{c}_t'\mathbf{x} \qquad t = 1, 2, \ldots, n^* \\ \text{subject to} \quad & A\mathbf{x} = \mathbf{b}_s \qquad s = 1, 2, \ldots, m^* \\ & \mathbf{x} \geq 0 \end{aligned} \tag{3.18}$$

where appropriate slack and surplus variables have been incorporated into the constraint equations.

Assume that we have solved the problem

$$\begin{aligned} \text{maximize} \quad & \mathbf{c}_1'\mathbf{x} \\ \text{subject to} \quad & A\mathbf{x} = \mathbf{b}_1 \\ & \mathbf{x} \geq 0 \end{aligned} \tag{3.19}$$

After solving (3.19), we know B^{-1} and $\mathbf{x}^0$.

Now can we quickly obtain an optimal solution to the following problem?

$$\begin{aligned} \text{maximize} \quad & \mathbf{c}_2'\mathbf{x} \\ \text{subject to} \quad & A\mathbf{x} = \mathbf{b}_1 \\ & \mathbf{x} \geq 0 \end{aligned} \tag{3.20}$$

In this situation, $\mathbf{x}^0$ is a feasible solution of (3.20), because if $\mathbf{x}^0$ is an optimal solution of (3.19) it must necessarily satisfy the constraints in (3.20).

The vector $\mathbf{x}^0$ is also an optimal solution of (3.20) if $z_j - c_j \geq 0$ for all j. Using $(B^{-1})_0$ we have

$$(B^{-1})_0 \begin{bmatrix} -c_{2j} \\ \overline{\mathbf{P}}_j \end{bmatrix} = z_j - c_{2j} \qquad \text{for all } j \tag{3.21}$$

The first entry in $\mathbf{P}_j$ is now $-c_{2j}$, the new objective function coefficient. After row 0 has been updated, it is not necessarily true that all $z_j - c_j = 0$ for the j basic variables (a basic vector may possess two nonzero entries), hence the tableau is not in canonical form. To obtain canonical form we multiply the appropriate basic row by the negative of this nonzero quantity and add this row to row 0, the objective function row. After the tableau is in canonical form, $\mathbf{x}^0$ is still an optimal solution if and only if all the updated coefficients in row 0 for all j nonbasic variables are greater than or equal to zero. Otherwise we must continue the iteration steps of the simplex procedure. This technique not only saves us from recalculating many unneeded tableaux but, as we shall see later, can give us almost immediately the optimal solution of (3.20).

Let us now consider the following situation. After (3.19) is solved, can one introduce a new resource vector $\mathbf{b}_2$ and in some way solve the new problem using B^{-1}? In this case, since the objective function coefficients have not changed, then necessarily $z_j - c_j \geq 0$ for all j. However, it can happen that premultiplying B^{-1} by $\mathbf{b}_2$ causes the solution to be infeasible; in particular, $B^{-1}\mathbf{b}_2 = \mathbf{x}^0 \ngeq 0$.

Many procedures can be used to obtain the proper form or, alternately, to solve any linear programming problem. Two of these methods are the dual simplex algorithm, and the pseudo-objective function technique described in Chapter 2.

3.2.1 *Dual Simplex Algorithm*

This algorithm is basically the work of C. E. Lemke (1954) and is very efficient when adding constraints to a linear programming problem in which one has already determined an optimal solution, or when one has introduced a new resource vector $\mathbf{b}^*$ in which $\mathbf{x}^0 = B_y^{-1}\mathbf{b}^* \ngeq 0$.

In the original simplex algorithm we start tableau 0 with a basic feasible solution, but where all $z_j - c_j$ are not necessarily nonnegative. We then make changes in the basis, one at a time, maintaining nonnegativity of the variables until all $z_j - c_j$ components are nonnegative, at which point we have an optimal feasible solution. In the dual simplex algorithm we start with a basic solution that is not feasible, but where all $z_j - c_j$ components are nonnegative. We make changes in the basis, keeping all $z_j - c_j$ nonnegative, until we have a feasible solution; at which point we have an optimal feasible solution.

The dual simplex algorithm suggests the following two operations:

Criterion I. If there are negative basic variables in the present solution, select the most negative to leave the basic solution (this determines the pivot row).

Criterion II. Select the nonbasic variable to enter the basic solution by taking the ratio of the cost row coefficients to the pivot row coefficients in each nonbasic vector (ignore positive and zero denominators). We then select the vector associated with the largest algebraic ratio.

EXAMPLE 3.3

Consider the following tableau

	$\mathbf{V}_0$	$\mathbf{P}_1$	$\mathbf{P}_2$	$\mathbf{P}_3$	$\mathbf{P}_4$	$\mathbf{P}_5$	$\mathbf{b}$
Row 0	1	2	5	0	0	0	20
Row 1	0	−1	1	1	0	0	−4
Row 2	0	−2	−3	0	1	0	−8
Row 3	0	1	0	0	0	1	5

In view of Criterion I, $\mathbf{P}_4$ leaves the basis since $b_2 = -8$. Taking ratios, we have that $-1 = \max\{2/-2, 5/-3\}$; hence we pivot on the element -2 and bring $\mathbf{P}_1$ into the solution. This gives us the following tableau

	$\mathbf{V}_0$	$\mathbf{P}_1$	$\mathbf{P}_2$	$\mathbf{P}_3$	$\mathbf{P}_4$	$\mathbf{P}_5$	$\mathbf{b}$
Row 0	1	0	2	0	1	0	12
Row 1	0	0	5/2	1	−1/2	0	0
Row 2	0	1	3/2	0	−1/2	0	4
Row 3	0	0	−3/2	0	1/2	1	1

Since all $z_j - c_j \geq 0$ and $\mathbf{x}^0 \geq 0$, our solution is feasible and moreover optimal.

3.3 Sensitivity Analysis

3.3.1 The Range on the Cost (profit) Vector

In the previous sections we discussed how we could solve similar problems using B^{-1} to adjust the new problem close to an optimal solution. In many cases we might ask how much a profit coefficient c_ℓ can vary without affecting our solution? In particular, over what range can c_ℓ vary without altering the present basic solution? This question is frequently asked since c_ℓ can be viewed as the profit per unit of a specific commodity that has a certain value this year and another value next year. Then we are asking if the optimal policy we determined last year is still the optimal policy this year.

Let us consider the ℓth component of $\mathbf{c}$ which is changed to $c_\ell + \alpha$ for some α, i.e.

$$c_j^* = c_j \qquad c_\ell^* = c_\ell + \alpha \qquad j \neq \ell \tag{3.22}$$

To determine the range on c_ℓ, let us consider two cases: ℓ corresponds to (1) a nonbasic variable or (2) a basic variable.

Case 1. The variable x_ℓ is nonbasic. In this situation

$$z_j - c_j^* \geq 0 \quad \text{for all } j \neq \ell \tag{3.23}$$

This is true since $z_j = \mathbf{c}_B^{*\prime} B_m^{-1} \overline{\mathbf{P}}_j = \mathbf{c}_B' B_m^{-1} \overline{\mathbf{P}}_j$ and $c_j^* = c_j$ for $j \neq \ell$. Hence $z_j - c_j^*$ in expression (3.23) has not been altered by changing c_ℓ.

If $c_\ell^* = c_\ell + \alpha$ is such that

$$z_\ell - c_\ell^* \geq 0 \tag{3.24}$$

then we still have an optimal solution. Now from (3.24), α must be such that

$$z_\ell - (c_\ell + \alpha) \geq 0$$

This implies that $-\alpha \geq -z_\ell + c_\ell$, or

$$\alpha \leq z_\ell - c_\ell \tag{3.25}$$

Thus if α is chosen such that

$$\alpha \in (-\infty, z_\ell - c_\ell] \tag{3.26}$$

our previous basic solution is still optimal.

It should be noted that x_ℓ will enter the solution when $\alpha \geq z_\ell - c_\ell$.

Case 2. The variable x_ℓ is basic. Changing c_ℓ to $c_\ell + \alpha$ in tableau 0 and then updating this tableau by multiplying it by B_{opt}^{-1}, we have

$$B_{\text{opt}}^{-1} \begin{bmatrix} -c_0 & -c_1 & \cdots & -c_\ell^* & \cdots & 0 \\ \overline{\mathbf{P}}_0 & \overline{\mathbf{P}}_1 & \cdots & \overline{\mathbf{P}}_\ell & \cdots & \mathbf{b} \end{bmatrix} \qquad B_{\text{opt}}^{-1} = \begin{bmatrix} 1 & \mathbf{c}_B' B_m^{-1} \\ 0 & B_m^{-1} \end{bmatrix}$$

Again, $z_j - c_j \geq 0$ for all $j \neq \ell$ since the only change from the original tableau is c_ℓ^*. For $j = \ell$, $z_j - c_j^* = z_j - c_j - \alpha = -\alpha$.

The updated tableau is not in canonical form since the ℓth column is now of the form

$$\begin{bmatrix} -\alpha \\ 0 \\ 0 \\ \vdots \\ 1 \\ 0 \\ \vdots \\ 0 \end{bmatrix}$$

To obtain canonical form we multiply the basic row by α and add this row to the objective function row. If all the new coefficients in row 0 are nonnegative, the previous basic solution is still optimal and the only change is the value of the optimum.

Let us determine the range of values that c^*_ℓ can assume while preserving optimality. For any given α and basic row i', the new updated $z_j - c_j$ coefficients for the nonbasic variables NB will be

$$(z_j - c_j) + \alpha(x_{i'j}) \tag{3.27}$$

The basic $z_j - c_j$ are still zero, hence it suffices to consider only the updated nonbasic columns.

If $x_{i'j} < 0$, then (3.27) remains nonnegative when

$$\alpha x_{i'j} \geq -(z_j - c_j) \quad \text{for } j \text{ nonbasic}$$
$$\alpha \leq -(z_j - c_j)/x_{i'j}$$

This implies that the maximum value α can have without changing the present basic solution is the greatest lower bound (glb) of the above inequalities, i.e.

$$\alpha_{\max} = \text{glb}\{-(z_j - c_j)/x_{i'j} \mid x_{i'j} < 0, j \in \text{NB}\} \tag{3.28}$$

If $x_{i'j} > 0$, then (3.27) remains nonnegative when

$$\alpha x_{i'j} \geq -(z_j - c_j) \quad \text{for } j \text{ nonbasic}$$

or $\alpha_{min} \geq -(z_j - c_j)/x_{i'j}$. Hence the maximum decrease in c_ℓ occurs when

$$\alpha_{min} = \text{lub}\{-(z_j - c_j)/x_{i'j} \mid x_{i'j} > 0, j \in \text{NB}\} \tag{3.29}$$

Here, lub is the least upper bound.

EXAMPLE 3.4

Consider again the problem

$$\begin{array}{ll} \text{maximize} & 3x_1 + 5x_2 \\ \text{subject to} & x_1 \leq 4 \\ & 3x_1 + 2x_2 \leq 18 \\ & x_1, \quad x_2 \geq 0 \end{array}$$

where in the final solution we have the equivalent formulation

$$\begin{array}{lll} \text{Row 0} & z_0 + (9/2)x_1 \qquad\qquad + (5/2)x_4 & = 45 \\ \text{Row 1} & x_1 \qquad + x_3 & = 4 \\ \text{Row 2} & (3/2)x_1 + x_2 \qquad + (1/2)x_4 & = 9 \end{array}$$

Assume we want to determine the range on c_1. Since x_1 is nonbasic, from (3.26) the maximum value α can assume is $z_1 - c_1 = 9/2$. Hence the range on c_1 is

$$c_1^* \in (-\infty, c_1 + 9/2] = (-\infty, 15/2]$$

To determine the range on c_2 we note that x_2 is basic. Now

$$\mathbf{x}_1 = \begin{bmatrix} x_{11} \\ x_{21} \end{bmatrix} = \begin{bmatrix} 1 \\ 3/2 \end{bmatrix} \qquad \mathbf{x}_4 = \begin{bmatrix} x_{14} \\ x_{24} \end{bmatrix} = \begin{bmatrix} 0 \\ 1/2 \end{bmatrix}$$

Since x_{21} and x_{24} are not negative, from (3.28) α can be any large value.

Next we determine the smallest value c_2^* can assume. From (3.29)

$$\begin{aligned} \alpha &= \text{lub}\{-(z_j - c_j)/x_{i'j} \mid x_{i'j} > 0, j = 1 \text{ or } 4\} \\ &= \text{lub}\{(-9/2)/(3/2), (-5/2)/(1/2), j = 1 \text{ or } 4\} = -3 \end{aligned}$$

Therefore, $c_2^* \in [c_2 - 3, \infty) = [2, \infty)$. Since $\alpha = -3$ for $j = 1$, this implies that x_1 will enter the solution for $c_2^* \leq 2$. To verify that x_1 will indeed enter the solution when $c_2^* \leq 2$, let us consider the updated tableau as specified by incorporating this lower limit. We have

	V_0	P_1	P_2	P_3	P_4	P_5
Row 0	1	9/2	3	0	5/2	45
Row 1	0	1	0	1	0	4
Row 2	0	3/2	1	0	1/2	9

where only the value of $z_2 - c_2$ has been changed. Putting this tableau into canonical form

	V_0	P_1	P_2	P_3	P_4	b
Row 0	1	0	0	0	1	18
Row 1	0	1	0	1	0	4
Row 2	0	3/2	1	0	1/2	9

Clearly, the optimal solution is still $x_2^0 = 9$, $x_3^0 = 4$, and $x_1^0 = 0 = x_4^0$. However, let us take a close look at x_1. Component $z_1 - c_1 = 0$, implying that if x_1 enters the solution, the optimum is still 18 with an optimal solution $x_1^0 = 4$, $x_2^0 = 3$, and $x_3^0 = 0 = x_4^0$. Hence when $c_2^* = 2$, we can identify an alternative optimal solution or, better yet, many alternative solutions since any convex combination of these two solutions is an optimal solution. Thus we see that if there exist more than m components $z_j - c_j = 0$ in the optimal tableau, there will exist multiple solutions, and when the number of $z_j - c_j = 0$ components is equal to m there is only one solution to our linear programming problem.

It is left as an exercise to show that when c_2^* takes a value below 2 its corresponding $z_j - c_j$ will be negative, implying that we must continue the simplex algorithm with x_1 entering the solution.

The discussion so far has been based on determining the range of the objective function coefficients so that the present *basic* solution remains optimal. One should not confuse this with determining the range on these coefficients so that the solution remains optimal. In particular, consider the following problem

$$\begin{array}{ll} \text{maximize} & -3x_1 + 3x_2 \\ \text{subject to} & 2x_1 + 3x_2 \leq 18 \\ & -2x_1 + 2x_2 \leq 12 \\ & x_1, \quad x_2 \geq 0 \end{array}$$

Our final tableau is

	$\mathbf{V}_0$	$\mathbf{P}_1$	$\mathbf{P}_2$	$\mathbf{P}_3$	$\mathbf{P}_4$	$\mathbf{b}$
Row 0	1	0	0	0	3/2	18
Row 1	0	5	0	1	−3/2	0
Row 2	0	−1	1	0	1/2	6

where $\mathbf{P}_2$ and $\mathbf{P}_3$ are in the basis. In this situation the ranges on c_1 and c_2 are $(-\infty, -3]$ and $[0, 3]$, respectively, as the reader can easily verify.

However, one could argue that c_2 can be any large value and x_2^0 will still be 6 with the other variables zero, which is true. The point is that when $c_2 \geq 3$ we could obtain a different basic solution even though the solution is the same in this case, i.e., x_1 and x_2 or x_2 and x_4 could be basic.

This example also illustrates another important point, namely, the range determined is not necessarily unique. In particular, when alternative basic solutions exist there may be different range analyses. For example, if $\mathbf{P}_4$ had been introduced in the basis so that $\mathbf{P}_3$ would not be in the final tableau, we would have

	$\mathbf{V}_0$	$\mathbf{P}_1$	$\mathbf{P}_2$	$\mathbf{P}_3$	$\mathbf{P}_4$	$\mathbf{b}$
Row 0	1	5	0	1	0	18
Row 1	0	2/3	1	1/3	0	6
Row 2	0	−10/3	0	−2/3	1	0

Here, the ranges on c_1 and c_2 are $(-\infty, 2]$ and $[0, \infty)$, respectively. Note that the solution is still the same as obtained previously. We see that a range analysis applies only to a specific basis, and if alternative solutions exist, the range analyses could differ.

3.3.2 A Change in the Resource Vector

Assume the resource vector has its rth component changed. In particular, let $b_r^* = b_r + \alpha$. To verify optimality we need only multiply this vector by B_{opt}^{-1}, where the subscript opt denotes optimal, in particular

$$B_{\text{opt}}^{-1}\mathbf{b}^* = B_{\text{opt}}^{-1}(\mathbf{b} + \begin{bmatrix} 0 \\ \vdots \\ \alpha \\ 0 \\ \vdots \\ 0 \end{bmatrix})$$

$$= \mathbf{x}^0 + \alpha(B_{\text{opt}}^{-1})^r$$

where superscript r denotes the rth column of B^{-1}.

To determine the range on b_r^*, we must determine when the updated right-hand vector remains greater than or equal to zero. With $\alpha(B_{\text{opt}}^{-1})^r = \alpha\mathbf{w}$, then $x_i^0 + \alpha w_i \geq 0$ for $i = 1, 2, \ldots, m$, or

$$\alpha w_i \geq -x_i^0 \quad \text{for } i = 1, 2, \ldots, m \tag{3.30}$$

Case 1. If $w_i < 0$, α must be such that $\alpha \leq -x_i^0/w_i$ for all i with $w_i < 0$. Thus, the maximum increase in $b_r^* = b_r + \alpha$ is restricted by

$$\alpha_{\max} = \text{glb}\{-x_i^0/w_i \mid w_i < 0\} \tag{3.31}$$

If this set is empty, the level can be increased without bound without changing the basic variables. Note the value of the optimal solution is changing since the right-hand vector is changing.

Case 2. If $w_i > 0$, then in view of (3.30), $\alpha \geq -x_i^0/w_i$. This implies that the maximum decrease in b_r occurs when

$$\alpha_{\min} = \text{lub}\{-x_i^0/w_i \mid w_i > 0\} \tag{3.32}$$

EXAMPLE 3.5

Again considering Example 3.4, let us determine the range of b_1 and b_2.

$$B_{\text{opt}}^{-1} = \begin{bmatrix} 1 & 0 & 5/2 \\ 0 & 1 & 0 \\ 0 & 0 & 1/2 \end{bmatrix} \qquad \mathbf{b} = \begin{bmatrix} 0 \\ 4 \\ 18 \end{bmatrix} \qquad B_{\text{opt}}^{-1}\mathbf{b} = \begin{bmatrix} 45 \\ 4 \\ 9 \end{bmatrix}$$

Concerning b_1, we have

$$B_{\text{opt}}^{-1}\begin{bmatrix} 0 \\ 4+\alpha \\ 18 \end{bmatrix} = \begin{bmatrix} 45 \\ 4 \\ 9 \end{bmatrix} + \alpha \begin{bmatrix} 0 \\ 1 \\ 0 \end{bmatrix}$$

Since $w_1 = 1$ and $w_2 = 0$, the set in (3.31) is empty and the upper bound on b_1 is ∞. In (3.32), $\alpha = -4$, so the lower bound on b_1 is $b_1 - 4 = 0$. Therefore the range on b_1 is $b_1^* \in [0, \infty)$.

As for b_2, we have

$$B^{-1}\begin{bmatrix} 0 \\ 4 \\ 18+\alpha \end{bmatrix} = \begin{bmatrix} 45 \\ 4 \\ 9 \end{bmatrix} + \alpha \begin{bmatrix} 5/2 \\ 0 \\ 1/2 \end{bmatrix}$$

Again the upper bound on b_2 is ∞ since $w_1 = 0$ and $w_2 = 1/2$. From (3.32), $\alpha = -9/(1/2) = -18$, and the lower bound on b_2 is therefore $b_2 - 18 = 0$. This implies that the same variables will remain basic for $b_2 \in [0, \infty)$.

Note that if $b_1 < 0$ or $b_2 < 0$, the basic variable associated with this right-hand change will leave the basis. With this change the coefficients in row 0 will change; this implies, for example, that the value of the objective function will decrease by 2.5 for every unit of resource 2 (row 2) eliminated from our original model until $b_2 = 0$. Beyond this level, the $z_4 - c_4$ coefficient will change.

3.3.3 Changes in the Constraint Coefficients

When the restriction matrix coefficients are changed, we are usually not interested in their range but rather in knowing if the problem still has the same optimal solution and, if not, how we can restart the simplex algorithm from the present optimal tableau.

Assume the coefficient $a_{i\ell}$ now has a new value so that

$$a_{i\ell}^* = a_{i\ell} + \alpha \tag{3.33}$$

Case 1. If x_ℓ is a nonbasic variable, then $\mathbf{x}^0 = B^{-1}\mathbf{b} \geq 0$ and $\mathbf{x}^0$ is still a feasible solution.

To verify optimality, one only needs to check the $z_j - c_j$ components. If $j \neq \ell$, then

$$z_j^* - c_j = \mathbf{c}_B' B_m^{-1}\overline{\mathbf{P}}_j - c_j = z_j - c_j$$

Thus for $j \neq \ell$ the value of $z_j - c_j$ remains the same. For $j = \ell$

$$z_\ell^* - c_\ell = \mathbf{c}'_B B_m^{-1} \overline{\mathbf{P}}_j^* - c_\ell \tag{3.34}$$

The only entry in row 0 that must be checked is (3.34). If (3.34) is nonnegative the solution is still optimal. However if (3.34) is negative, then we must continue the simplex algorithm with x_ℓ entering the solution.

Case 2. If x_ℓ is basic, the situation is more complicated. The basic matrix B^* is different and hence $(B^*)^{-1}$ is different from B_{opt}^{-1} or it may not even exist.

Assume $(B^*)^{-1}$ does not exist (the columns of B^* are not linearly independent); then one way to get back into canonical form is to decompose the ℓ th column. Letting

$$\mathbf{P}_\ell^* = \mathbf{P}_\ell + \begin{bmatrix} 0 \\ \vdots \\ \alpha \\ 0 \\ \vdots \\ 0 \end{bmatrix} = \mathbf{P}_\ell + \mathbf{P}_\ell$$

we have added to the original tableau a new artificial vector, $\mathbf{P}_\ell$. In the final tableau the updated vector is $B^{-1}\mathbf{P}_\ell$, which corresponds to a nonbasic variable. Hence one can proceed as outlined in Case 1.

If $(B^*)^{-1}$ does exist, three conditions can arise:

(1) $(\mathbf{x}^*)^0 = (B^*)^{-1}\mathbf{b} \ngeq 0$
(2) $(\mathbf{x}^*)^0 \geq 0$, but $z_j^* - c_j \ngeq 0$ for all j
(3) $(\mathbf{x}^*)^0 \geq 0$ and $z_j^* - c_j \geq 0$ for all j

In (1) use the dual simplex or introduce a pseudo-objective function to obtain canonical form. In (2) continue the simplex algorithm; (3) implies that the solution is optimal.

3.4 Parametric Programming

In this section we investigate what happens to the optimal solution if we systematically change the cost vector or the resource vector.

3.4.1 Cost Vector

Let us consider the series of objective functions defined as

$$\mathbf{c}(\theta) = \mathbf{c} + \theta\mathbf{d} \tag{3.35}$$

where $\mathbf{d}$ is an n-dimensional vector. When θ takes any positive value the original objective function is denoted by (3.35). As noted in sensitivity analysis (Section 3.3), there could be a value of $\theta > 0$ in which the optimal solution is exactly the same as when we considered only $\mathbf{c}$. Therefore when we investigate what happens with a systematic change in (3.35), we actually ask for what values or range of $\theta > 0$ the present basic solution will remain optimal and what activity will come into the solution if $\theta \geq \theta^*$.

First we solve the problem with $\theta = 0$. We obtain the optimal solution $\mathbf{x}^0$ and B^{-1}. Now if $\mathbf{c}(\theta) = \mathbf{c} + \theta\mathbf{d}$ for any θ, then the $(z_j - c_j)^*$ for the basic variables are

$$\begin{aligned}(z_j - c_j)^* &= (B^{-1})_0 \begin{bmatrix} -c(\theta)_j \\ \overline{\mathbf{P}}_j \end{bmatrix} \\ &= \mathbf{c}_B' B_m^{-1} \overline{\mathbf{P}}_j - c(\theta)_j \\ &= z_j - (c_j + \theta d_j) \\ &= (z_j - c_j) - \theta d_j \\ &= -\theta d_j \end{aligned} \tag{3.36}$$

Recall that (3.36) is like $-\alpha$ (in Case 2, p. 71) where the only change in the objective function occurred as if $c_\ell^* = c_\ell + \alpha_\ell$. Hence, to obtain canonical form we multiply each basic row by θd_j and add it to the objective function row. We shall denote θd_j by θd_{Bj} for the jth basic variable; for each basic variable $(z_j - c_j)^* = -\theta d_{Bj}$. Without loss of generality assume the basic variables are $j = 1, 2, \ldots, m$. Now for $j = m + 1, m + 2, \ldots, n$ initially

$$(z_j - c_j)^* = (z_j - c_j) - \theta d_j \tag{3.37}$$

Suppose we systematically eliminate θd_{B1} for the first basic variable, θd_{B2} for the second, etc. Upon eliminating θd_{B1}, (3.37) is modified to

$$(z_j - c_j) - \theta d_j + \theta d_{B1}(x_{1j}) \tag{3.38}$$

Then eliminating θd_{B2}, (3.38) is updated to

$$(z_j - c_j) - \theta d_j + \theta d_{B1}(x_{1j}) + \theta d_{B2}(x_{2j})$$

Continuing in this manner, we finally modify the cost coefficients to

$$(z_j - c_j) - \theta d_j + \theta d'_B B_m^{-1}\overline{\mathbf{P}}_j \tag{3.39}$$

for any nonbasic j.

Now let us determine how far θ can be increased while still preserving the present solution. In particular, for what values of θ will (3.39) be nonnegative for all j nonbasic? This will happen if θ is such that

$$\theta(\mathbf{d}'_B B_m^{-1}\overline{\mathbf{P}}_j - d_j) \geq -(z_j - c_j)$$

Hence the maximum increase in θ occurs when

$$\theta \leq -(z_j - c_j)/(\mathbf{d}'_B B_m^{-1}\overline{\mathbf{P}}_j - d_j) \quad \text{for } \mathbf{d}'_B B_m^{-1}\overline{\mathbf{P}}_j - d_j < 0$$

or

$$\theta^* = \text{glb}\{-(z_j - c_j)/(\mathbf{d}'_B B_m^{-1}\overline{\mathbf{P}}_j - d_j) \mid \mathbf{d}'_B B_m^{-1}\overline{\mathbf{P}}_j - d_j < 0, \; j \in NB\} \tag{3.40}$$

For $\theta \in [0, \theta^*]$ as defined in (3.40), the same activities are in the solution as when $\theta = 0$; that is, the basic solution is preserved and these activities are at the same level, but clearly the value of the objective function has been changed.

When $\theta > \theta^*$, the basis will change and the vector which gave the minimum ratio will enter the basis.

3.4.2 *Resource Vector*

Assume the resource vector $\mathbf{b}$ will systematically be changed by adding the vector $\theta\mathbf{b}_0$, i.e., $\mathbf{b}(\theta) = \mathbf{b} + \theta\mathbf{b}_0$. For some value of θ, $(\mathbf{x}^*)^0 = B_m^{-1}(\mathbf{b} + \theta\mathbf{b}_0)$ will no longer be a feasible solution. To determine the maximum increase in θ before the solution becomes infeasible, it suffices to determine θ^* such that $B_m^{-1}(\mathbf{b} + \theta^*\mathbf{b}_0) \geq 0$. Thus θ^* must be such that $\theta^* B_m^{-1}\mathbf{b}_0 \geq -\mathbf{x}^0$. Hence

$$\theta^* = \text{glb}\{-x_i^0/(B_m^{-1}\mathbf{b}_0)_i \mid (B_m^{-1}\mathbf{b}_0)_i < 0\}$$

and for $\theta \in [0, \theta^*]$ the tableau is still optimal since all the $z_j - c_j$ components are nonnegative. Moreover, only the level of these activities and the optimum will change.

EXAMPLE 3.6

Consider again the problem

$$\begin{aligned} \text{maximize} \quad & 3x_1 + 5x_2 \\ \text{subject to} \quad & x_1 \leq 4 \\ & 3x_1 + 2x_2 \leq 18 \\ & x_1, \quad x_2 \geq 0 \end{aligned}$$

The final equivalent formulation is

$$\begin{array}{llcl} \text{Row 0} & z_0 + (9/2)x_1 & + (5/2)x_4 & = 45 \\ \text{Row 1} & x_1 + x_3 & & = 4 \\ \text{Row 2} & (3/2)x_1 + x_2 & + (1/2)x_4 & = 9 \end{array}$$

Suppose that one wishes to investigate a trade-off between the two activities x_1 and x_2, whereby the unit profit of activity x_1 can be increased by decreasing the unit profit of activity x_2 by half. The objective function will take the form

$$\mathbf{c}(\theta)'\mathbf{x} = (3 + 2\theta)x_1 + (5 - \theta)x_2 = \mathbf{x}'(\mathbf{c} + \theta\mathbf{d})$$

where $\mathbf{d}' = (2, -1, 0, 0)$.

In the final tableau

$$B_m^{-1} = \begin{bmatrix} 1 & 0 \\ 0 & 1/2 \end{bmatrix} \qquad \mathbf{d}_B = \begin{bmatrix} d_3 \\ d_2 \end{bmatrix} = \begin{bmatrix} 0 \\ -1 \end{bmatrix}$$

The nonbasic variables are x_1 and x_4. Therefore for $j = 1$ we have

$$\begin{aligned} &(z_1 - c_1) + \theta(\mathbf{d}_B' B_m^{-1}\overline{\mathbf{P}}_1 - d_1) \\ &\quad = 9/2 + \theta\left((0, -1)\begin{bmatrix} 1 & 0 \\ 0 & 1/2 \end{bmatrix}\begin{bmatrix} 1 \\ 3 \end{bmatrix} - 2\right) \\ &\quad = 9/2 + \theta\left((0, -1/2)\begin{bmatrix} 1 \\ 3 \end{bmatrix} - 2\right) \\ &\quad = 9/2 + \theta(-7/2) \end{aligned}$$

Hence for $j = 1, \theta_1 = 9/7$.

For $j = 4$

$$(z_4 - c_4) + \theta(\mathbf{d}'_B B_m^{-1}\overline{\mathbf{P}}_4 - d_4) = 5/2 + \theta(-1/2)$$

Hence, $\theta_4 = 5$ and $\theta^* = \min\{\theta_1, \theta_4\} = 9/7$. This implies that the corresponding basic solution remains optimal for $0 \leq \theta \leq 9/7$; for $\theta > 9/7$, x_1 enters the solution since the minimum ratio occurs when $j = 1$. When $\theta > 9/7$, x_3 will leave the solution since

$$\min\{x_i^0/x_{i1} \mid x_{i1} > 0\} = \min\{4, 6\} = 4$$

We therefore pivot on $x_{i1} = 1$ and

$$\eta_2 = \begin{bmatrix} -(z_j - c_j)^* \\ 1 \\ -3/2 \end{bmatrix} = \begin{bmatrix} 0 \\ 1 \\ -3/2 \end{bmatrix}$$

Note that $(z_j - c_j)^* = 0$ when $\theta^* = 9/7$.

To obtain B_Y^{-1}, the new B^{-1}, we recall that

$$B_Y^{-1} = EB_y^{-1}$$

However, the first row of B_y^{-1} is altered since $\theta^* = 9/7$. An easy way to find B_Y^{-1} is to calculate EB_y^{-1}, and then change the first row of B_Y^{-1}, since $(EB_y^{-1})_0 = [1, \mathbf{c}'_B B_m^{-1}]$ where $\mathbf{c}'_B$ is now equal to $[3 + 2\theta^*, 5 - \theta^*] = (39/7, 26/7)$.

Therefore

$$B_Y^{-1} = EB_y^{-1}$$

$$= \begin{bmatrix} 1 & 0 & 0 \\ 0 & 1 & 0 \\ 0 & -3/2 & 1 \end{bmatrix} \begin{bmatrix} 1 & 0 & 5/2 \\ 0 & 1 & 0 \\ 0 & 0 & 1/2 \end{bmatrix} = \begin{bmatrix} 1 & 0 & 5/2 \\ 0 & 1 & 0 \\ 0 & -3/2 & 1/2 \end{bmatrix} \tag{3.41}$$

Now modifying the last m components of the first row of (3.41) we have

$$\mathbf{c}_B{}' B_m^{-1} = (1/7)(39, 26)\begin{bmatrix} 1 & 0 \\ -3/2 & 1/2 \end{bmatrix} = (0, 13/7)$$

Also, the new value of the objective function can be easily computed, i.e.

$$(B_Y^{-1})_0 \begin{bmatrix} 0 \\ \mathbf{b} \end{bmatrix} = 234/7$$

An alternate way to update this value is to modify each component in row 0 by $\theta^*(4d_3 + 9d_2)$. Hence

$$\begin{aligned}\mathbf{c}'\mathbf{x}_Y^0 &= \mathbf{c}'\mathbf{x}_y^0 + \theta^*(4d_3 + 9d_2)\\ &= 45 + \theta^*(-9) = 234/7\end{aligned}$$

The same argument applies if one wishes to modify the first row of B_y^{-1} before calculating B_Y^{-1}. At this point, when $\theta^* = 9/7$, the new objective function is $\mathbf{c}(\theta^*)' = (39/7, 26/7, 0, 0)$ and the optimal solution is $x_1^0 = 4$ and $x_2^0 = 3$ (basic variables) and $x_3^0 = 0 = x_4^0$ (nonbasic variables).

Continuing in our investigation, for what value of θ greater than 9/7 do we encounter another change in the basis? Consider

$$\bar{\theta} = \theta + 9/7 \qquad \theta = \bar{\theta} - 9/7$$

Now with x_3 and x_4 nonbasic and $\mathbf{d}' = (2, -1, 0, 0)$, then $\mathbf{d}'_B = (d_1, d_2) = (2, -1)$. We have for $j = 3$

$$\begin{aligned}&(z_3 - c_3) + \theta(\mathbf{d}'_B B_m^{-1}\bar{\mathbf{P}}_3 - d_3)\\ &\quad = 0 + \theta\left((2, -1)\begin{bmatrix}1 & 0\\ -3/2 & 1/2\end{bmatrix}\begin{bmatrix}1\\0\end{bmatrix} - 0\right) = (7/2)\theta\end{aligned} \tag{3.42}$$

and for $j = 4$

$$(z_4 - c_4) + \theta\left((7/2, -1/2)\begin{bmatrix}0\\1\end{bmatrix} - 0\right) = 13/7 + \theta(-1/2) \tag{3.43}$$

Thus $\theta_4 = 26/7$, and at $\bar{\theta} = 26/7 + 9/7 = 5$ the next basic change will occur with $\mathbf{P}_4$ entering the basis. Also

$$B_m^{-1}\bar{\mathbf{P}}_4 = \begin{bmatrix}1 & 0\\ -3/2 & 1/2\end{bmatrix}\begin{bmatrix}0\\1\end{bmatrix} = \begin{bmatrix}0\\1/2\end{bmatrix}$$

$$B_m^{-1}\mathbf{b} = \begin{bmatrix}4\\3\end{bmatrix}$$

Therefore $\mathbf{P}_2$ leaves the basis and we pivot on 1/2.

To determine B_Y^{-1}, we have

$$\eta_3 = \begin{bmatrix}0\\0\\2\end{bmatrix}$$

$$B_Y^{-1} = \begin{bmatrix}1 & 0 & 0\\ 0 & 1 & 0\\ 0 & 0 & 2\end{bmatrix}\begin{bmatrix}1 & 0 & 13/7\\ 0 & 1 & 0\\ 0 & -3/2 & 1/2\end{bmatrix} = \left[\begin{array}{c|cc}1 & & \\ \hline 0 & 1 & 0\\ 0 & -3 & 1\end{array}\right]$$

The remaining elements in the first row of this matrix are the $z_j - c_j$ coefficients for x_3 and x_4. Hence for $j = 3$ we have from (3.42) that

$$z_j - c_j = (7/2)\bar{\theta}^* = (\bar{\theta} - 9/7)7/2 = (26/7)(7/2) = 13$$

From (3.43) for $j = 4$

$$z_j - c_j = 13/7 + \bar{\theta}^*(-1/2) = 13/7 + (\theta - 9/7)(-1/2) = 0$$

Thus

$$B_Y^{-1} = \begin{bmatrix} 1 & 13 & 0 \\ 0 & 1 & 0 \\ 0 & -3 & 1 \end{bmatrix}$$

Updating $\mathbf{c}(\bar{\theta})'$, the new objective function is now the vector $(3 + 2\bar{\theta}^*, 5 - \bar{\theta}^*, 0, 0) = (13, 0, 0, 0)$.

It can be shown by the above procedure that for any $\theta \geq 5$ the present solution remains optimal.

3.5 Bounded Variable Algorithm

In linear programming problems, constraints of the following type often occur

$$x_j \leq u_j$$
$$x_j \geq \ell_j$$

where u_j and ℓ_j are the upper and lower bounds on the jth activity. In our previous models $\ell_j = 0$ and $u_j \rightarrow \infty$.

These constraints can be incorporated into the problem as ordinary constraints; then the number of constraints in the problem is $m + 2n$ if every activity has finite upper and lower bounds and m is the number of constraints originally in A. Instead, the algorithm to be described keeps m constraints, reducing the size of the tableau and the number of numerical computations. The number of tableaux is also reduced.

We first restrict our attention to problems with lower bounds of zero, so we only need consider problems with upper bounds on the activities.

Let us consider the following system

$$\begin{aligned} z_0 - c_1 x_1 - c_2 x_2 \qquad\qquad &= v \\ a_{11}x_1 + a_{12}x_2 + x_3 \qquad &= b_1 \\ a_{21}x_1 + a_{22}x_2 \qquad + x_4 &= b_2 \end{aligned} \tag{3.44}$$

where each activity has an upper bound u_j.

The upper bound constraints have not been incorporated into (3.44), but we shall modify the simplex method to account for the upper bound restrictions.

Suppose in (3.44) that $-c_1$ is the most negative element in the objective function. Then according to the simplex procedure x_1 is the entering variable and x_1 should be increased from zero to a positive value. The maximum value which x_1 can take in the ordinary simplex method is

$$\min \{b_i/a_{i1} \mid a_{i1} > 0\} \tag{3.45}$$

Now take into account u_1, the upper bound on x_1. Hence the maximum value that x_1 can take is

$$\min \{\min \{b_i/a_{i1} \mid a_{i1} > 0\}, u_1\} \tag{3.46}$$

Two cases can arise; either the minimum is $b_{i'}/a_{i'1}$ for some i' or x_1 must assume its upper bound u_1. In the first situation the system (3.44) is transformed as usual with $a_{i'1}$ as a pivot.

In the second case we must modify the tableau by letting x_1 assume its upper bound u_1. This can be done by replacing x_1 in (3.44) by $x_1 = u_1 - x_1^*$. The new variable x_1^* is a slack variable incorporated to make the inequality $x_1 \leq u_1$ an equality. Thus when $x_1^* = 0$ then $x_1 = u_1$, and when $x_1^* = u_1$ then $x_1 = 0$. Substituting

$$x_1 = u_1 - x_1^* \tag{3.47}$$

into (3.44), we have

$$\begin{aligned} z_0 + c_1 x_1^* - c_2 x_2 \qquad\qquad &= v + c_1 u_1 \\ - a_{11} x_1^* + a_{12} x_2 + x_3 \qquad &= b_1 - a_{11} u_1 \\ - a_{21} x_1^* + a_{22} x_2 \qquad + x_4 &= b_2 - a_{21} u_1 \end{aligned} \tag{3.48}$$

The tableau has been changed, therefore, by

(i) subtracting u_1 times the coefficient of x_1 from the values of the basic variables and the current value of the objective function
(ii) multiplying all coefficients of x_1 by -1

It should be noted that the new variable x_1^* is nonbasic, and therefore $x_1^* = 0$ and $x_1 = u_1$. It is quite apparent that in a later tableau x_1^* could become basic or, more important, attain its upper bound u_1. When this happens we again substitute according to (3.47), or follow steps (i) and (ii), with x_1 now a nonbasic variable.

In the final tableau any combination of x_j or x_j^* could possibly be in the solution. If x_j is in the final tableau, then x_j^0 will equal the right-hand side corresponding to x_j^0 basic row. If x_j^* is in the final tableau, then $x_j^0 = u_1 - x_j^*$.

In (3.48), it should also be noted that the updated right-hand side is still greater than or equal to zero. This is easy to verify since

$$b_i - a_{i1}u_1 = a_{i1}[(b_i/a_{i1}) - u_1] \tag{3.49}$$

which is nonnegative for $a_{i1} > 0$ since u_1 is a minimum in (3.46). For $a_{i1} < 0$, $[(b_i/a_{i1}) - u_1]$ is negative and hence (3.49) is nonnegative.

We have not considered the possibility that, due to an increase of the new basic variable, the present basic variables may exceed their upper bounds. In other words, in (3.44) with x_2 nonbasic we have

$$x_3 = b_1 - a_{11}x_1$$
$$x_4 = b_2 - a_{21}x_1$$

Hence the upper bounds u_3 or u_4 can be reached if $b_1 - a_{11}x_1 = u_3$ or $b_2 - a_{21}x_1 = u_4$. This implies that x_1 cannot be increased above

$$\min \{(b_i - u^i)/a_{i1} \mid a_{i1} < 0\}$$

where u^i is the upper bound of the basic variable in row i. This implies the maximum that x_1 can be increased is

$$\min \{\min \{(b_i - u^i)/a_{i1} \mid a_{i1} < 0\}, \min \{b_i/a_{i1} \mid a_{i1} > 0\}, u_1\} \tag{3.50}$$

Assuming that the minimum occurs at u_3, the upper bound of x_3, first put x_3 at its upper bound by substituting in (3.44) for x_3 according to (3.47), i.e., $x_3 = u_3 - x_3^*$, and then transform the tableau with a_{11} as pivot.

The rules for modifying the simplex algorithm to handle upper bound constraints are summarized as follows

I. Section of the new basic variable. Select the column with the minimum $z_j - c_j$, i.e., select min $\{a_{0j} \mid a_{0j} < 0\}$. Let k be the column of the new basic variable. If all $a_{0j} \geq 0$, the optimal solution has been obtained.

II. Section of the leaving basic variable. Select the row connected with min $\{\theta_1, \theta_2, \theta_3\}$ where

$$\theta_1 = \min_i \{b_i/a_{ik} \mid a_{ik} > 0\}$$

$$\theta_2 = \min_i \{(b_i - u^i)/a_{ik} \mid a_{ik} < 0\}$$

$$\theta_3 = u_k$$

Let the row connected with this minimum be row r.

III. Transformation of the tableau.

A. $\theta = \theta_1$

1. Perform a normal simplex iteration with a_{rk} as pivot

B. $\theta = \theta_2$

1. Change the upper bound switch (u.b.s.) on the vector that will leave the basis as described under (3.50) (initially this switch is set at zero)
2. Change the right-hand side of the rth row to $b_r - u^r$
3. Multiply the basic vector which is to leave the basis by -1
4. Transform the tableau with a_{rk} as pivot

C. $\theta = \theta_3$

1. Change the upper bound switch on the kth vector
2. Update the right-hand sides, i.e., $b_i - a_{ik}u_k \quad (i = 1, 2, \ldots, m)$
3. Multiply all coefficients of the kth vector by -1

Let us consider the following problem

$$\begin{aligned} \text{maximize} \quad & 2x_1 + x_2 - x_3 \\ \text{subject to} \quad & x_1 - 2x_2 \leq 3 \\ & -x_1 + x_2 + 2x_3 \leq 4 \\ & 0 \leq x_1 \leq 1 \\ & 0 \leq x_2 \leq 2 \\ & 0 \leq x_3 \leq 3 \\ & 0 \leq x_4 \leq 4 \\ & 0 \leq x_5 \leq 5 \end{aligned}$$

Then our initial tableau is

	$\mathbf{P}_1$	$\mathbf{P}_2$	$\mathbf{P}_3$	$\mathbf{P}_4$	$\mathbf{P}_5$	$\mathbf{b}$	θ_1
u.b.s.	0	0	0	0	0		
u.b.	1	2	3	4	5		
$z_j - c_j$	-2	-1	1	0	0	0	
$\mathbf{P}_4$	1	-2	0	1	0	3	3
$\mathbf{P}_5$	-1	1	2	0	1	4	

Since $\mathbf{P}_1$ has the most negative $z_j - c_j$, it will enter the basis; and $a_{11} = 1$ is the only positive entry so that $\theta_1 = 3$. Now $\theta_2 = (b_2 - u^2)/a_{21} = (4 - 5)/(-1) = 1$, and $\theta_3 = 1$, i.e., the bound on the incoming vector $\mathbf{P}_k$ is 1. Hence $\theta = 1 = \min \{\theta_1, \theta_2, \theta_3\}$. We shall adopt the rule that in case of ties involving θ_3 we shall select θ_3 since the computations are relatively easier than the other options. Therefore $\theta = \theta_3$ and tableau 0 can be updated quite quickly by considering only the kth vector and the right-hand side. The updated tableau is

	$\mathbf{P}_1$	$\mathbf{P}_2$	$\mathbf{P}_3$	$\mathbf{P}_4$	$\mathbf{P}_5$	$\mathbf{b}$	θ_1
u.b.s.	1	0	0	0	0		
u.b.	1	2	3	4	5		
$z_j - c_j$	2	-1	1	0	0	2	
$\mathbf{P}_4$	-1	-2	0	1	0	2	
$\mathbf{P}_5$	1	1	2	0	1	5	5

In this tableau $\mathbf{P}_2$ will be the entering vector.

$\theta_1 = 5$

$\theta_2 = (2 - 4)/(-2) = 1$ for the 4th variable

$\theta_3 = 2$

Hence, $\theta = \theta_2 = 1$, and applying the transformation procedures for $\dot{\theta}_2$ we have before pivoting

	$\mathbf{P}_1$	$\mathbf{P}_2$	$\mathbf{P}_3$	$\mathbf{P}_4$	$\mathbf{P}_5$	$\mathbf{b}$
u.b.s.	1	0	0	1	0	
u.b.	1	2	3	4	5	
$z_j - c_j$	2	-1	1	0	0	2
$\mathbf{P}_4$	-1	-2	0	-1	0	-2
$\mathbf{P}_5$	1	1	2	0	1	5

Now, applying rule (4) of (B) we see that the pivot element corresponds to $a_{12} = -2$. Therefore the next tableau will be

	$\mathbf{P}_1$	$\mathbf{P}_2$	$\mathbf{P}_3$	$\mathbf{P}_4$	$\mathbf{P}_5$	$\mathbf{b}$
u.b.s.	1	0	0	1	0	
u.b.	1	2	3	4	5	
$z_j - c_j$	5/2	0	1	1/2	0	3
$\mathbf{P}_2$	1/2	1	0	1/2	0	1
$\mathbf{P}_5$	1/2	0	2	−1/2	1	4

Since all $z_j - c_j$ are nonnegative the solution is optimal; it is

$$x_1^0 = \bar{x}_1 + I_1 u_1 = 1$$
$$x_2^0 = \bar{x}_2 + I_2 u_2 = 1$$
$$x_3^0 = \bar{x}_3 + I_3 u_3 = 0$$
$$x_4^0 = \bar{x}_4 + I_4 u_4 = 4$$
$$x_5^0 = \bar{x}_5 + I_5 u_5 = 4$$

where $\bar{x}_j$ is the corresponding solution in the final tableau and I_j is the upper bound switch for $j = 1, 2, \ldots, 5$.

In our discussion we have considered only upper bounds on the activities. Problems with finite lower bounds can be transformed so the lower bound on each activity is zero. Then the above algorithm is still valid.

Consider $\ell_j \leq x_j \leq u_j$ for any j. Then defining $\bar{x}_j = x_j - \ell_j$, we have that $\bar{x}_j \geq 0$. Moreover, $0 \leq \bar{x}_j \leq u_j - \ell_j = \bar{u}_j$. In this manner problems with lower bounds can be reformulated as problems with lower bounds of zero.

Exercises

1. Consider the problem

$$\begin{array}{ll} \text{maximize} & (-7/2)x_1 + 3x_2 \\ \text{subject to} & 2x_1 + 3x_2 \leq 18 \\ & -2x_1 + 2x_2 \leq 10 \\ & x_1, \quad x_2 \geq 0 \end{array}$$

In the optimal tableau

$$B^{-1} = \begin{bmatrix} 1 & 0 & 3/2 \\ 0 & 1 & -3/2 \\ 0 & 0 & 1/2 \end{bmatrix}$$

a) Determine the final tableau and identify its corresponding optimal solution and optimum.

b) Determine the range on c_1; c_2; b_1; b_2.

c) If c_2 is moved to its upper limit, determine the set of all alternate optimal solutions.

d) If b_1 is at its lower limit, determine its corresponding optimal solution.

2. Assume we wish to investigate the trade-off between activities x_1 and x_2 in Exercise 1, whereby the unit profit from activity 1 can be increased by the same amount as the unit profit decrease from activity 2; i.e., $\mathbf{c}(\theta) = \mathbf{c} + \theta\mathbf{d}$ for $\theta \geq 0$. Determine θ^* such that for $0 \leq \theta \leq \theta^*$ the present solution will remain optimal. If $\theta > \theta^*$, which vector will enter the solution?
3. Assume the unit profit of the second activity in Exercise 1 cannot be determined exactly from one year to the next, but from previous information it follows a uniform distribution with a lower value of 2 and an upper value of 4. What is the probability that the solution we identified in part *b*) of Exercise 1 will be an optimal solution, say, next year?
4. Consider Exercise 13 in Chapter 2. Determine the range on the profit coefficients of wheat and cattle, respectively. Determine the range on each resource.
5. Consider the problem

$$\begin{aligned} \text{maximize} \quad & -3x_1 + 3x_2 \\ \text{subject to} \quad & 2x_1 + 3x_2 \leq 18 \\ & -2x_1 + 2x_2 \leq 12 \\ & x_1, \quad x_2 \geq 0 \end{aligned}$$

Note this problem is similar to Exercise 1. Answer the following questions using B_{opt}^{-1} obtained in Exercise 1:

a) Determine the final tableau and associated basic solution.

b) Determine the range on c_2. How does it correspond to the range determined for c_2 in Exercise 1?

c) If $c_2 = 4$, show that the final tableau will yield the same extreme point and solution as determined in *a*). How can you account for this inasmuch as 4 lies above the upper limit of the range on c_2?

6. Suppose a variable x_j is unrestricted in some linear programming model.

a) After transforming this model to standard form using the transformation $x_j = x_j^{(1)} - x_j^{(2)}$ where $x_j^{(1)}$ and $x_j^{(2)}$ are nonnegative, show that if $x_j^{(1)}$ is in the basis at some stage of the simplex algorithm its counterpart $x_j^{(2)}$ must be nonbasic.

b) If $x_j^{(1)}$ is in the final basis, what is the corresponding $z_j - c_j$ for $x_j^{(2)}$? What does this imply?

7. Consider the problem

$$\begin{array}{ll} \text{maximize} & x \\ \text{subject to} & x \leq 2 \end{array}$$

a) Solve this problem via the simplex algorithm. (Use the transformation $x = x_1 - x_2; x_1, x_2 \geq 0$.)

b) Show that the set of all optimal solutions is $\{x \in E^2 \mid x_1 - x_2 = 2\}$. Discuss why one can only generate one extreme point solution in the final tableau.

8. Solve the following problems via the simple upper bounding procedure

a)
$$\begin{array}{ll} \text{maximize} & x_1 + 2x_2 \\ \text{subject to} & x_1 + x_2 \leq 1 \\ & 1 \leq x_1 \leq 2 \\ & x_2 \geq 3 \end{array}$$

b)
$$\begin{array}{ll} \text{minimize} & x_1 - x_2 \\ \text{subject to} & 2 \leq x_1 \leq 3 \\ & x_2 \geq 0 \end{array}$$

References

Charnes, A. 1952. Optimality and degeneracy in linear programming. *Econometrica* 20:160–70.

Charnes, A., and Cooper, W. W. 1961. *Management Models and Industrial Applications of Linear Programming*, vols. 1, 2. Wiley, New York.

Cooper, L., and Steinberg, D. 1970. *Introduction to Methods of Optimization.* Saunders, Philadelphia.

Dantzig, G. B. 1963. *Linear Programming and Extensions.* Princeton Univ. Press, Princeton, N.J.

Dantzig, G. B., and Van Slyke, R. M. 1967. Generalized upper bounding techniques. *J. Comput. Syst. Sci.* 1:213–26.

Gass, S. L. 1969. *Linear Programming: Methods and Applications.* McGraw-Hill, New York.

Hadley, G. 1962. *Linear Programming.* Addison-Wesley, Reading, Mass.

Hillier, F. S., and Lieberman, G. J. 1967. *Introduction to Operations Research.* Holden-Day, San Francisco.

Hirshfeld, David S. 1972. Very large linear programming models and how to solve them professionally. Paper presented at 41st Natl. ORSA Meet. New Orleans.

Lasdon, Leon S. 1970. *Optimization Theory for Large Systems.* Macmillan, New York.

Lemke, C. E. 1954. The dual method of solving the linear programming problem. Nav. Res. Logist. Q. 1:36–47.

Orchard-Hays, W. 1968. *Advanced Linear Programming Computing Techniques.* McGraw-Hill, New York.

Sposito, V. A. 1973. Solutions to a special class of linear programming problems. *J. Oper. Res.* 21:386–88.

Van de Panne, C. 1971. *Linear Programming and Related Techniques.* American Elsevier, New York.

Wagner, H. M. 1969. *Principles of Operations Research.* Prentice-Hall, Englewood Cliffs, N.J.

CHAPTER FOUR

LINEAR DUALITY THEORY

One of the most important early discoveries in linear programming was the notion of duality. Associated with every linear programming problem is a *dual* linear programming problem which bears an important and special relationship to the original problem. This chapter discusses significant aspects of and the relationship between the original linear programming problem and its dual.

4.1 Linear Duality

Let the original or *primal* linear programming problem be defined as

Problem I

$$\begin{aligned} &\text{maximize} && \mathbf{c}'\mathbf{x} \\ &\text{subject to} && A\mathbf{x} \leq \mathbf{b} \\ &&& \mathbf{x} \geq 0 \end{aligned}$$

where A is an $m \times n$ matrix and a_{ij}, b_i, and c_j are given constants ($i = 1, 2, \ldots, m; j = 1, 2, \ldots, n$).

We shall define Problem II to be the *dual* problem.

Problem II

$$\begin{aligned} &\text{minimize} && \mathbf{b}'\mathbf{y} \\ &\text{subject to} && A'\mathbf{y} \geq \mathbf{c} \\ &&& \mathbf{y} \geq 0 \end{aligned}$$

The dual problem is obtained by

(i) transposing the matrix A
(ii) changing maximization to minimization
(iii) interchanging **b** and **c**
(iv) reversing the inequality sign in the matrix constraint

With the above transformation we can construct a dual of the primal which is again a linear programming problem. This raises an obvious question: if Problem II is the dual of Problem I, what is the dual of Problem II?

LEMMA 4.1. *The dual of the dual is the primal.*

Proof. The dual problem can be equivalently expressed as

$$\begin{aligned} -\text{maximize} \quad & (-\mathbf{b}'\mathbf{y}) \\ \text{subject to} \quad & -A'\mathbf{y} \le -\mathbf{c} \\ & \mathbf{y} \ge 0 \end{aligned} \tag{4.1}$$

Hence (4.1) is in the form of a primal problem with a signed objective function and its dual is

$$\begin{aligned} -\text{minimize} \quad & (-\mathbf{c}'\mathbf{x}) \\ \text{subject to} \quad & -A\mathbf{x} \ge -\mathbf{b} \\ & \mathbf{x} \ge 0 \end{aligned}$$

which can be written

$$\begin{aligned} \text{maximize} \quad & \mathbf{c}'\mathbf{x} \\ \text{subject to} \quad & A\mathbf{x} \le \mathbf{b} \\ & \mathbf{x} \ge 0 \end{aligned}$$

i.e., the dual of Problem II is the primal.

Thus there is complete symmetry between primal and dual problems and the problem specified as the primal or dual is arbitrary. We could have just as well called Problem I the dual and Problem II the primal.

To show why such problems are of interest, consider the following lemma..

LEMMA 4.2. *If* $\bar{\mathbf{x}}$ *and* $\bar{\mathbf{y}}$ *are feasible solutions of Problems I and II, respectively, then* $\mathbf{c}'\bar{\mathbf{x}} \le \mathbf{b}'\bar{\mathbf{y}}$.

Proof. If $\bar{\mathbf{x}}$ and $\bar{\mathbf{y}}$ are feasible solutions of the primal and dual, respectively, then

$$A\bar{\mathbf{x}} \le \mathbf{b} \tag{4.2}$$

$$A'\bar{\mathbf{y}} \ge \mathbf{c} \tag{4.3}$$

Since $\bar{\mathbf{x}} \geq 0$ and $\bar{\mathbf{y}} \geq 0$, we multiply (4.2) by $\bar{\mathbf{y}}$ and (4.3) by $\bar{\mathbf{x}}$ to obtain

$$\bar{\mathbf{y}}'A\bar{\mathbf{x}} \leq \bar{\mathbf{y}}'\mathbf{b} = \mathbf{b}'\bar{\mathbf{y}} \tag{4.4}$$

$$\bar{\mathbf{x}}'A'\bar{\mathbf{y}} \geq \bar{\mathbf{x}}'\mathbf{c} = \mathbf{c}'\bar{\mathbf{x}} \tag{4.5}$$

From (4.4) and (4.5) we have $\mathbf{b}'\bar{\mathbf{y}} \geq \bar{\mathbf{x}}'A'\bar{\mathbf{y}} \geq \mathbf{c}'\bar{\mathbf{x}}$. Therefore, $\mathbf{c}'\bar{\mathbf{x}} \leq \mathbf{b}'\bar{\mathbf{y}}$.

COROLLARY 4.2.1. *If $\bar{\mathbf{x}}$ and $\bar{\mathbf{y}}$ are feasible solutions of Problems I and II, respectively, and $\mathbf{c}'\bar{\mathbf{x}} = \mathbf{b}'\bar{\mathbf{y}}$, then $\bar{\mathbf{x}}$ is an optimal solution of Problem I and $\bar{\mathbf{y}}$ is an optimal solution of Problem II.*

Proof. From Lemma 4.2, for any feasible solution $\mathbf{x}$, $\mathbf{c}'\mathbf{x} \leq \mathbf{b}'\bar{\mathbf{y}}$, and hence $\mathbf{c}'\mathbf{x} \leq \mathbf{b}'\bar{\mathbf{y}} = \mathbf{c}'\bar{\mathbf{x}}$. Thus $\mathbf{c}'\mathbf{x} \leq \mathbf{c}'\bar{\mathbf{x}}$ for any $\mathbf{x} \in \{\mathbf{x} \mid A\mathbf{x} \leq \mathbf{b}, \mathbf{x} \geq 0\}$; i.e., $\bar{\mathbf{x}}$ is an optimal solution of Problem I.

Similarly, $\bar{\mathbf{y}}$ solves the dual.

COROLLARY 4.2.2. *If the primal has an unbounded solution, then the dual problem is infeasible.*

Proof. Assume that the dual problem has a feasible solution $\bar{\mathbf{y}}$; then from Lemma 4.2, $\mathbf{c}'\mathbf{x}$ is bounded above by $\mathbf{b}'\bar{\mathbf{y}}$ for all $\mathbf{x} \in \mathcal{A} = \{\mathbf{x} \mid A\mathbf{x} \leq \mathbf{b}, \mathbf{x} \geq 0\}$. But by hypothesis there exists $\bar{\mathbf{x}} \in \mathcal{A}$ such that $\mathbf{c}'\bar{\mathbf{x}} = \infty$, a contradiction, and hence the dual problem is infeasible.

The converse of Corollary 4.2.2 is not true as shown by Example 4.1 below. Hence if a problem is infeasible its dual may be unbounded or it may be infeasible. We shall prove later that these are the only two possibilities.

EXAMPLE 4.1

$$\begin{array}{ll} \text{maximize} & x_1 + x_2 \\ \text{subject to} & -x_1 + x_2 \leq -1 \\ & x_1 - x_2 \leq 0 \\ & x_1, \quad x_2 \geq 0 \end{array}$$

and its dual problem

$$\begin{array}{ll} \text{minimize} & -y_1 \\ \text{subject to} & -y_1 + y_2 \geq 1 \\ & y_1 - y_2 \geq 1 \\ & y_1, \quad y_2 \geq 0 \end{array}$$

Both problems have infeasible solutions which can be verified graphically by the reader.

The dual problems discussed so far are symmetric in form because all constraints are expressed as inequalities. It may happen that a linear programming model is constructed with equality as well as inequality constraints; in this case, Problem II is expressed in an asymmetric dual representation as shown in the following lemma:

LEMMA 4.3. *If the ith row of the primal is an equality constraint, then the ith variable in the dual problem is unrestricted in sign.*

Proof. Without loss of generality, assume the primal has the form

$$\begin{aligned} &\text{maximize} \quad \mathbf{c}'\mathbf{x} \\ &\text{subject to} \quad \textstyle\sum_{j=1}^{n} a_{1j}x_j = b_1 \\ &\qquad \textstyle\sum_{j=1}^{n} a_{ij}x_j \leq b_i \qquad i = 2, 3, \ldots, m \\ &\qquad x_j \geq 0 \qquad j = 1, 2, \ldots, n \end{aligned} \tag{4.6}$$

Since (4.6) can be expressed equivalently as

$$\sum_{j=1}^{n} a_{1j}x_j \geq b_1 \qquad \sum_{j=1}^{n} a_{1j}x_j \leq b_1$$

the primal problem can be rewritten as

$$\begin{aligned} &\text{maximize} \quad \mathbf{c}'\mathbf{x} \\ &\text{subject to} \quad \bar{A}\mathbf{x} \leq \bar{\mathbf{b}} \\ &\qquad \mathbf{x} \geq 0 \end{aligned} \tag{4.7}$$

where

$$\bar{A}' = \begin{bmatrix} -a_{11} & \\ -a_{12} & \\ \vdots & A' \\ -a_{1n} & \end{bmatrix} \qquad \bar{\mathbf{b}} = \begin{bmatrix} -b_1 \\ \vdots \\ b_m \end{bmatrix}$$

Then the dual of (4.7) is

$$\begin{aligned} &\text{minimize} \quad \bar{\mathbf{b}}'\mathbf{y} \\ &\text{subject to} \quad \bar{A}'\mathbf{y} \geq \mathbf{c} \\ &\qquad \mathbf{y} \geq 0 \end{aligned}$$

or

$$\begin{aligned} &\text{minimize} \quad [b_1(-y_0 + y_1) + b_2y_2 + \cdots + b_my_m] \\ &\text{subject to} \quad A'\hat{\mathbf{y}} \geq \mathbf{c} \\ &\qquad y_0, y_1, \ldots, y_m \geq 0 \end{aligned}$$

$$\hat{\mathbf{y}} = \begin{bmatrix} -y_0 + y_1 \\ y_2 \\ \vdots \\ y_m \end{bmatrix}$$

Hence, letting $\hat{y}_1 = -y_0 + y_1$, then $\hat{y}_1$ can be either positive or negative.

In solving a linear programming problem it is customary to examine a special function, the Lagrangian function. When some constraints are expressed as inequalities the analysis of the corresponding Lagrangian form is more difficult. The theory of convex cones and sets overcomes these difficulties and leads to the development of the Existence Theorem and Duality Theorem of linear programming. These are known as the two fundamental theorems of linear programming.

We shall define

$$\psi(\mathbf{x}, \mathbf{y}) = \mathbf{c}'\mathbf{x} + \mathbf{y}'(\mathbf{b} - A\mathbf{x}) \qquad \mathbf{x} \geq 0 \tag{4.8}$$

as the Lagrangian of Problem I. The Lagrangian form associated with Problem II consistent with (4.8) is defined as

$$\phi(\mathbf{y}, \mathbf{x}) = \mathbf{b}'\mathbf{y} + \mathbf{x}'(\mathbf{c} - A'\mathbf{y}) \qquad \mathbf{y} \geq 0$$

The following lemma is an immediate consequence.

LEMMA 4.4. $\phi(\mathbf{y}, \mathbf{x}) = \psi(\mathbf{x}, \mathbf{y})$ *for any* $\mathbf{x} \geq 0$ *and* $\mathbf{y} \geq 0$.

To achieve our objective in the remainder of this chapter, it is necessary to consider the following *saddle value problem.*

find a pair of vectors $(\mathbf{x}^0, \mathbf{y}^0)$
such that $\psi(\mathbf{x}, \mathbf{y}^0) \leq \psi(\mathbf{x}^0, \mathbf{y}^0) \leq \psi(\mathbf{x}^0, \mathbf{y}) \qquad \forall \mathbf{x} \geq 0 \qquad \forall \mathbf{y} \geq 0$

By considering the saddle value problem we can show that if either the primal or the dual problem has an optimal solution, then so does the other; and moreover, the optima are equal, i.e., $\mathbf{c}'\mathbf{x}^0 = \mathbf{b}'\mathbf{y}^0$. The solution of the saddle value problem, the *saddle point solution,* is the combined solutions of Problems I and II. To see why the saddle value solution is the exact ingredient needed to prove the duality result, let us consider the following lemma.

LEMMA 4.5. *If* $(\mathbf{x}^0, \mathbf{y}^0)$ *is a saddle point solution of* $\psi(\mathbf{x}, \mathbf{y})$, *then*

(i) $\mathbf{y}^{0\prime}(\mathbf{b} - A\mathbf{x}^0) = 0$
(ii) $\mathbf{x}^{0\prime}(\mathbf{c} - A'\mathbf{y}^0) = 0$

Proof. Since $(\mathbf{x}^0, \mathbf{y}^0)$ is a saddle point solution of $\psi(\mathbf{x}, \mathbf{y})$, then

$$\psi(\mathbf{x}^0, \mathbf{y}^0) \leq \psi(\mathbf{x}^0, \mathbf{y}) \quad \text{for all } \mathbf{y} \geq 0$$
$$\mathbf{y}^{0\prime}(\mathbf{b} - A\mathbf{x}^0) \leq \mathbf{y}'(\mathbf{b} - A\mathbf{x}^0) \quad \text{for all } \mathbf{y} \geq 0 \tag{4.9}$$

(a) To now prove that $\mathbf{b} - A\mathbf{x}^0 \geq 0$. Assume this is not true; then there exists $\bar{\mathbf{y}} > 0$ such that $\bar{\mathbf{y}}'(\mathbf{b} - A\mathbf{x}^0) < 0$. In particular, if the ith row of $\mathbf{b} - A\mathbf{x}^0 \ngeq 0$, then take $\bar{y}_i > 0$ and all the other $\mathbf{y}$ components equal to zero. Hence there exists $k > 0$ such that

$$\mathbf{y}^{0\prime}(\mathbf{b} - A\mathbf{x}^0) > k\bar{\mathbf{y}}'(\mathbf{b} - A\mathbf{x}^0)$$

which violates (4.9). Therefore $\mathbf{b} - A\mathbf{x}^0 \geq 0$.

(b) To now show that *(i)* holds. If not, then since $\mathbf{y}^0 \geq 0$ and $\mathbf{b} - A\mathbf{x}^0 \geq 0$, we have $\mathbf{y}^{0\prime}(\mathbf{b} - A\mathbf{x}^0) > 0$. But this implies that in (4.9)

$$0 < \mathbf{y}'(\mathbf{b} - A\mathbf{x}^0) \quad \text{for all } \mathbf{y} \geq 0$$

which is contradicted by $\bar{\mathbf{y}} = 0$. Hence conclusion *(i)* holds.

By a similar argument, conclusion *(ii)* holds.

COROLLARY 4.5.1. *If* $(\mathbf{y}^0, \mathbf{x}^0)$ *is a saddle point solution of* $\phi(\mathbf{y}, \mathbf{x})$, *then*

(i) $\phi(\mathbf{y}^0, \mathbf{x}) \leq \phi(\mathbf{y}^0, \mathbf{x}^0) \leq \phi(\mathbf{y}, \mathbf{x}^0) \quad \forall \mathbf{x} \geq 0 \quad \forall \mathbf{y} \geq 0$
(ii) $\mathbf{y}^{0\prime}(\mathbf{b} - A\mathbf{x}^0) = 0$
(iii) $\mathbf{x}^{0\prime}(\mathbf{c} - A'\mathbf{y}^0) = 0$

Proof. It follows from Lemmas 4.4 and 4.5 that *(i)*, *(ii)*, and *(iii)* are true.

COROLLARY 4.5.2. *If* $(\mathbf{x}^0, \mathbf{y}^0)$ *is a saddle point solution of*

$$\psi(\mathbf{x}, \mathbf{y}) = \mathbf{c}'\mathbf{x} + \mathbf{y}'(\mathbf{b} - A\mathbf{x})$$

then $\mathbf{x}^0$ *solves Problem I,* $\mathbf{y}^0$ *solves Problem II, and* $\mathbf{c}'\mathbf{x}^0 = \mathbf{b}'\mathbf{y}^0$.

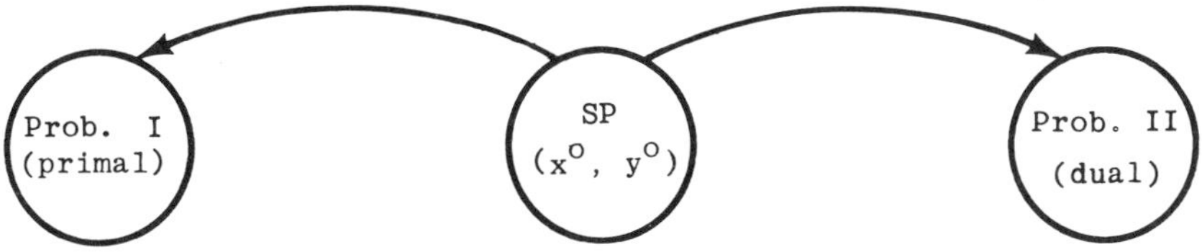

Figure 4.1. Implication of saddle value solution.

Proof. From *(i)* and *(ii)* in Lemma 4.5, $\mathbf{c}'\mathbf{x}^0 = \mathbf{b}'\mathbf{y}^0$. Thus from Corollary 4.2.1, $\mathbf{x}^0$ and $\mathbf{y}^0$ are optimal solutions of Problems I and II, respectively, if $\mathbf{x}^0$ and $\mathbf{y}^0$ are appropriate feasible solutions. However, since $\mathbf{b} - A\mathbf{x}^0 \geq 0$ from the proof of Lemma 4.5, $\mathbf{x}^0 \in \{\mathbf{x} \mid A\mathbf{x} \leq \mathbf{b}, \mathbf{x} \geq 0\}$, which implies that $\mathbf{x}^0$ solves Problem I. Likewise $\mathbf{y}^0$ solves Problem II.

Figure 4.1 shows an important relationship between the saddle value problem and the primal or dual linear programming problem as given by the above results. If it is possible to establish an equivalence between the primal and the saddle value problems, then in view of Lemma 4.5 and Corollary 4.5.2, if one problem has a solution, so will the other. The clarification of the relationships among Problem I, the saddle value problem, and Problem II is given in Theorem 4.1.

The following lemma (Karlin 1959) is of basic importance in deriving the main results of this chapter. Its proof is based on some results on convex cones, given in Appendix B.

LEMMA 4.6. *Let S be a closed convex polyhedral set in E^{m+1} which satisfies the following two conditions:*

(1) if $\begin{bmatrix} \mathbf{w} \\ z \end{bmatrix} \in S$ *and* $\mathbf{w} \geq 0$ $(\mathbf{w} \in E^m)$, *then the scalar* $z \leq 0$

(2) S contains at least one point $\begin{bmatrix} \mathbf{w}^0 \\ 0 \end{bmatrix}$ *where* $w_i^0 \geq 0$ $(i = 1, 2, \ldots, m)$

Then there exists a vector $\mathbf{u}^0$ of m components and a scalar $v^0 > 0$ satisfying the following two properties:

(i) $\mathbf{u}^0 \geq 0$

(ii) $\mathbf{u}^{0\prime}\mathbf{w} + v^0 z \leq 0$ *for all* $\begin{bmatrix} \mathbf{w} \\ z \end{bmatrix} \in S$

Remark: If S contains no point of the positive orthant except on the face $z = 0$ and does have at least one point on this face, then there exists a hyperplane through the origin which separates S from the interior of the positive orthant and strictly away from the face $w = 0$.

Before giving the proof we present an example.

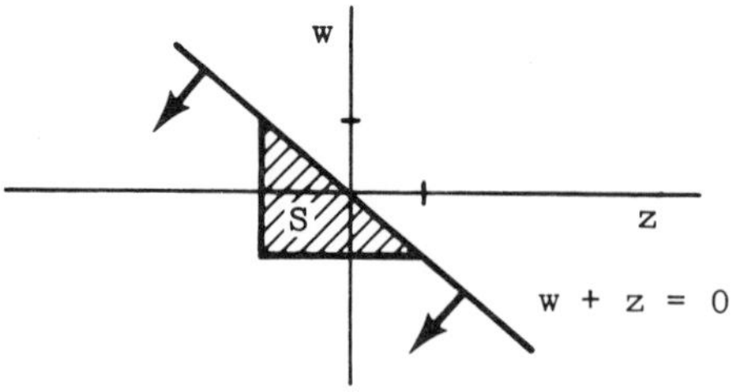

Figure 4.2. Example of Lemma 4.6.

Example 4.2

Let $S = \{(w, z) \in E^2 \mid w + z \leq 0,\ w \geq -1,\ z \geq -1\}$, then an appropriate hyperplane separating S from the interior of the positive orthant is $w + z \leq 0$, i.e., $u^0 = 1 = v^0$. This is illustrated in Fig. 4.2.

Proof of Lemma 4.6. Define T as

$$T = \left\{ \begin{bmatrix} \mathbf{r} \\ s \end{bmatrix} \middle| \begin{bmatrix} \mathbf{r} \\ s \end{bmatrix} \leq \begin{bmatrix} \mathbf{w} \\ z \end{bmatrix} \text{ for some } \begin{bmatrix} \mathbf{w} \\ z \end{bmatrix} \in S \right\}.$$

The set S is enlarged to T where T is a closed convex polyhedral set which from (2) contains the negative orthant.

Let T^0 be the dual cone of T expressed as

$$T^0 = \left\{ \begin{bmatrix} \alpha \\ \beta \end{bmatrix} \middle| \begin{bmatrix} \alpha \\ \beta \end{bmatrix}' \begin{bmatrix} \mathbf{r} \\ s \end{bmatrix} \leq 0 \quad \text{for} \begin{bmatrix} \mathbf{r} \\ s \end{bmatrix} \in T \right\}$$

$$= \left\{ \begin{bmatrix} \alpha \\ \beta \end{bmatrix} \geq 0 \middle| \begin{bmatrix} \alpha \\ \beta \end{bmatrix}' \begin{bmatrix} \mathbf{r} \\ s \end{bmatrix} \leq 0 \quad \text{for} \begin{bmatrix} \mathbf{r} \\ s \end{bmatrix} \in T \right\} \tag{4.10}$$

since the set T contains the negative orthant.

In particular, suppose that $\begin{bmatrix} \bar{\alpha} \\ \bar{\beta} \end{bmatrix} < 0$. Then since T contains the negative orthant $\bar{\alpha}'\mathbf{r} + \bar{\beta}s \nleq 0$ for some $\begin{bmatrix} \mathbf{r} \\ s \end{bmatrix} \in T$. Hence $\begin{bmatrix} \bar{\alpha} \\ \bar{\beta} \end{bmatrix} \notin T^0$.

We show first that (4.11) is equivalent to (4.10).

$$\left\{ \begin{bmatrix} \alpha \\ \beta \end{bmatrix} \geq 0 \middle| \begin{bmatrix} \alpha \\ \beta \end{bmatrix}' \begin{bmatrix} \mathbf{w} \\ z \end{bmatrix} \leq 0 \quad \text{for all} \begin{bmatrix} \mathbf{w} \\ z \end{bmatrix} \in S \right\} \tag{4.11}$$

Since $\begin{bmatrix} \mathbf{r} \\ s \end{bmatrix} \leq \begin{bmatrix} \mathbf{w} \\ z \end{bmatrix}$ from the definition of T, then

$$\alpha'\mathbf{r} + \beta s \leq \alpha'\mathbf{w} + \beta z \quad \text{for all} \begin{bmatrix} \alpha \\ \beta \end{bmatrix} \geq 0$$

Hence if (4.11) holds, then $\alpha'\mathbf{r} + \beta s \leq \alpha'\mathbf{w} + \beta z \leq 0$; and (4.11) implies (4.10).

Conversely, if $\begin{bmatrix} \alpha \\ \beta \end{bmatrix} \in T^0$, then $\alpha'\mathbf{r} + \beta s \leq 0$ for $\begin{bmatrix} \mathbf{r} \\ s \end{bmatrix} \in S$ since $S \subset T$. Hence T^0 can be equivalently expressed as (4.11).

Now define

$$H = \left\{ \begin{bmatrix} \mathbf{u} \\ v \end{bmatrix} \middle| \mathbf{u} \geq 0, v > 0 \right\} \subset E^{m+1}$$

If $T^0 \cap H \neq \phi$, then there exists $\begin{bmatrix} \mathbf{u}^0 \\ v^0 \end{bmatrix} \in H$ such that $\begin{bmatrix} \mathbf{u}^0 \\ v^0 \end{bmatrix} \in T^0$, or

$$\mathbf{u}^{0\prime}\mathbf{w} + v^0 z \leq 0 \quad \text{for all} \begin{bmatrix} \mathbf{w} \\ z \end{bmatrix} \in S,$$

and the results of the lemma hold.

Suppose on the contrary that $T^0 \cap H = \phi$. Let $H_0 = \left\{ \begin{bmatrix} \mathbf{u} \\ 0 \end{bmatrix} \middle| \mathbf{u} \geq 0 \right\}$. Then

$$H_0 \cap H = \phi \tag{4.12}$$

In view of (4.12), if $T^0 \cap H = \phi$, then $T^0 \subset H_0$, which from Lemma B.2 in Appendix B implies that

$$T^{00} \supset H_0^0 \tag{4.13}$$

where

$$H_0^0 = \left\{ \begin{bmatrix} \mathbf{r} \\ s \end{bmatrix} \middle| \begin{bmatrix} \mathbf{r} \\ s \end{bmatrix}' \begin{bmatrix} \mathbf{u} \\ 0 \end{bmatrix} \leq 0 \quad \text{for all } \mathbf{u} \geq 0 \right\}$$

$$= \left\{ \begin{bmatrix} \mathbf{r} \\ s \end{bmatrix} \middle| \mathbf{r} \leq 0, s \text{ arbitrary} \right\}$$

$$T^{00} = \mathcal{P}(T) = \left\{ \lambda \begin{bmatrix} \mathbf{r} \\ s \end{bmatrix} \middle| \lambda \geq 0, \begin{bmatrix} \mathbf{r} \\ s \end{bmatrix} \in T \right\} \tag{4.14}$$

Thus T^{00} is a closed convex cone spanned by T with vertex at the origin. Hence

$$T \subset T^{00} \tag{4.15}$$

Now consider the vector $\mathbf{c} = \begin{bmatrix} 0 \\ \delta \end{bmatrix}$ $(\delta > 0)$. Then

$$\mathbf{c} \in H_0^0 \tag{4.16}$$

This implies, from (4.13) and (4.14), that there exists $\begin{bmatrix} 0 \\ \bar{\eta} \end{bmatrix} \in T$ such that $\lambda \begin{bmatrix} 0 \\ \bar{\eta} \end{bmatrix} = \begin{bmatrix} 0 \\ \delta \end{bmatrix}$. However, $\mathbf{c} \in T^{00}$ and $T \subset T^{00}$ imply that $\lambda > 0$ and $\bar{\eta} > 0$. But if $\begin{bmatrix} 0 \\ \bar{\eta} \end{bmatrix} \in T$, then there exists

$$\begin{bmatrix} \bar{\mathbf{w}} \\ \bar{z} \end{bmatrix} \geq \begin{bmatrix} 0 \\ \bar{\eta} \end{bmatrix} > \begin{bmatrix} 0 \\ 0 \end{bmatrix}$$

which contradicts assumption (1). Therefore $T^0 \not\subset H_0$ and *(i)* and *(ii)* of the lemma follow.

Now, applying Lemma 4.6, we can establish Theorem 4.1.

THEOREM 4.1 *(Equivalence Theorem). A vector* $\mathbf{x}^0$ *is an optimal solution to Problem I if and only if there exists* $\mathbf{y}^0 \geq 0$ *such that* $(\mathbf{x}^0, \mathbf{y}^0)$ *is a saddle point solution of* $\psi(\mathbf{x}, \mathbf{y})$.

Proof. Let $\mathbf{x}^0$ be an optimal solution to Problem I, and define

$$SS = \left\{ \begin{bmatrix} \mathbf{b} - A\mathbf{x} \\ \mathbf{c}'(\mathbf{x} - \mathbf{x}^0) \end{bmatrix} \middle| \; \mathbf{x} \geq 0 \right\}$$

The set SS is a polyhedral convex set which satisfies conditions (1) and (2) of Lemma 4.6. In particular,

(i) $\mathbf{b} - A\mathbf{x} \geq 0$ and $\mathbf{x} \geq 0$ imply that $\mathbf{c}'\mathbf{x} - \mathbf{c}'\mathbf{x}^0 \leq 0$, since $\mathbf{x}^0$ solves Problem I

(ii) if $\mathbf{c}'(\mathbf{x} - \mathbf{x}^0) = 0$ for $\mathbf{x} \in \{\mathbf{x} \mid \mathbf{b} - A\mathbf{x} \geq 0, \mathbf{x} \geq 0\}$, then

$$\mathbf{x}^0 \in \{\mathbf{x} \mid \mathbf{b} - A\mathbf{x} \geq 0, \mathbf{x} \geq 0\},$$

i.e., $\mathbf{b} - A\mathbf{x}^0 \geq 0$. Therefore the results of Lemma 4.6 hold and there exist $\mathbf{u}^0 \geq 0$, $v^0 > 0$ such that

$$\mathbf{u}^{0\prime}\mathbf{w} + v^0 z \leq 0 \quad \text{for all} \quad \begin{bmatrix} \mathbf{w} \\ z \end{bmatrix} \in SS \tag{4.17}$$

Thus with $\mathbf{w} = \mathbf{b} - A\mathbf{x}$ in (4.17) and dividing by v^0 we have

$$(1/v^0)\mathbf{u}^{0\prime}(\mathbf{b} - A\mathbf{x}) + z \leq 0 \tag{4.18}$$

Letting $\mathbf{y}^0 = (1/v^0)\mathbf{u}^0 \geq 0$ and with $z = \mathbf{c}'(\mathbf{x} - \mathbf{x}^0)$ in (4.18)

$$\mathbf{y}^{0\prime}(\mathbf{b} - A\mathbf{x}) + \mathbf{c}'(\mathbf{x} - \mathbf{x}^0) \leq 0 \quad \text{for all} \quad \mathbf{x} \geq 0 \tag{4.19}$$

To show that $\mathbf{y}^{0\prime}(\mathbf{b} - A\mathbf{x}^0) = 0$, let $\mathbf{x} = \mathbf{x}^0$ in (4.19); then

$$\mathbf{y}^{0\prime}(\mathbf{b} - A\mathbf{x}^0) \leq 0$$

However, $\mathbf{y}^0 \geq 0$ and $\mathbf{b} - A\mathbf{x}^0 \geq 0$ imply that $\mathbf{y}^{0\prime}(\mathbf{b} - A\mathbf{x}^0) \geq 0$, hence

$$\mathbf{y}^{0\prime}(\mathbf{b} - A\mathbf{x}^0) = 0$$

Adding this zero term to the right side of (4.19) and rearranging terms

$$\mathbf{c}'\mathbf{x} + \mathbf{y}^{0\prime}(\mathbf{b} - A\mathbf{x}) \leq \mathbf{c}'\mathbf{x}^0 + \mathbf{y}^{0\prime}(\mathbf{b} - A\mathbf{x}^0) \quad \text{for all } \mathbf{x} \geq 0$$

Therefore

$$\psi(\mathbf{x}, \mathbf{y}^0) \leq \psi(\mathbf{x}^0, \mathbf{y}^0) \quad \text{for all } \mathbf{x} \geq 0$$

Likewise, $\psi(\mathbf{x}^0, \mathbf{y}^0) \leq \psi(\mathbf{x}^0, \mathbf{y}) \quad$ for all $\mathbf{y} \geq 0$ since

$$\mathbf{c}'\mathbf{x}^0 \leq \mathbf{c}'\mathbf{x}^0 + \mathbf{y}'(\mathbf{b} - A\mathbf{x}^0) \quad \text{for all } \mathbf{y} \geq 0$$
$$\mathbf{c}'\mathbf{x}^0 + \mathbf{y}^{0\prime}(\mathbf{b} - A\mathbf{x}^0) \leq \mathbf{c}'\mathbf{x}^0 + \mathbf{y}'(\mathbf{b} - A\mathbf{x}^0)$$
$$\psi(\mathbf{x}^0, \mathbf{y}^0) \leq \psi(\mathbf{x}^0, \mathbf{y}) \quad \text{for all } \mathbf{y} \geq 0$$

Thus $(\mathbf{x}^0, \mathbf{y}^0)$ is a saddle point of $\psi(\mathbf{x}, \mathbf{y})$.

To show next that if $(\mathbf{x}^0, \mathbf{y}^0)$ is a saddle point solution of $\psi(\mathbf{x}, \mathbf{y})$, then $\mathbf{x}^0$ solves Problem I. However, this is a direct consequence of Lemma 4.5 and Corollary 4.5.2.

THEOREM 4.2 *(Duality Theorem). If Problem I has an optimal solution* $\mathbf{x}^0$, *then Problem II has an optimal solution* $\mathbf{y}^0$. Moreover, $\mathbf{c}'\mathbf{x}^0 = \mathbf{b}'\mathbf{y}^0$.

Proof. From Theorem 4.1, if $\mathbf{x}^0$ solves Problem I, then there exists $\mathbf{y}^0 \geq 0$ such that $(\mathbf{x}^0, \mathbf{y}^0)$ is a saddle point solution of $\psi(\mathbf{x}, \mathbf{y})$. Again by Corollary 4.5.2, $\mathbf{y}^0$ solves Problem II and $\mathbf{c}'\mathbf{x}^0 = \mathbf{b}'\mathbf{y}^0$.

The Equivalence Theorem and Duality Theorem of linear programming establish that if $(\mathbf{x}^0, \mathbf{y}^0)$ is a saddle point solution of the Lagrangian $\psi(\mathbf{x}, \mathbf{y})$, then $\mathbf{x}^0$ solves Problem I and $\mathbf{y}^0$ solves Problem II. Moreover, if either Problem I or Problem II has an optimal solution, this is a sufficient condition that its dual problem has an optimal solution. These results hold free from any regularity conditions or constraint qualifications; this is not necessarily the case for nonlinear programming problems. The equivalence relationship between the primal and dual problems is illustrated in Fig. 4.3.

The Equivalence Theorem was established using Lemma 4.6 under the assumption that a certain set T was closed and convex. Now let us establish an equivalence theorem for certain problems over linear constraints with an arbitrary objective function $F(\mathbf{x})$.

Consider *Condition A*. The set

$$T = \left\{ \begin{bmatrix} \mathbf{r} \\ s \end{bmatrix} \middle| \begin{bmatrix} \mathbf{r} \\ s \end{bmatrix} \leq \begin{bmatrix} \mathbf{w} \\ z \end{bmatrix} \quad \text{for some} \begin{bmatrix} \mathbf{w} \\ z \end{bmatrix} \in S \right\}$$

is a closed convex set where

$$S = \left\{ \begin{bmatrix} \mathbf{b} - A\mathbf{x} \\ F(\mathbf{x}) - F(\mathbf{x}^0) \end{bmatrix} \middle| \mathbf{x} \geq 0 \right\}$$

and $\mathbf{x}^0$ is an optimal solution of Problem I.

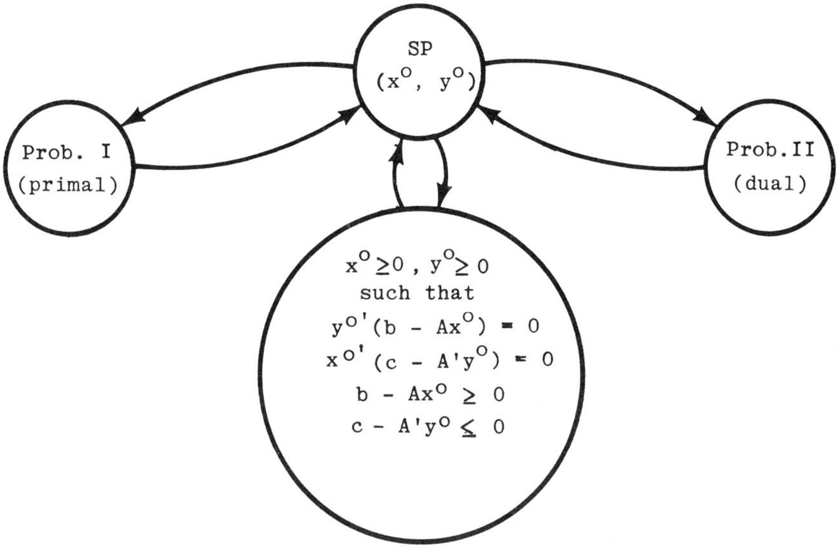

Figure 4.3. Equivalence relationship of linear programming problems.

If $F(\mathbf{x})$ is a concave function defined for all $\mathbf{x} \geq 0$, then Condition A is satisfied. In particular, for any $\alpha \in [0, 1]$ and any two vectors $\mathbf{x}_1$ and $\mathbf{x}_2$, $F(\mathbf{x})$ is said to be a concave function if $\alpha F(\mathbf{x}_1) + (1 - \alpha)F(\mathbf{x}_2) \leq F[\alpha\mathbf{x}_1 + (1 - \alpha)\mathbf{x}_2]$. Now consider any $\alpha \in [0, 1]$ and

$$\begin{bmatrix} \mathbf{r}_1 \\ s_1 \end{bmatrix}, \begin{bmatrix} \mathbf{r}_2 \\ s_2 \end{bmatrix} \in T$$

We have

$$\begin{aligned} \bar{\mathbf{r}} &= \alpha\mathbf{r}_1 + (1 - \alpha)\mathbf{r}_2 \leq \mathbf{b} - A[\alpha\mathbf{x}_1 + (1 - \alpha)\mathbf{x}_2] = \mathbf{b} - A\bar{\mathbf{x}} \\ \bar{s} &= \alpha s_1 + (1 - \alpha)s_2 \leq \alpha[F(\mathbf{x}_1) - F(\mathbf{x}^0)] + (1 - \alpha)[F(\mathbf{x}_2) - F(\mathbf{x}^0)] \\ &\leq F[\alpha\mathbf{x}_1 + (1 - \alpha)\mathbf{x}_2] - F(\mathbf{x}^0) \\ &= F(\bar{\mathbf{x}}) - F(\mathbf{x}^0) \end{aligned}$$

from the concavity of F. Therefore $\begin{pmatrix} \bar{\mathbf{r}} \\ \bar{s} \end{pmatrix} \in T$.

Hence the above results are valid when the constraints are linear and $F(\mathbf{x})$ is concave, of the form $\mathbf{c}'\mathbf{x} + \mathbf{x}'D\mathbf{x}$ where D is negative semidefinite, i.e., $\mathbf{x}'D\mathbf{x} \le 0$ for all $\mathbf{x}$. These problems are classified as quadratic programming problems and are discussed in Chapter 7.

THEOREM 4.3 *(Existence Theorem). If Problems I and II have at least one feasible solution* $\bar{\mathbf{x}}$ *and* $\bar{\mathbf{y}}$, *respectively, then both problems have optimal solutions. Moreover, the optima are equal.*

Proof. From Lemma 4.2 the set $R = \{\mathbf{c}'\mathbf{x} \mid \mathbf{b} - A\mathbf{x} \ge 0,\ \mathbf{x} \ge 0\}$ is bounded above by $\mathbf{b}'\bar{\mathbf{y}}$, and since R is a closed convex interval in E^1, the set R has a maximal element.

Similarly, the set $S = \{\mathbf{b}'\mathbf{y} \mid A'\mathbf{y} - \mathbf{c} \ge 0,\ \mathbf{y} \ge 0\}$ has a minimal element, since it is bounded below by $\mathbf{c}'\bar{\mathbf{x}}$.

Therefore, applying the Duality Theorem, Problems I and II both have optimal solutions and the optima are equal.

4.2 Fractional Programming

This section illustrates how some of the results of this chapter can be applied to obtain optimal solutions to a class of nonlinear problems called fractional programming problems (see Charnes and Cooper 1962). These problems are characterized by linear constraints and an objective function composed of a quotient of two linear functions $(\mathbf{b}'\mathbf{w} + b_0)/(\mathbf{d}'\mathbf{w} + d_0)$ where b_0 and d_0 are scalars. The general form of a fractional programming problem is given by the following problem

Problem III

$$\begin{aligned} \text{minimize} \quad & (\mathbf{b}'\mathbf{w} + b_0)/(\mathbf{d}'\mathbf{w} + d_0) \\ \text{subject to} \quad & A'\mathbf{w} - \mathbf{c} \ge 0 \\ & \mathbf{w} \ge 0 \end{aligned}$$

where $\mathbf{d}'\mathbf{w} + d_0 > 0$ for all $\mathbf{w} \in \Omega_{\text{III}} = \{\mathbf{w} \mid A'\mathbf{w} - \mathbf{c} \ge 0, \mathbf{w} \ge 0\}$.

Before discussing how to obtain an optimal solution of Problem III, consider the following linear programming problem

$$\begin{aligned} \text{minimize} \quad & \mathbf{b}'\mathbf{y} + b_0 y_0 \\ \text{subject to} \quad & A'\mathbf{y} - \mathbf{c}y_0 \ge 0 \\ & \mathbf{d}'\mathbf{y} + d_0 y_0 = 1 \\ & \mathbf{y} \ge 0 \\ & y_0 \ge 0 \end{aligned} \tag{4.20}$$

Then its dual problem is

$$\begin{aligned} \text{maximize} \quad & x_0 \\ \text{subject to} \quad & A\mathbf{x} + \mathbf{d}x_0 \le \mathbf{b} \\ & -\mathbf{c}'\mathbf{x} + d_0 x_0 \le b_0 \\ & \mathbf{x} \ge 0 \\ & x_0 \in E^1 \end{aligned} \tag{4.21}$$

By the Duality Theorem, if either (4.20) or (4.21) has an optimal solution then so does the other and, moreover, the optima are equal, i.e.

$$\mathbf{b}'\mathbf{y}^0 + b_0 y_0^0 = x_0^0 \tag{4.22}$$

If we select the easiest to solve computationally, say (4.21), then its optimal solution can be expressed as

$$(\mathbf{x}^0, x_0^0) = B^{-1}\begin{bmatrix} \mathbf{b} \\ b_0 \end{bmatrix}$$

and from Chapter 2, Lemma 2.5, $(\mathbf{y}^0, y_0^0) = \bar{\mathbf{c}}_B' B^{-1}$ solves (4.20).

Before showing that an optimal solution of (4.20) can yield a solution to Problem III, consider Problem III and (4.21) with $b_0 = 0$, $d_0 = 1$, and $\mathbf{d} = 0$. Then we have

$$\begin{aligned} \text{minimize} \quad & \mathbf{b}'\mathbf{w} \\ \text{subject to} \quad & A'\mathbf{w} - \mathbf{c} \ge 0 \\ & \mathbf{w} \ge 0 \end{aligned}$$

and

$$\begin{aligned} \text{maximize} \quad & x_0 \\ \text{subject to} \quad & A\mathbf{x} \le \mathbf{b} \\ & -\mathbf{c}'\mathbf{x} + x_0 \le 0 \\ & \mathbf{x} \ge 0 \\ & x_0 \in E^1 \end{aligned} \tag{4.23}$$

But under these conditions (4.23) is the dual of Problem III, since (4.23) can be equivalently expressed as

$$\begin{aligned} \text{maximize} \quad & \mathbf{c}'\mathbf{x} \\ \text{subject to} \quad & A\mathbf{x} \le \mathbf{b} \\ & \mathbf{x} \ge 0 \end{aligned}$$

At first glance this may seem surprising. However, we shall now show how Problem III can be transformed to (4.21). Let

$$v \;=\; \mathbf{d}'\mathbf{w} + d_0 > 0 \quad \text{for all } \mathbf{w} \in \Omega_{\text{III}} \tag{4.24}$$

$$y_0 = 1/v \qquad > 0 \tag{4.25}$$

$$\mathbf{y} \;=\; y_0\mathbf{w} \qquad \geq 0 \tag{4.26}$$

Then the objective function can equivalently be expressed as

$$\begin{aligned}(\mathbf{b}'\mathbf{w} + b_0)/(\mathbf{d}'\mathbf{w} + d_0) &= \mathbf{b}'\mathbf{w}/v + b_0(1/v)\\ &= \mathbf{b}'\mathbf{y} + b_0 y_0\end{aligned} \tag{4.27}$$

Also, the constraints can be rewritten as

$$\begin{aligned}A'\mathbf{y}v - \mathbf{c} &\geq 0\\ \mathbf{d}'\mathbf{w} + d_0 &= v\\ v &> 0\\ \mathbf{y} &\geq 0\end{aligned}$$

or

$$\begin{aligned}A'\mathbf{y} - \mathbf{c}y_0 &\geq 0\\ \mathbf{d}'\mathbf{y} + d_0 y_0 &= 1\\ \mathbf{y} &\geq 0\\ y_0 &> 0\end{aligned} \tag{4.28}$$

Under (4.24), (4.25), and (4.26) all feasible solutions in Problem III are in (4.20), and conversely, except when $y_0 = 0$. Hence if $(\mathbf{y}^0, y_0^0)$ is an optimal solution of (4.20) and $y_0^0 \neq 0$, then

$$\min\,(\mathbf{b}'\mathbf{w} + b_0)/(\mathbf{d}'\mathbf{w} + d_0) = \mathbf{b}'\mathbf{y}^0 + b_0 y_0^0 \qquad \mathbf{w} \in \Omega_{\text{III}}$$

and $\mathbf{w}^0 = (1/y_0^0)\mathbf{y}^0$ solves Problem III.

It is left as an exercise to show that if one has an optimal solution of (4.20) with $y_0^0 = 0$, then glb $(\mathbf{b}'\mathbf{w} + b_0)/(\mathbf{d}'\mathbf{w} + d_0) = \mathbf{b}'\mathbf{y}^0\,(\mathbf{w} \in \Omega_{\text{III}})$.

EXAMPLE 4.3

$$\begin{aligned}\text{minimize}\quad & (2w_1 + w_2)/(w_1 + 1)\\ \text{subject to}\quad & w_1 + w_2 \geq 1\\ & w_1 - w_2 \leq 2\\ & w_1,\quad w_2 \geq 0\end{aligned} \tag{4.29}$$

then

$$\mathbf{b} = \begin{bmatrix} 2 \\ 1 \end{bmatrix}, \mathbf{d} = \begin{bmatrix} 1 \\ 0 \end{bmatrix}, b_0 = 0, d_0 = 1, A' = \begin{bmatrix} 1 & 1 \\ -1 & 1 \end{bmatrix}, \mathbf{c} = \begin{bmatrix} 1 \\ -2 \end{bmatrix}$$

and (4.20) and its dual problem are

$$\begin{array}{ll} \text{minimize} & 2y_1 + y_2 \\ \text{subject to} & y_1 + y_2 - y_0 \geq 0 \\ & -y_1 + y_2 + 2y_0 \geq 0 \\ & y_1 + y_0 = 1 \\ & y_1, \ y_2, \ y_0 \geq 0 \end{array} \tag{4.30}$$

$$\begin{array}{ll} \text{maximize} & x_0 \\ \text{subject to} & x_1 - x_2 + x_0 \leq 2 \\ & x_1 + x_2 \leq 1 \\ & -x_1 + 2x_2 + x_0 \leq 0 \\ & x_1, \ x_2 \geq 0 \\ & x_0 \in E^1 \end{array} \tag{4.31}$$

Letting $x_0 = x_3 - x_4$, $x_3 \geq 0$, and $x_4 \geq 0$ in (4.31), we can obtain an optimal solution of (4.31) using the simplex algorithm. However, by noting that $y_0 = 1 - y_1$, (4.30) can be expressed as

$$\begin{array}{ll} \text{minimize} & 2y_1 + y_2 \\ \text{subject to} & 2y_1 + y_2 \geq 1 \\ & -3y_1 + y_2 \geq -2 \\ & y_1, \ y_2 \geq 0 \end{array}$$

But $(y_1^0, y_2^0) = (1/2, 0)$ is a feasible solution, and since $2y_1^0 + y_2^0 = 1$, (y_1^0, y_2^0) is an optimal solution. Therefore the solution of (4.29) is

$$\mathbf{w}^0 = \frac{1}{1/2}\begin{bmatrix} 1/2 \\ 0 \end{bmatrix} = \begin{bmatrix} 1 \\ 0 \end{bmatrix}$$

In this example, there exist alternative optimal solutions. In particular, $\mathbf{y}^{0\prime} = (0, 1)$ is an optimal solution and any convex combination of $(1/2, 0)$ and $(0, 1)$ is an optimal solution. This implies that the above fractional programming problem has multiple optimal solutions, namely, any convex combination of $(1/2, 0)$ and $(0, 1)$.

4.3 Economic Interpretation of Duality

In Chapter 2, Lemma 2.5, the reader was introduced to the notion of duality, in which the optimal solution of the dual problem could be expressed as $\mathbf{y}^{0\prime} = \mathbf{c}_B'B^{-1}$ where B is the optimal basis associated with the primal problem. Also, Section 2.4 discussed an economic interpretation of the so-called $z_j - c_j$ criterion. We shall now consider the above notion in the spirit of linear duality and relate this interpretation to the concept of range analysis.

A typical interpretation of the primal problem, Problem I of this chapter or Problem I after introducing slack variables in Chapter 2, is that these problems involve allocating limited resources among competing processes or activities. Here, as in Example 2.1, c_j is the unit profit, say in dollars, of process x_j and a_{ij} is the amount of b_i consumed by each unit of process x_j. In other words, the economic interpretation of the *primal* problem is: If c_j is the profit of each unit of activity x_j and b_i is the limiting amount of each input resource, then how much of each activity should be produced to maximize the total profit of these activities?

Let us now interpret the corresponding meaning of the dual problem. The constraint restrictions are bound below by the profit vector, i.e., $A'\mathbf{y} \geq \mathbf{c}$. Therefore, for any constraint i, the units associated with each y_i must be units of dollars per unit of b_i, since the units of each a_{ij} are units of b_i per unit of x_j and the units of c_j are dollars per unit of x_j. Thus each dual variable can be interpreted as the marginal value of resource i. The economic interpretation of the *dual* problem is: If b_i is the assumed supply of each input and c_j is the limiting bound on the unit value of each output, then what marginal or unit values should be allocated to each y_i to minimize the total value of these inputs?

Suppose now that $\mathbf{y}^0$ is the optimal solution of the dual problem. Then since there exists some $\mathbf{x}^0$ which solves the primal problem we have from the Duality Theorem that

$$\mathbf{b}'\mathbf{y}^0 = \mathbf{c}'\mathbf{x}^0$$

This implies that total prices of the resources to the user are equal to the total profit, and these prices should be set so they are a minimum. Also, if the availability of b_i increased (decreased) one unit, then the optimal value of $\mathbf{c}'\mathbf{x}$ would increase (decrease) by y_i^0; increasing (decreasing) the total income for the user. Hence y_i^0 (or $z_i - c_i$) is the maximum price that the user should be willing to pay if additional units of this resource were made available on the open market. Moreover, by the law of supply and demand, whenever the available supply of b_i is not exhausted then, in view of the fact that $\mathbf{y}^{0\prime}(\mathbf{b} - A\mathbf{x}^0) = 0$, it must be that $y_i^0 = 0$; that is, the value of additional units of resource i is zero and has no value on the open market.

Let us now reconsider each y_i^0 and relate the increase (decrease) of b_i

to sensitivity analysis or the range on b_i. Recall in Chapters 2 and 3 that each z_i was introduced by updating row 0 at each iteration until an optimal tableau was obtained. This implies that if b_i is increased (decreased) by n units, the new value of the optimum would change by nb_i provided the present basic solution remains the same; and if the availability of resource i does not move out of its range, y_i^0 indicates the rate of change at which profit would increase (decrease).

4.4 Extensions of Linear Duality

Variants of the linear duality have replaced the two orthant domains of the problem

$$\begin{aligned} &\text{maximize} \quad \mathbf{c}'\mathbf{x} \\ &\text{subject to} \quad \mathbf{b} - A\mathbf{x} \geq 0 \\ &\qquad\qquad\qquad\quad \mathbf{x} \geq 0 \end{aligned} \tag{4.32}$$

by cone domains L and C of the form

$$\mathbf{b} - A\mathbf{x} \in L \qquad \mathbf{x} \in C \tag{4.33}$$

Programming problems over cone domains have been extensively studied by Ben-Israel, Charnes, and Kortanek (1969); Berman and Ben-Israel (1971); Duffin (1965); Kretschmer (1961); Sposito (1974); and Sposito and David (1971, 1972). Other pertinent studies are given in the references.

In general, the dual of (4.32) over cone domains (4.33) is

$$\begin{aligned} &\text{minimize} \quad \mathbf{b}'\mathbf{y} \\ &\text{subject to} \quad A'\mathbf{y} - \mathbf{c} \in C^* \\ &\qquad\qquad\qquad\quad \mathbf{y} \in L^* \end{aligned} \tag{4.34}$$

where C^* and L^* denote the polar cones of C and L, respectively.

In view of the above, and the fact that $\{\mathbf{x} \mid \mathbf{x} \geq 0\}$ can be expressed as a closed convex cone Q_n^+, it follows that Problems I and II of this chapter could have been equivalently expressed as

$$\begin{aligned} &\text{maximize} \quad \mathbf{c}'\mathbf{x} \\ &\text{subject to} \quad \mathbf{b} - A\mathbf{x} \in Q_m^+ \\ &\qquad\qquad\qquad\quad \mathbf{x} \in Q_n^+ \end{aligned}$$

and

$$\begin{array}{ll} \text{minimize} & \mathbf{b}'\mathbf{y} \\ \text{subject to} & A'\mathbf{y} - \mathbf{c} \in Q_n^+ \\ & \qquad\ \mathbf{y} \in Q_m^+ \end{array}$$

An interesting observation concerning Problems I and II over polyhedral cone domains is that the two problems are symmetric in form even if $\mathbf{x}$ is unrestricted; for instance, when $C = E^n$ and $C^* = 0$. In our earlier discussion these problems over inequality and equality constraints had an asymmetric representation.

When the cone domains are not necessarily polyhedral, it can be shown under a certain regularity assumption that the dual of (4.32), restricted by (4.33), will again take the form of (4.34). This is covered by Sposito and David (1972).

Exercises

1. Formulate the dual problems of the following linear programming problems

a) $$\begin{array}{ll} \text{maximize} & x_1 + 2x_2 \\ \text{subject to} & x_1 + 3x_2 \le 2 \\ & x_1 + 2x_2 \le 1 \\ & -x_1 - x_2 \ge 0 \\ & x_1, \quad x_2 \ge 0 \end{array}$$

b) $$\begin{array}{ll} \text{maximize} & 2x_1 - x_2 \\ \text{subject to} & x_1 - x_2 \ge 1 \\ & x_1 \qquad \le 1 \\ & x_1, \quad x_2 \ge 0 \end{array}$$

c) $$\begin{array}{ll} \text{minimize} & 20x_1 + 30x_2 \\ \text{subject to} & x_1 + 2x_2 \le 20 \\ & x_1 + 3x_2 \ge 4 \\ & x_1, \quad x_2 \ge 0 \end{array}$$

2. Formulate the dual of the following problem so the dual variables are nonnegative

$$\begin{array}{ll} \text{maximize} & -3x_1 - 5x_2 - 2x_3 \\ \text{subject to} & 2x_1 + 3x_2 + x_3 = 21 \\ & 3x_1 + x_2 + 4x_3 = 38 \\ & x_1, \quad x_2, \quad x_3 \ge 0 \end{array}$$

3. Reformulate the following problem so all the variables are nonnegative

$$\begin{array}{ll} \text{maximize} & x_1 + x_2 \\ \text{subject to} & x_1 + x_2 \leq 2 \\ & 2x_1 + x_2 \geq 1 \\ & x_1 \geq 0 \\ & x_2 \text{ unrestricted} \end{array}$$

4. Consider Problems I and II.
   a) Prove that if **b** is a vector with all nonnegative components, then Problem I has a feasible solution.
   b) Prove that if the vector **b** and matrix A possess all positive components, then both Problems I and II have optimal solutions.
   c) Prove that if all the components of **c** and $-\mathbf{b}$ are nonpositive, then both Problems I and II have optimal solutions.
   d) Prove that if all the components of **c** and **b** and $-A$ are positive, then Problem I is unbounded.
5. Find a dual problem for the following linear programming problem

$$\begin{array}{ll} \text{minimize} & \mathbf{c}'\mathbf{x} \\ \text{subject to} & \mathbf{a} \leq A\mathbf{x} \leq \mathbf{b} \\ & \mathbf{x} \text{ unrestricted} \end{array}$$

6. Using the Duality Theorem and the two linear programming problems given below, prove the following well-known lemma from Farkas: A vector **b** will satisfy $\mathbf{b}'\mathbf{y} \geq 0$ for all $\mathbf{y} \geq 0$ with $A'\mathbf{y} \geq 0$ if and only if there exists $\mathbf{x} \geq 0$ such that $\mathbf{b} - A\mathbf{x} \geq 0$.

$$\begin{array}{ll} \text{minimize} & \mathbf{b}'\mathbf{y} \\ \text{subject to} & \mathbf{A}'\mathbf{y} \geq 0 \\ & \mathbf{y} \geq 0 \end{array}$$

and its dual

$$\begin{array}{ll} \text{maximize} & 0'\mathbf{x} \\ \text{subject to} & A\mathbf{x} \leq \mathbf{b} \\ & \mathbf{x} \geq 0 \end{array}$$

7. Prove that the system $A\mathbf{x} \in \tilde{L}, \mathbf{x} \in \tilde{C}$ is consistent if and only if $-\mathbf{y} \in L^*$, $A'\mathbf{y} \in C^*$ implies $\mathbf{y} = 0$. L and C are closed, convex cones and $\tilde{L}$ and $\tilde{C}$ denote the interiors of L and C.
8. Consider the linear model $\mathbf{y} = X\beta + \epsilon$ where $E(\epsilon) = 0$ and $E(\epsilon\epsilon') = \sigma^2$.
   (a) Show that the problem of minimizing the sum of squares; i.e., minimize $(\mathbf{Y} - X\beta)'(\mathbf{Y} - X\beta)$, can be solved by the simplex algorithm. (Hint: Consider the normal equations.)
   b) Find the dual problem of the linear programming problem established in *a*).
   c) Show that the normal equations are always consistent by considering only the dual problem.
   d) Assume the following restriction is imposed on the above least squares problem in *a*)

$$A\beta = \mathbf{b}$$

where A is an $m \times p$ matrix of arbitrary rank. Show that this problem can also be solved by the simplex algorithm by deriving an equivalent linear programming problem.

9. Consider Example 2.1 in Chapter 2.
   a) Establish its dual problem.
   b) What is the range associated with each resource and profit coefficient?
   c) Give an economic interpretation of Example 2.1 and its dual.
10. Consider Exercise 13 in Chapter 2. Give an economic interpretation of this problem and its dual.
11. Let L be $\{\mathbf{x} \in E^2 \mid x_1 - x_2 \leq 0, x_2 \geq 0\}$
   a) Show that L is a convex cone.
   b) Consider the problem

$$\begin{aligned} \text{maximize} \quad & -x_1 + 2x_2 \\ \text{subject to} \quad & x_1 + x_2 - 1 \in Q_1^+ \\ & \mathbf{x} \in L \end{aligned}$$

   Establish the dual of this problem over appropriate cone domains.
   c) Convert the above primal problem to constraints of the classical inequality type; do the same for the dual problem.
   d) Are the two problems in *c*) still mutually dual? Also, are they still symmetric?

References

Bector, C. R. 1968. Duality in fractional and indefinite programming. *Z. Angew. Math. Mech.* 48:418–20.

Ben-Israel, Adi. 1971. Linear equations and inequalities or finite dimensional, real or complex, vector spaces: a unified theory. *J. Math. Anal. Appl.* 27:367–89.

Ben-Israel, A.; Charnes, A.; and Kortanek, K. O. 1969. Duality and asymptotic solvability over cones. *Bull. Am. Math. Soc.* 75:318–24.

Berman, A., and Ben-Israel, A. 1971. Linear inequalities, mathematical programming and matrix theory. *J. Math. Program.* 1:291–300.

Charnes, A., and Cooper, W. W. 1962. Programming with linear fractional functionals. *Nav. Res. Logist. Q.* 9:181–86.

Duffin, R. J. 1965. Infinite programs. In *Linear Inequalities and Related Systems*. Eds. H. W. Kuhn and A. W. Tucker, pp. 157–70. Princeton Univ. Press, Princeton, N. J.

Fan, Ky. 1965. A generalization of the Alaoglu-Bourbaki theorem and its applications. *Math. Z.* 88:48–60.

Farkas, J. 1902. Uber die Theorie der einfachen Ungleichungen. *J. Reine Angew. Math.* 124:1–24.

Karlin, S. 1959. *Mathematical Methods and Theory of Games, Programming, and Economics*. Addison-Wesley, Reading, Mass.

Kretschmer, K. S. 1961. Programmes in paired spaces. *Can. J. Math.* 13:221–38.

Kuhn, H. W., and Tucker, A. W. 1950. Nonlinear programming. 2nd Berkeley Symp. Proc. Math. Stat. Probab., pp. 481–92. Univ. Calif. Press, Berkeley.

Kunzi, H., and Krelle, W. 1966. *Nonlinear Programming*. Blaisdell, Waltham, Mass.

Mangasarian, O. L. 1969. *Nonlinear Programming*. McGraw-Hill, New York.

Sposito, V. A. 1974. Modified regularity conditions for nonlinear programming problems over mixed cone domains. *J. Math. Program.* 6:167–79.

Sposito, V. A., and David, H. T. 1971. Saddle point optimality criteria of nonlinear programming problems over cones without differentiability. *SIAM J. Appl. Math.* 20:698–702.

______. 1972. A note on Farkas lemmas over cone domains. *SIAM J. Appl. Math.* 22:356–58.

Van Slyke, R. M., and Wets, R. J. 1968. A duality theory for abstract mathematical programs with applications to optimal control theory. *J. Math. Anal. Appl.* 22:679–706.

Varaiya, P. P. 1967. Nonlinear programming in Banach space. *SIAM J. Appl. Math.* 15:284–93.

Whittle, P. 1971. *Optimization under Constraints: Theory and Applications of Nonlinear Programming*. Wiley, New York.

CHAPTER FIVE

TOPICS IN LINEAR PROGRAMMING AND STATISTICS

The general area of linear and nonlinear programming has been applied to many types of problems in the last decade. This chapter focuses on a class of problems commonly found in the field of statistics, in particular, on statistical problems viewed as linear programming problems. These topics include estimating regression parameters in linear models, determining the solutions of Tchebycheff inequalities, the generalized Neyman-Pearson problem, and a special class of stochastic programming often called chance-constrained programming (see Francis and Meeks 1972, Karst 1958, Kempthorne and Folks 1971, Mood and Graybill 1963, and Whittle 1971).

In Chapter 3 it is noted that often a model builder may not have perfect information on the coefficients in the model, but there may be some basis to assume that these coefficients follow certain distributions, i.e., normal, uniform, etc. The model builder must determine a way of solving this probabilistic or stochastic model. A common approach is to convert it to a deterministic model with fixed coefficients that approximate the real situation. In this chapter, we discuss converting stochastic linear programming models to deterministic models, and in this vein, chance-constrained programming is introduced (see Charnes and Cooper 1959).

5.1 Linear Programming Techniques in Regression Analysis

As pointed out in Chapter 1, least squares theory is usually used as a criterion of "best fit." If the regression parameters are restricted to certain intervals, then the problem is a quadratic programming problem since the objective function can be expressed as a quadratic function. Otherwise we can solve the normal equations, i.e., solve the system $X'X\boldsymbol{\beta} = X'\mathbf{y}$. In the proper form this problem can be solved by linear programming techniques; namely, in the linear programming model

$$\begin{aligned} &\text{maximize} \quad \mathbf{c}'\mathbf{x} \\ &\text{subject to} \quad A\mathbf{x} = \mathbf{b} \end{aligned}$$

let $\mathbf{c} = 0$, $\mathbf{b} = X'\mathbf{y}$, and $A = X'X$, and we have the linear programming problem

$$\begin{array}{ll} \text{maximize} & 0'\beta \\ \text{subject to} & X'X\beta = X'\mathbf{y} \\ & \beta \text{ unrestricted} \end{array}$$

In regression problems alternatives to least squares as criteria of best fit are *least absolute deviations* and *least maximum deviation.* Using the duality theory of linear programming we can solve these latter two problems easily by linear programming techniques as shown below.

5.1.1 Minimizing the Maximum Absolute Deviation

Consider the model $\mathbf{y} = X\beta + \mathbf{e}$, where x_{ij} $(i = 1, 2, \ldots, k,\ j = 1, 2, \ldots, p)$ denotes a set of k observational measurements on p independent variables, and y_i $(i = 1, 2, \ldots, k)$ denotes the associated measurement on the dependent variable. Note that in the case of curvilinear regression we may have $x_{ij} = z_i^j$, $\log z_{ij}$, or $w_{ij}v_{ij}$, etc. We wish to find regression coefficients b_j that solve the following minimax problem

$$\underset{b_j}{\text{minimize}} \left\{ \underset{i}{\text{maximum}} \left| \sum_j x_{ij} b_j - y_i \right| \right\} \tag{5.1}$$

This problem is solved by linear programming techniques considering the following unknown bound on $(X\beta)_i - y_i$, namely, a variable e such that

$$-e \leq (X\beta)_i - y_i \leq e \quad \text{for all } i \tag{5.2}$$

where e denotes the value of the maximum absolute deviation we wish to minimize. Note from (5.2) that $e \geq 0$. We therefore have the following linear programming problem, i.e., a system with $2k$ equations in $p + 1$ unknowns.

$$\begin{array}{lll} \text{minimize} & e & \\ \text{subject to} & -(X\beta)_i + e \geq -y_i & \\ & (X\beta)_i + e \geq y_i & i = 1, 2, \ldots, k \end{array} \tag{5.3}$$

For this particular formulation, it is usually better computationally to solve the dual problem if $2k \gg p + 1$.

$$\begin{array}{ll} \text{maximize} & -\mathbf{y}'\mathbf{d}_1 + \mathbf{y}'\mathbf{d}_2 \\ \text{subject to} & -X'\mathbf{d}_1 + X'\mathbf{d}_2 = 0 \\ & \sum_i (d_{1i} + d_{2i}) = 1 \\ & d_{1i},\ d_{2i} \geq 0 \end{array} \tag{5.4}$$

If d_{1i} (or d_{2i}) is strictly positive in the optimal solution of (5.4), the maximum deviation occurs at the ith sample point and, furthermore, this point will lie below (or above) the fitted line.

5.1.2 *Minimizing the Sum of Absolute Deviations*

This criterion is similar to least squares except that deviations from the fitted line are measured in terms of absolute value rather than the square of deviations.

To convert this problem into a linear programming problem let

e_{1i} denote positive deviations

e_{2i} denote negative deviations

$$i = 1, 2, \ldots, k \qquad (5.5)$$

The e_{1i} and e_{2i} are vertical deviations above and below the fitted line for the ith set of observations; therefore, any fitted equation can be represented in terms of e_{1i} and e_{2i}. In particular, for any i

$$\sum_j x_{ij} b_j + e_{1i} - e_{2i} = y_i \qquad (5.6)$$

Hence, by the nature of the problem, at most one deviation can be nonzero. Also, one should note that for any ith set of observations, $e_{1i} + e_{2i}$ is the absolute deviation between the fit $\sum_j x_{ij} b_j$ and y_i.

Thus an appropriate objective function for a linear programming problem would be $\sum_{i=1}^{k}(e_{1i} + e_{2i})$. With (5.5) and (5.6) we have the following linear programming problem with k equations in $p + 2k$ unknowns.

$$\begin{aligned} &\text{minimize} \quad \textstyle\sum_{i=1}^{k} e_{1i} + \sum_{i=1}^{k} e_{2i} \\ &\text{subject to} \quad X\beta + I\mathbf{e}_1 - I\mathbf{e}_2 = \mathbf{y} \\ &\qquad\qquad\qquad \mathbf{e}_1, \quad \mathbf{e}_2 \geq 0 \end{aligned} \qquad (5.7)$$

where I denotes the identity matrix.

If the number of observations k is large, then it is easier to solve the dual of (5.7), i.e., $p + 2k$ equations in k unknowns

$$\begin{aligned} &\text{maximize} \quad \mathbf{y}'\mathbf{d} \\ &\text{subject to} \quad X'\mathbf{d} = 0 \\ &\qquad\qquad\quad d_i \leq 1 \\ &\qquad\qquad -d_i \leq 1 \quad \text{for all } i \end{aligned} \qquad (5.8)$$

This problem is reduced further by letting $w_i = d_i + 1$ $(i = 1, 2, \ldots, k)$. Then (5.8) becomes a system of p equations in k unknowns and k bounded variables.

$$\begin{aligned}
&\text{maximize} \quad \mathbf{y}'\mathbf{w} - \Sigma y_i \\
&\text{subject to} \\
&\qquad X'\mathbf{w} = X' \begin{bmatrix} 1 \\ \vdots \\ 1 \end{bmatrix} \\
&\qquad 0 \leq w_i \leq 2 \quad \text{for all } i
\end{aligned}$$

This problem can be solved readily by special simplex algorithms for bounded variable problems or with computer systems such as the IBM Mathematical Programming System in which one can input a bound section independent of the linear constraint system.

5.2 Necessary and Sufficient Conditions for L_1 Estimation

Minimizing the sum of absolute deviations is often called L_1 estimation.

As noted earlier, either a nonnegative primal variable takes value zero or the corresponding dual constraint is satisfied as an equality. For our problem, (5.7) and (5.8), this implies that

(i) either $e_{1i} = 0$ or $d_i = 1$, and
(ii) either $e_{2i} = 0$ or $d_i = -1$

Hence if the ith point is above a certain hyperplane, then $e_{1i} > 0$ so that $d_i = 1$; if the point is below the hyperplane, then $e_{2i} > 0$ so that $d_i = -1$.

This lemma, the Lemma of Complementary Slackness, is used to establish a sufficient condition under L_1.

Appa and Smith (1973) have shown that

(1) at least one hyperplane giving minimum sum of absolute deviations passes through p of the N points, and
(2) under the assumption that no set of $p + 1$ observations lies on one hyperplane in p dimensions, if one denotes N_1 as the number of points above a certain hyperplane and N_2 as the number of points below this hyperplane, then the hyperplane cannot be optimal under L_1 unless $|N_1 - N_2| \leq p$.

The proof of these properties is left to the reader.

These two conditions imply that the search for the best hyperplane is restricted to a subset of $\binom{N}{p}$ hyperplanes. The following lemma and corollary allow us to identify an optimal hyperplane from this subset of $\binom{N}{p}$ hyperplanes.

LEMMA 5.1. *Any feasible extreme point solution of the dual problem associated with an optimal solution of the primal problem is an optimal solution.*

Proof. Let $\mathbf{d}^*$ be any feasible extreme point solution of the dual problem associated with an optimal solution $(\mathbf{e}^*, \beta^*)$ of the primal problem.

Multiplying the primal constraints by $\mathbf{d}^*$ we have

$$\mathbf{d}^{*\prime}(X\beta^* + I\mathbf{e}_1^* - I\mathbf{e}_2^*) = \mathbf{d}^{*\prime}\mathbf{y}$$

Since $\mathbf{d}^*$ is a feasible solution of the dual problem, then $X'\mathbf{d}^* = 0$. Therefore we have

$$\mathbf{d}^{*\prime}(I\mathbf{e}_1^* - I\mathbf{e}_2^*) = \mathbf{d}^{*\prime}\mathbf{y}$$

Now N_1 of the e_{1i} and N_2 of the e_{2i} are positive; the remaining $e_{1i} = 0 = e_{2i}$. Hence by the nature of the problem, if $e_{2i}^* > 0$, then $e_{1i}^* = 0$ and $d_i^* = -1$. In view of the above and by the Lemma of Complementary Slackness, we have

$$\mathbf{d}^{*\prime}(I\mathbf{e}_1^* - I\mathbf{e}_2^*) = \mathbf{e}'I\mathbf{e}_1^* + \mathbf{e}'I\mathbf{e}_2^* = \sum_{i=1}^{k} (e_{1i}^* + e_{2i}^*)$$

where $\mathbf{e}$ is a unit vector. This implies that the optima of the primal and dual problems are equal, and by the Duality Theorem of linear programming, $\mathbf{d}^*$ is an optimal solution.

COROLLARY 5.1.1. *If $\mathbf{d}^*$ is any feasible dual solution associated with any feasible primal solution $(\beta^*, \mathbf{e}_1^*, \mathbf{e}_2^*)$ which satisfies the following conditions*

(1) $\mathbf{d}^{*\prime}(I\mathbf{e}_1^* - I\mathbf{e}_2^* - \mathbf{y}) = 0$

(2) $\mathbf{e}_1^{*\prime}(I\mathbf{d}^* - 1) = 0$, *and*

(3) $\mathbf{e}_2^{*\prime}(-I\mathbf{d}^* - 1) = 0$,

then the respective vectors are optimal.

Exercises 12, 13, and 14 exhibit additional properties of the L_1 estimator.

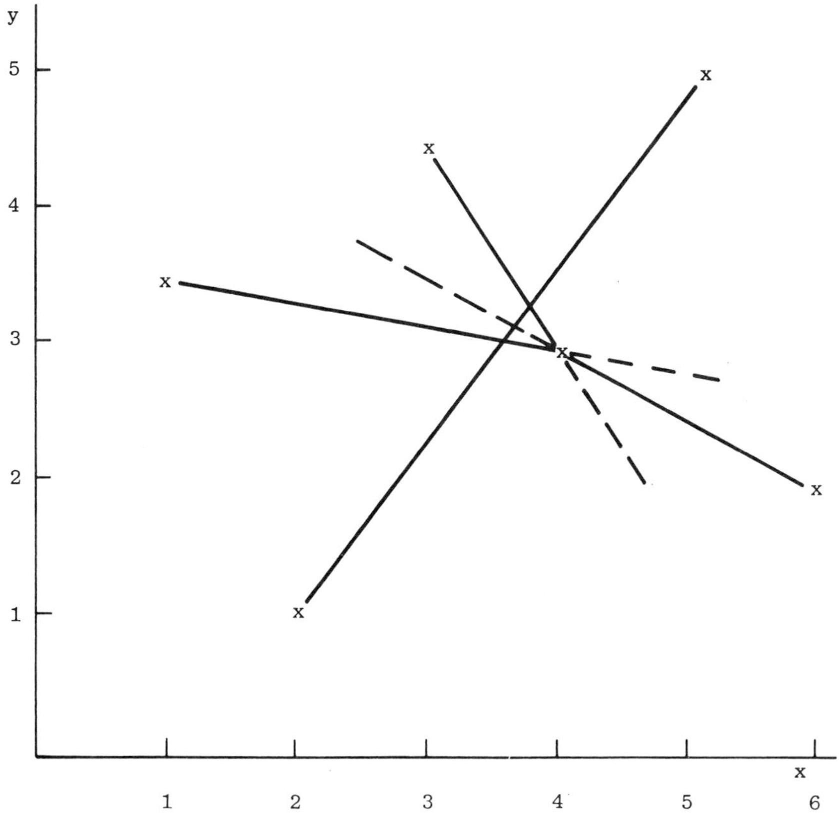

Figure 5.1. Example 5.1.

EXAMPLE 5.1

Consider the following points (x, y): $(1, 7/2)$, $(2, 1)$, $(3, 9/2)$, $(4, 3)$, $(5, 5)$, and $(6, 2)$ (see Fig. 5.1). In view of Appa and Smith's necessary conditions, with $p = 2$ one can restrict the search of the best linear model $y = b_0 + b_1 x$ under L_1 to four linear equations assuming the optimal equation is unique. In particular, since N is even, only lines that bipartition the points need be considered.

Consider the candidate (line) that passes through points $(1, 7/2)$ and $(4, 3)$. Then

$$e_{22} > 0 \text{ so that } d_2 = -1$$
$$e_{13} > 0 \text{ so that } d_3 = 1$$
$$e_{15} > 0 \text{ so that } d_5 = 1$$
$$e_{26} > 0 \text{ so that } d_6 = -1$$

If this line is to be optimal under L_1, then d_1 and d_4 must satisfy the following

$$\begin{aligned} d_1 + d_4 &= 0 \\ d_1 + 4d_4 &= 0 \\ d_1, \quad d_4 &\in [-1, 1] \end{aligned}$$

This is clearly satisfied by $d_1 = 0 = d_4$. Therefore, in view of Lemma 5.1, $\mathbf{d}^* = (0, -1, 1, 0, 1, -1)$ is an optimal solution of the dual problem and consequently the model $y = 11/3 - (1/6)x$ is optimal under L_1.

5.2.1 *Unbiased Estimators*

The following algorithms (Sielken and Hartley 1973) yield unbiased estimators under L_1 or L_∞ (minimizing the maximum deviation).

Consider an estimator such that

$$\beta - \beta_0(\mathbf{e}) = -[\beta - \beta_0(\mathbf{e})]$$

This estimator is an antisymmetrical estimator of β. The least squares estimator $\hat{\beta} = (X'X)^{-1}X'\mathbf{y}$ is an antisymmetrical estimator since

$$\begin{aligned} \hat{\beta}(\mathbf{e}) &= (X'X)^{-1}X'(X\beta + \mathbf{e}) = \beta + (X'X)^{-1}X'\mathbf{e} \\ \hat{\beta}(-\mathbf{e}) &= (X'X)^{-1}X'(X\beta - \mathbf{e}) = \beta - (X'X)^{-1}X'\mathbf{e} \end{aligned}$$

Hence it follows that

$$\begin{aligned} &\beta - \hat{\beta}(\mathbf{e}) = -[\beta - \hat{\beta}(-\mathbf{e})] \\ &E[\hat{\beta}(\mathbf{e})] = \beta = E[\hat{\beta}(-\mathbf{e})] \end{aligned}$$

The feasible region associated with the primal problem under L_1 is expressed as

$$\begin{aligned} \mathbf{y} - X\mathbf{b} + I\mathbf{e}_1 + I\mathbf{e}_2 &= 2I\mathbf{e}_1 \geq 0 \\ \mathbf{y} - X\mathbf{b} - I\mathbf{e}_1 - I\mathbf{e}_2 &= -2I\mathbf{e}_2 \leq 0 \end{aligned}$$

Letting $u_i = e_{1i} + e_{2i}$, we have the following equivalent linear programming problem

$$\begin{aligned} \text{minimize} \quad & \textstyle\sum_{i=1}^{n} u_i \\ \text{subject to} \quad & -I\mathbf{u} \leq \mathbf{y} - X\mathbf{b} \leq I\mathbf{u} \\ & \mathbf{u} \geq 0 \end{aligned}$$

Rewriting $\mathbf{b}$ and $\hat{\beta}$ as the difference of two nonnegative quantities, $\mathbf{b} = \mathbf{b}_1 - \mathbf{b}_2$ and $\hat{\beta} = \beta^{(1)} - \beta^{(2)}$, we can rewrite the constraints as

$$\begin{aligned} -\mathbf{X}_i(\mathbf{b}_1 + \beta^{(2)}) + \mathbf{X}_i(\mathbf{b}_2 + \beta^{(1)}) - u_i &\le -y_i + \mathbf{X}_i\hat{\beta} \\ \mathbf{X}_i(\mathbf{b}_1 + \beta^{(2)}) - \mathbf{X}_i(\mathbf{b}_2 + \beta^{(1)}) - u_i &\le y_i - \mathbf{X}_i\hat{\beta} \\ \mathbf{b}_j, \beta^{(j)}, \mathbf{u} &\ge 0 \qquad j = 1, 2 \end{aligned}$$

where $\mathbf{X}_i$ denotes the ith row of X.

Let $\mathbf{B}_1 = \mathbf{b}_1 + \beta^{(2)}$ and $\mathbf{B}_2 = \mathbf{b}_2 + \beta^{(1)}$. If $(\bar{\mathbf{B}}_1, \bar{\mathbf{B}}_2, \bar{\mathbf{u}})$ solves the following problem P_1

$$\begin{aligned} &\text{minimize} \quad \Sigma u_i \\ &\text{subject to} \quad \begin{bmatrix} -X & X & -I \\ X & -X & -I \end{bmatrix} \begin{bmatrix} \mathbf{B}_1 \\ \mathbf{B}_2 \\ \mathbf{u} \end{bmatrix} \le \begin{bmatrix} -\mathbf{y} + X\hat{\beta} \\ \mathbf{y} - X\hat{\beta} \end{bmatrix} \\ &\qquad\qquad \mathbf{B}_1, \mathbf{B}_2, \mathbf{u} \ge 0 \end{aligned}$$

then $\bar{\mathbf{B}}_1 - \bar{\mathbf{B}}_2 + \hat{\beta}$ is an estimator of $\mathbf{b}$ that minimizes the sum of absolute deviations. This program can be written equivalently as P_2

$$\begin{aligned} &\text{minimize} \quad \Sigma u_i \\ &\text{subject to} \quad \begin{bmatrix} -X & X & -I \\ X & -X & -I \end{bmatrix} \begin{bmatrix} \mathbf{B}_2 \\ \mathbf{B}_1 \\ \mathbf{u} \end{bmatrix} \le \begin{bmatrix} \mathbf{y} - X\hat{\beta} \\ -\mathbf{y} + X\hat{\beta} \end{bmatrix} \\ &\qquad\qquad \mathbf{B}_1, \mathbf{B}_2, \mathbf{u} \ge 0 \end{aligned}$$

Algorithm:

(i) select one of the problems, P_1 or P_2, with probability 1/2
(ii) estimate $\mathbf{b}$ by $\bar{\beta} = \bar{\mathbf{B}}_1 - \bar{\mathbf{B}}_2 + \hat{\beta}$

We now show that $\bar{\beta}$ is an unbiased estimator of $\mathbf{b}$; $P_1(\mathbf{e})$ and $P_2(-\mathbf{e})$ are identical, as are $P_2(\mathbf{e})$ and $P_1(-\mathbf{e})$, since

$$\begin{aligned} \mathbf{y}(-\mathbf{e}) - X\hat{\beta}(-\mathbf{e}) &= X\mathbf{b} - \mathbf{e} - X\hat{\beta}(-\mathbf{e}) \\ &= -\mathbf{e} + X[\beta - \hat{\beta}(-\mathbf{e})] \\ &= -\mathbf{e} - X[\beta - \hat{\beta}(\mathbf{e})] \\ &= -\mathbf{y}(\mathbf{e}) + X\hat{\beta}(\mathbf{e}) \end{aligned}$$

Hence

$$\bar{\mathbf{B}}_1[P_1(\mathbf{e})] - \bar{\mathbf{B}}_2[P_1(\mathbf{e})] = \bar{\mathbf{B}}_2[P_2(-\mathbf{e})] - \bar{\mathbf{B}}_1[P_2(-\mathbf{e})]$$
$$\bar{\mathbf{B}}_1[P_2(\mathbf{e})] - \bar{\mathbf{B}}_2[P_2(\mathbf{e})] = \bar{\mathbf{B}}_2[P_1(-\mathbf{e})] - \bar{\mathbf{B}}_1[P_1(-\mathbf{e})]$$

Using the above we have

$$\begin{aligned} E[\beta - \bar{\beta}] &= E[\beta - (\bar{\mathbf{B}}_1 - \bar{\mathbf{B}}_2 + \hat{\beta})] \\ &= E[\beta - \hat{\beta}] - E[\bar{\mathbf{B}}_1 - \bar{\mathbf{B}}_2] \\ &= 0 - E\{[\bar{\mathbf{B}}_1 - \bar{\mathbf{B}}_2 \mid \mathbf{e}] + [\bar{\mathbf{B}}_1 - \bar{\mathbf{B}}_2 \mid -\mathbf{e}]\} \\ &= -E\{[\bar{\mathbf{B}}_1 - \bar{\mathbf{B}}_2 \mid P_1(\mathbf{e})]\,1/2 \\ &\quad + [\bar{\mathbf{B}}_1 - \bar{\mathbf{B}}_2 \mid P_2(\mathbf{e})]\,1/2 \\ &\quad + [\bar{\mathbf{B}}_1 - \bar{\mathbf{B}}_2 \mid P_1(-\mathbf{e})]\,1/2 \\ &\quad + [\bar{\mathbf{B}}_1 - \bar{\mathbf{B}}_2 \mid P_2(-\mathbf{e})]\,1/2\} = 0 \end{aligned}$$

i.e., $\bar{\beta} = \bar{\mathbf{B}}_1 - \bar{\mathbf{B}}_2 + \hat{\beta}$ is an unbiased estimator of $\mathbf{b}$.

As an exercise show that one can also use the above procedure to find an unbiased estimator under L_∞, i.e., minimize the maximum deviation.

5.3 Generalized Tchebycheff Inequalities

Suppose, given a bounded random variable X $(-\infty < a \le X \le b < \infty)$ for which the distribution function is unknown, we want to determine a sharp upper bound U for the expected value of some function of X, say $f(X)$, where the expected values of some $n + 1$ functions of X, say $g_0(X)$, $g_1(X), \ldots, g_n(X)$ are known; i.e., we want to determine a sharp constant U such that

$$E[f(X)] \le U$$

subject to the constraints $E[g_i(X)] = c_i$ $(i = 0, 1, \ldots, n)$. By "sharp" we mean that there exists a random variable X_0 with values in $[a, b]$ such that the maximum of $E[f(X_0)] = U$.

First, suppose we "linearize" by assuming there exists y_i $(i = 0, 1, \ldots, n)$ such that

$$f(X) \le \sum_{i=0}^{n} y_i g_i(X) \quad \text{for all } a \le X \le b$$

then we would have

$$E[f(X)] \leq \sum_{i=0}^{n} y_i E[g_i(X)] = \mathbf{y}'\mathbf{c}$$

To obtain the best upper bound subject to this assumption, we must determine a solution to the following minimization problem LP_1

find a vector $\mathbf{y}' = (y_0, y_1, \ldots, y_n)$
which minimizes $\mathbf{y}'\mathbf{c}$
subject to

$$f(x) \leq \sum_{i=0}^{n} y_i g_i(x) \quad \text{for all } x \in [a, b] \tag{5.9}$$

Note the transition from X to x. We shall show that this linear approach does indeed yield a sharp upper bound. Very often the dual of a primal programming problem either is easier to solve than the primal or the mere formulation of the dual sheds insight into the solution of the primal; both of these statements are true for the primal programming problem LP_1.

Suppose instead of demanding that the side condition (5.9) in LP_1 be satisfied for all $x \in [a, b]$, we require only that it be satisfied on some finite set of points $a = x_1 < x_2 < x_3 < \cdots < x_{m-1} < x_m = b$. Then the modified minimization problem is a linear programming problem LP

$$\begin{aligned} &\text{minimize} && \mathbf{y}'\mathbf{c} \\ &\text{subject to} && A\mathbf{y} \geq \mathbf{b} \\ &&& A = [a_{ji}] = [g_i(x_j)] \\ &&& \mathbf{b}' = (f(x_1), f(x_2), \ldots, f(x_m)) \end{aligned}$$

Let us now formulate the dual LD

$$\begin{aligned} &\text{maximize} && \mathbf{w}'\mathbf{b} \\ &\text{subject to} && A'\mathbf{w} = \mathbf{c} \\ &&& \mathbf{w} \geq 0 \end{aligned}$$

or

$$\begin{aligned} &\text{maximize} && \textstyle\sum_{j=1}^{m} w_j f(x_j) \\ &\text{subject to} && \textstyle\sum_{j=1}^{m} g_0(x_j)\, w_j = c_0 \\ &&& \textstyle\sum_{j=1}^{m} g_1(x_j)\, w_j = c_1 \\ &&& \vdots \\ &&& \textstyle\sum_{j=1}^{m} g_n(x_j)\, w_j = c_n \\ &&& w_j \geq 0 \qquad j = 1, 2, \ldots, m \end{aligned}$$

In some (unspecified) sense we can visualize LP_1 as a limiting form of LP where we let m, the number of values of x in $[a, b]$ for which (5.9) holds, increase to infinity. If this concept is valid, then we would expect the dual of LP_1 to take a limiting form of LD, and consequently the objective function and the constraints become limiting forms of the sums in LD, or integrals; i.e., we expect the dual to be of the form

$$\begin{aligned}
\text{maximize} \quad & \int_a^b f(x)\,dw(x) \quad \text{over all } w(x) \\
\text{subject to} \quad & \int_a^b g_0(x)\,dw(x) = c_0 \\
& \int_a^b g_1(x)\,dw(x) = c_1 \\
& \quad \vdots \\
& \int_a^b g_n(x)\,dw(x) = c_n \\
& \qquad\qquad w(x) \geq 0 \quad \text{for all } x \in [a, b]
\end{aligned}$$

Let $g_i(x) = x^i$ $(i = 0, 1, \ldots, n)$ then the primal LP_1 is

$$\begin{aligned}
\text{minimize} \quad & \mathbf{y}'\mathbf{c} \quad \text{over all } \mathbf{y} \\
\text{subject to} \quad & f(x) \leq \textstyle\sum_{i=0}^{n} y_i x^i \quad \text{for all } x \in [a, b]
\end{aligned}$$

and its dual LD_1 is

$$\begin{aligned}
\text{maximize} \quad & \int_a^b f(x)\,dw(x) \quad \text{over all } w(x) \geq 0 \\
\text{subject to} \quad & \int_a^b dw(x) = 1 \\
& \int_a^b x\,dw(x) = c_1 \\
& \int_a^b x^2\,dw(x) = c_2 \\
& \quad \vdots \\
& \int_a^b x^n\,dw(x) = c_n
\end{aligned}$$

Thus the dual becomes a problem in maximizing a Stieltjes integral of $f(x)$ with respect to a distribution function satisfying n moment constraints.

Now suppose that an optimal finite solution $\mathbf{y}^0$ of LP_1 exists and LD_1 is its dual. Then letting $w_0(x)$ be the optimal finite solution of LD_1 we have

$$\mathbf{y}^{0\prime}\mathbf{c} = \int_a^b f(x)\,dw_0(x) \leq \int_a^b \left[\sum_{i=0}^{n} y_i^0 g_i(x)\right] dw_0(x) = \mathbf{y}^{0\prime}\mathbf{c}$$

so that

$$\int_a^b \left[f(x) - \sum_{i=0}^{n} y_i^0 g_i(x) \right] dw_0(x) = 0 \ .$$

which implies that $w_0(x) = 0$ whenever $f(x) \neq \sum_{i=0}^{n} y_i^0 g_i(x)$ and that w_0 concentrates all its positive mass on the zeroes of the function $h(x)$, where

$$h(x) = f(x) - \sum_{i=0}^{n} y_i^0 g_i(x)$$

since $h(x)$ is always nonpositive. Thus the vector $\mathbf{y}^0$ is obtained by ascertaining all possible schemes in which the nonpositive function

$$h(x) = f(x) - \sum_{i=0}^{n} y_i g_i(x) = 0$$

and then assigning mass points for w in accordance with the side conditions for LD_1. We then pick out that scheme and its corresponding w_0 which gives a maximum for $\int_a^b f(x)\, dw(x) = I$. If s_0 is that set of zeroes for $h(x)$ which maximizes I and w_0 is its corresponding probability measure, then we have

$$\mathbf{y}^{0\prime}\mathbf{c} = \int_a^b f(x)\, dw_0(x) = \sum_{x_j \in s_0} (w_0)_j f(x_j) = U$$

The following theorem (a modification of Theorems 1 and 2 of Kingman 1963) shows that LD_1 is indeed the dual of LP_1 and LP_1 possesses a finite optimal solution under some mild regularity conditions.

We shall define

(i) $\mathcal{P}(\mathbf{c})$ to be the set of random variables X taking values in $\Omega \subset E^1$ satisfying $E[g_i(X)] = c_i \ (i = 0, 1, \ldots, n)$ with $g_0(X) = c_0 = 1$

(ii) $I = \{\mathbf{z} \in E^{n+1} \mid z_1 = c_1, z_2 = c_2, \ldots, z_n = c_n\}$

(iii) $H = \{\mathbf{z} \in E^{n+1} \mid z_0 = E[f(X)], z_1 = E[g_0(X)], \ldots, z_n = E[g_n(X)]\}$

THEOREM 5.1 (*Kingman*). *If $I \cap H$ contains at least one point and $\mathcal{P}(\mathbf{c})$ is nonempty, then either*

(i) there exist $n + 1$ numbers $y_0^0, y_1^0, \ldots, y_n^0$ such that

$$f(t) \leq \sum_{i=0}^{n} y_i^0 g_i(t) \ \textit{for all } t \tag{5.10}$$

and lub $E[f(X)] = \mathbf{y}^{0\prime}\mathbf{c}$ *for* $X \in \mathcal{P}(\mathbf{c})$, *or*

(ii) no vector $\mathbf{y}$ *satisfies (5.10) and* lub $E[f(X)] = +\infty$ *for* $X \in \mathcal{P}(\mathbf{c})$.

LEMMA 5.2 (*Weak Duality*). *If there exists* $\mathbf{y}^0 \in E^{n+1}$ *such that* $f(t) \leq \sum_{i=0}^{n} y_i^0 g_i(t)$ *for all* t, *then*

$$\text{lub } E[f(X)] \leq \text{glb } \mathbf{y}'\mathbf{c} \qquad X \in \mathcal{P}(\mathbf{c}) \quad \mathbf{y} \textit{ restricted by } (5.10)$$

Proof. Consider all $\mathbf{y}$ such that

$$0 \leq \sum_{i=0}^{n} y_i g_i(t) - f(t) \quad \text{for all } t$$

then $0 \leq \mathbf{y}'\mathbf{c} - E[f(X)]$ or $E[f(X)] \leq \mathbf{y}'\mathbf{c}$. Therefore

$$\text{lub } E[f(X)] \leq \mathbf{y}'\mathbf{c} \qquad X \in \mathcal{P}(\mathbf{c})$$

$$\text{lub } E[f(X)] \leq \text{glb } \mathbf{y}'\mathbf{c} \qquad X \in \mathcal{P}(\mathbf{c}) \quad \mathbf{y} \text{ restricted by } (5.10)$$

In view of Lemma 5.2, condition (*i*) of Theorem 5.1 is stated equivalently: there exists a vector $\mathbf{y}^0 \in E^{n+1}$ such that

$$\mathbf{y}^{0\prime}\mathbf{c} = \text{glb } \mathbf{y}'\mathbf{c} = \text{lub } E[f(X)] \quad \mathbf{y} \text{ restricted by } (5.10)$$

Thus, letting $\Omega = [a, b]$, the theorem tells us if $\mathbf{y}^0$ exists such that the primal problem LP_1 has a solution, then its corresponding dual problem LD_1 has an optimal solution

$$\begin{aligned} &\text{find lub}_w \int_a^b f(x)\,dw(x) \\ &\text{subject to} \quad \int_a^b g_i(x)\,dw(x) = c_i \qquad i = 0, 1, \ldots, n \\ &\qquad\qquad\qquad\qquad w(x) \geq 0 \end{aligned}$$

Moreover, the greatest lower bound for the primal equals the least upper bound for the dual or, since their values are attained, the minimum for the primal equals the maximum for the dual, i.e.

$$\mathbf{y}^{0\prime}\mathbf{c} = \int_a^b f(x)\,dw_0(x)$$

We therefore have the following corollary.

COROLLARY 5.2.1. *Suppose LP_1 is feasible and (5.10) holds. Then either LP_1 and LD_1 have optimal solutions and their optima are equal or LP_1 is unbounded and LD_1 is infeasible.*

EXAMPLE 5.2

Suppose X is a random variable bounded between $-a$ and a with an average value or mean 1, and we want a sharp upper bound on $E[X^2]$. An easy way to solve this problem is to consider the following dual problem

$$\begin{aligned} &\text{minimize} \quad y_0 + \mu y_1 = y_0 + y_1 \\ &\text{subject to} \quad x^2 \leq y_0 + y_1 \quad \text{for all } x \in [-a, a] \end{aligned}$$

or subject to the restriction that $a^2 \leq y_0 + y_1$. Therefore $E[X^2] = a^2$ and is achieved for $\bar{\mathbf{y}}^{0\prime} = (0, a^2)$, $\bar{\bar{\mathbf{y}}}^{0\prime} = (a^2, 0)$, or a convex combination of $\bar{\mathbf{y}}^0$ and $\bar{\bar{\mathbf{y}}}^0$.

EXAMPLE 5.3

Suppose we have a bounded random variable X $(a \leq X \leq b)$ whose mean μ is known, and we want a sharp Tchebycheff upper bound for $E[X^2]$. We have $h(x) = x^2 - (y_0 + y_1 x) \leq 0$ on $[a, b]$, and thus $h(x)$ can have at most two zeroes.

The dual problem is

$$\begin{aligned} &\text{maximize} \quad \int_a^b x^2\, dw(x) \text{ over all probability measures } w(x) \\ &\text{subject to} \quad \int_a^b x\, dw(x) = \mu \\ &\qquad\qquad\quad \int_a^b dw(x) = 1 \end{aligned}$$

Since $h(x)$ is a quadratic function that is nonpositive for all $x \in [a, b]$, then $h(x)$ must have two zeroes in $[a, b]$ at the points $x = a$ and $x = b$.

To determine $w_0(x)$, we must solve the following two equations

$$\begin{aligned} w_0(a) + w_0(b) &= 1 \\ a w_0(a) + b w_0(b) &= \mu \end{aligned}$$

or $a w_0(a) + b[1 - w_0(a)] = \mu$. Therefore

$$\begin{aligned} w_0(a) &= (b - \mu)/(b - a) \\ w_0(b) &= 1 - w_0(a) = (\mu - a)/(b - a) \end{aligned}$$

Finally we have

$$\underset{w}{\text{lub}} \int_a^b x^2\,dw(x) = \int_a^b x^2\,dw_0(x)$$

$$= a^2(b - \mu)/(b - a) + b^2(\mu - a)/(b - a)$$
$$= [\mu(b^2 - a^2) + ab(a - b)]/(b - a)$$
$$= \mu(b + a) - ab = E[X^2]$$

It should be noted that the variance of X is $E[X^2] - E[X]^2 = \mu(b + a) - ab - \mu^2 = -(a - \mu)(b - \mu)$.

5.4 Tchebycheff Inequality

Suppose we have an unknown distribution function with mean μ and finite variance σ^2. Suppose we wish to estimate the probability of drawing an observation at random which is a distance as much as $k\sigma$ away from the mean. The well-known Tchebycheff inequality is a popular tool to estimate this probability, namely

$$P\{|X - \mu| \geq k\sigma\} \leq 1/k^2 \tag{5.11}$$

For example, the probability of drawing a sample value of more than 3 standard deviations (3σ) away from the mean is equal to or less than 1/9 regardless of the shape of the population distribution.

Often we have a random sample of size n from some unknown distribution and wish to obtain some conservative confidence intervals on $\bar{x}$, the sample mean. This is easily obtained as $[\mu - k\sigma/n, \mu + k\sigma/n]$ by considering the following lemma.

LEMMA 5.3 (*Tchebycheff's Inequality for samples of size n*). *Let $f(x)$ be an unknown distribution with mean μ and finite variance σ^2, k be any positive number, and $\bar{x}$ be the sample mean. Then*

$$P\{|\bar{x} - \mu| \geq k\sigma/n\} \leq 1/k^2 \tag{5.12}$$

Without loss of generality, let $n = 1$ and $k\sigma = \alpha$; then (5.12) is written equivalently

$$P\{|X - \mu| \geq \alpha\} \leq \sigma^2/\alpha^2 \tag{5.13}$$

In view of Kingman's Theorem (Theorem 5.1), when would (5.13) yield a sharp inequality?

Consider the random variable X_0 that concentrates its mass at

$(\mu - \alpha, \mu, \mu + \alpha)$ with probabilities $\sigma^2/2\alpha^2$, $1 - \sigma^2/\alpha^2$, and $\sigma^2/2\alpha^2$, respectively. Then

$$\begin{aligned} P\{\,|X_0 - \mu| \geq \alpha\} &= \sigma^2/\alpha^2 \quad \text{if } \sigma^2/\alpha^2 \leq 1 \\ &= 1 \quad \text{if } \sigma^2/\alpha^2 \geq 1 \end{aligned}$$

Hence, we have the following lemma in view of Theorem 5.1.

LEMMA 5.4. *If* $\sigma^2/\alpha^2 \leq 1$, *then*

$$\text{lub } E[b(X)] = E[b(X_0)] = \sigma^2/\alpha^2 \qquad X \in \mathcal{P}(\mathbf{c})$$

If $\sigma^2/\alpha^2 \geq 1$, *then*

$$\text{lub } E[b(X)] = \text{glb } \mathbf{y}'\mathbf{c} = 1 \qquad X \in \mathcal{P}(\mathbf{c}) \quad \mathbf{y} \textit{ restricted by } (5.9)$$

EXAMPLE 5.4

Suppose we have a bounded random variable, $-\alpha \leq X \leq \alpha$, with mean zero and finite variance σ^2. Then

$$\begin{aligned} &\text{lub } P\{\,|X| \geq \alpha\} = \text{glb } (y_0 + y_2\sigma^2) \quad \mathbf{y} \text{ restricted by } (5.9) \\ &E[X] = 0 \\ &E[X^2] = \sigma^2 \end{aligned}$$

and $\mathbf{y}$ is restricted so that

$$b(t) \leq y_0 + ty_1 + t^2y_2 \quad \text{for all } t$$

or $E[b(t)] \leq y_0 + \sigma^2 y_2$. Thus $\mathbf{y}^{0\prime} = (0, 0, 1/\alpha^2)$ is the least upper bound solution and $\mathbf{y}^{0\prime}\mathbf{c} = \sigma^2/\alpha^2$ if $\sigma^2/\alpha^2 \leq 1$; otherwise $\mathbf{y}^{0\prime}\mathbf{c} = 1$.

5.5 Generalized Neyman-Pearson Lemma

Let F_{x_θ} be a continuous family of distributions and assume we are testing some hypothesis H_0 against alternative H_a.

$$H_0 : F = F_{x_{\theta_0}} \qquad H_a : F = F_{x_{\theta_1}}$$

For each random sample from our sample space Ω, the experimenter must construct or determine some decision function based on the sample; i.e., the experimenter must decide either to accept or reject H_0.

In hypothesis testing we usually define a decision function by means of a real-valued statistic $\phi(x) \in [0, 1]$ called the test function which is defined over Ω. In particular, for a given outcome x, the probability that the experimenter accepts the alternative is taken to be $\phi(x)$, and the probability that the experimenter accepts the hypothesis is $1 - \phi(x)$.

Usually we want a test function that gives a test of size $\alpha \in (0, 1)$, the probability of rejecting a true hypothesis. We also desire a test which will maximize the probability of rejecting a hypothesis when it is in fact false. In other words, we are looking for a test that has maximum power for a fixed size α. The first solution of this problem was given by Neyman and Pearson for a simple hypothesis versus a simple alternative; it is often referred to as the Neyman-Pearson Lemma (see Lehmann 1959). It is written as

$$\begin{aligned} &\text{maximize} \quad \int_{E^1} \phi(t)\, dF_{x_{\theta_1}}(t) \qquad \text{for all } \phi(t) \in [0, 1] \\ &\text{subject to} \quad \int \phi(t)\, dF_{x_{\theta_0}}(t) = \alpha \end{aligned} \tag{5.14}$$

or generalized to

$$\int_{E^1} \phi(t) f_{i0}(t)\, d(t) = \alpha_i \qquad i = 1, 2, \ldots, m$$

For simplicity consider (5.14); then following the same approach as in Section 5.3, the above problem can be approximated by

$$\begin{aligned} &\text{maximize} \quad \textstyle\sum_{j=1}^{n} \phi(t_j) f_1(t_j)[t_j - t_{j-1}] \\ &\text{subject to} \quad \textstyle\sum_{j=1}^{n} \phi(t_j) f_0(t_j)[t_j - t_{j-1}] = \alpha \\ &\qquad \phi(t_j) + \psi(t_j) = 1 \\ &\qquad \phi(t_j) \geq 0 \\ &\qquad \psi(t_j) \geq 0 \qquad j = 1, 2, \ldots, n \end{aligned} \tag{5.15}$$

Letting

$$\mathbf{x} = \begin{bmatrix} \phi(t_1) \\ \vdots \\ \phi(t_n) \\ \psi(t_1) \\ \vdots \\ \psi(t_n) \end{bmatrix} \qquad \mathbf{b} = \begin{bmatrix} f_1(t_1)[t_1 - t_0] \\ \vdots \\ f_1(t_n)[t_n - t_{n-1}] \\ 0 \\ \vdots \\ 0 \end{bmatrix} \qquad \mathbf{c} = \begin{bmatrix} \alpha \\ 1 \\ \vdots \\ 1 \end{bmatrix}$$

$$A_{(n+1)\times(n+n)} = \left[\begin{array}{ccc|ccc} f_0(t_1)[t_1 - t_0] & \cdots & f_0(t_n)[t_n - t_{n-1}] & 0 & \cdots & 0 \\ \hline & I & & & I & \end{array}\right]$$

we have the dual problem

$$\begin{aligned} \text{minimize} \quad & \mathbf{y'c} \\ \text{subject to} \quad & f_0(t_1)[t_1 - t_0]\, y_1 + u_1 \geq f_1(t_1)[t_1 - t_0] \\ & \vdots \qquad\qquad\qquad\qquad \vdots \\ & f_0(t_n)[t_n - t_{n-1}]\, y_1 + u_n \geq f_1(t_n)[t_n - t_{n-1}] \\ & u_1, \ldots, u_n \geq 0 \end{aligned} \tag{5.16}$$

where $\mathbf{y}' = (y_1, u_1, u_2, \ldots, u_n)$. If we let $u_j = z(t_j)[t_j - t_{j-1}]$, then (5.16) can be written as

$$\begin{aligned} \text{minimize} \quad & \alpha y_1 + \textstyle\sum_{j=1}^{n} z(t_j)[t_j - t_{j-1}] \\ \text{subject to} \quad & f_0(t_j) y_1 + z(t_j) \geq f_1(t_j) \\ & z(t_j) \geq 0 \qquad j = 1, 2, \ldots, n \end{aligned}$$

Letting $n \to \infty$, we have

$$\begin{aligned} \text{minimize} \quad & \alpha y_1 + \textstyle\int z(t)\, d(t) \\ \text{subject to} \quad & z(t) \geq f_1(t) - y_1 f_0(t) \\ & z(t) \geq 0 \quad \text{for all } t \end{aligned}$$

Since we are minimizing, $z(t)$ will be taken as small as possible, namely,

$$\begin{aligned} z(t) &= 0 \quad \text{if } f_1(t) - y_1 f_0(t) \leq 0 \\ &= f_1(t) - y_1 f_0(t) \quad \text{otherwise} \end{aligned}$$

hence $z(t) = [f_1(t) - y_1 f_0(t)]^+ = L(y_1, t)^+ \geq 0$. Our problem then becomes an unconstrained minimization problem, i.e.

$$\text{minimize} \quad \alpha y_1 + \int [L(y_1, t)]^+\, dt$$

For fixed y_1 and $\{t \mid L(\bar{y}_1, t) \geq 0\} = A$, we have

$$\alpha\bar{y}_1 + \int_A L(\bar{y}_1, t)\,dt = \alpha\bar{y}_1 + \int_A [f_1(t) - \bar{y}_1 f_0(t)]\,dt$$

$$= \int_A f_1(t)\,dt + \bar{y}_1[\alpha - \int_A f_0(t)\,dt]$$

$$= \int_A f_1(t)\,dt$$

by the size α requirement, and the latter is equal to the optimum of the primal problem. Thus we have the following lemma.

LEMMA 5.5. *If y_1 is chosen such that $\int f_0(t)\,dt = \alpha$ over the set $\{t \mid L(y_1, t) \geq 0\}$, then $\alpha y_1 + \int [L(y_1, t)]^+\,dt$ is minimized.*

5.6 Chance-Constrained Programming

5.6.1 Constraint Requirements Random

In some linear programming problems the b_i cannot be determined exactly from one period to the next but behave like a random variable from some distribution. Under this condition a reasonable criterion is that each constraint must hold for most combinations of the parameter values.

In particular, our constraints are constructed so that

$$P\left\{\sum_{j=1}^{n} a_{ij}x_j \leq b_i\right\} \geq \alpha_i \qquad i = 1, 2, \ldots, m \qquad \alpha_i \in [0, 1]$$

where $1 - \alpha_i$ denotes the allowable "risk" that a random variable will be chosen such that $\sum_{j=1}^{n} a_{ij}x_j > b_i$. Now

$$P\left\{\sum_{j=1}^{n} a_{ij}x_j \leq b_i\right\} = P\left\{\frac{\sum_{j=1}^{n} a_{ij}x_j - E[b_i]}{\sigma_{b_i}} \leq \frac{b_i - E[b_i]}{\sigma_{b_i}}\right\} \tag{5.17}$$

where $E[b_i]$ and σ_{b_i} are the mean and standard deviations of b_i, respectively.

Assume that $b_i \sim N(\mu_{b_i}, \sigma_{b_i}^2)$, then

$$\frac{b_i - E[b_i]}{\sigma_{b_i}} \sim N(0, 1)$$

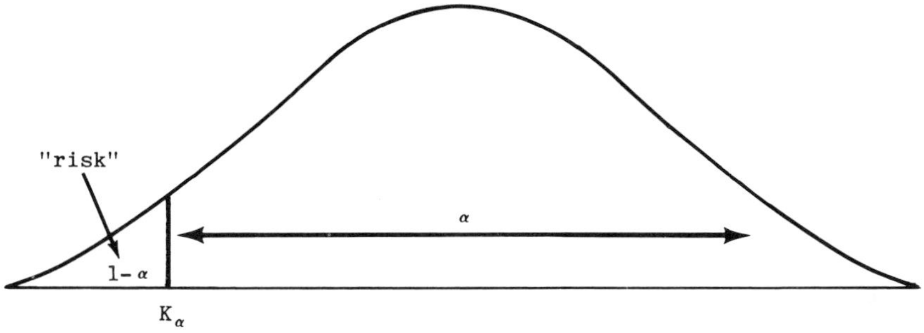

Figure 5.2. Area of allowable risk.

Therefore, since $P\{Z \geq K_\alpha\} = \alpha$ (see Fig. 5.2)

$$P\left\{K_{\alpha_i} \leq \frac{b_i - E[b_i]}{\sigma_{b_i}}\right\} = \alpha_i$$

It should be noted that this probability will increase if K_{α_i} is replaced by a smaller value and decrease if K_{α_i} is replaced by a larger value. Hence

$$P\left\{\frac{\Sigma a_{ij}x_j - E[b_i]}{\sigma_{b_i}} \leq \frac{b_i - E[b_i]}{\sigma_{b_i}}\right\} \geq \alpha_i \tag{5.18}$$

if and only if

$$\frac{\Sigma a_{ij}x_j - E[b_i]}{\sigma_{b_i}} \leq K_{\alpha_i} \tag{5.19}$$

or $\Sigma a_{ij}x_j \leq E[b_i] + K_{\alpha_i}\sigma_{b_i}$. Hence by solving the problem

$$\begin{array}{lll} \text{maximize} & \mathbf{c}'\mathbf{x} & \\ \text{subject to} & \Sigma a_{ij}x_j \leq E[b_i] + K_{\alpha_i}\sigma_{b_i} & i = 1, 2, \ldots, m \\ & \mathbf{x} \geq 0 & \end{array}$$

we obtain an approximate solution to our original problem.

EXAMPLE 5.5

$$\begin{array}{lrcrcrl} \text{maximize} & 20x_1 & + & 30x_2 & + & 25x_3 & \\ \text{subject to} & 3x_1 & + & 2x_2 & + & x_3 & \leq b_1 \\ & 2x_1 & + & 4x_2 & + & 2x_3 & \leq b_2 \\ & x_1 & + & 3x_2 & + & 5x_3 & \leq b_3 \\ & & & & & \mathbf{x} & \geq 0' \end{array}$$

where each b_i is normally and independently distributed as

$$b_1 \sim N(30, 1) \qquad b_2 \sim N(50, 4) \qquad b_3 \sim N(60, 9)$$

The first, second, and third constraints are required to hold with probability .975, .95, and .90, respectively.

Consider $\bar{\mathbf{x}}' = (7/3, 22/3, 19/3)$ and let us determine the probabilities that the three equations, respectively, are satisfied by $\bar{\mathbf{x}}$ and the probability that all three equations are satisfied. Considering each constraint, we have

$$\begin{aligned}
P\{3(7/3) + 2(22/3) + 19/3 \le b_1\} &= P\{(84/3 - 30)/1 \le (b_1 - 30)/1\} \\
&= P\{-2 \le Z\} = .9772 \\
P\{2(7/3) + 4(22/3) + 2(19/3) \le b_2\} &= P\{(140/3 - 50)/2 \le (b_2 - 50)/2\} \\
&= P\{-5/3 \le Z\} = .9515 \\
P\{7/3 + 3(22/3) + 5(19/3) \le b_3\} &= P\{-4/3 \le Z\} = .9082
\end{aligned}$$

Thus we have the probability that each constraint is satisfied. Therefore, the probability that all constraints are satisfied is .8450. Moreover, since .9772, .9515, and .9082 are greater than α_1, α_2, and α_3, respectively, then $\bar{\mathbf{x}}$ is a feasible solution. Reformulating the above probabilistic model into a deterministic model, we have

$$\begin{aligned}
E[b_1] + K_{.975}\,1 &= 30 + (-1.96) = 28.04 \\
E[b_2] + K_{.95}\,2 &= 50 + (-1.645)(2) = 46.71 \\
E[b_3] + K_{.90}\,3 &= 60 + (-1.28)(3) = 56.16
\end{aligned}$$

Hence our approximate solution is obtained by solving the problem

$$\begin{aligned}
\text{maximize}\quad & 20x_1 + 30x_2 + 25x_3 \\
\text{subject to}\quad & 3x_1 + 2x_2 + x_3 \le 28.04 \\
& 2x_1 + 4x_2 + 2x_3 \le 46.71 \\
& x_1 + 3x_2 + 5x_3 \le 56.16 \\
& x_1,\ x_2,\ x_3 \ge 0
\end{aligned}$$

Note that $\bar{\mathbf{x}}$ is not the optimal solution to this problem, but only a feasible solution.

In the above discussion we assumed that each b_i followed a normal distribution; we can consider other distributions. First consider what transpired above. Note that b_i was changed to $E[b_i] + K_{\alpha_i}\sigma_i = B_{\alpha_i}$. This implies that the new coefficient B_{α_i} is determined such that $F_b(B_{\alpha_i}) = 1 - \alpha$ where F_b denotes the cumulative distribution function of b, i.e., $F_b(z) = P\{b \le z\}$. Hence, we can consider the cumulative distribution function of b without specifically requiring that b_i follow a normal distribution; then $P\{\mathbf{a}'\mathbf{x} \leq b\} \geq \alpha$ is converted to a deterministic constraint by determining B_α such that $1 - F_b(B_\alpha) = \alpha$. Hence $B_\alpha = F_b^{-1}(1 - \alpha)$ and we have

$$P\{\mathbf{a}'\mathbf{x} \leq b\} \geq \alpha \text{ is equivalent to } \mathbf{a}'\mathbf{x} \leq B_\alpha$$

By an analogous argument we can determine the following equivalences

$$P\{\mathbf{a}'\mathbf{x} \geq b\} \geq \alpha \text{ is equivalent to } \mathbf{a}'\mathbf{x} \geq B_{1-\alpha}$$
$$P\{\mathbf{a}'\mathbf{x} \leq b\} \leq \alpha \text{ is equivalent to } \mathbf{a}'\mathbf{x} \geq B_\alpha$$
$$P\{\mathbf{a}'\mathbf{x} \geq b\} \leq \alpha \text{ is equivalent to } \mathbf{a}'\mathbf{x} \leq B_{1-\alpha}$$

EXAMPLE 5.6

Assume $b_1 \sim N(30, 1)$ and $b_2 \sim U(50,100)$ and require that

$$P\{\mathbf{a}_1'\mathbf{x} \leq b_1\} \geq .025$$
$$P\{\mathbf{a}_2'\mathbf{x} \geq b_2\} \geq .20$$

We can easily convert these probabilities to

$$\mathbf{a}_1'\mathbf{x} \leq B_{\alpha_1} \equiv 28.04$$
$$\mathbf{a}_2'\mathbf{x} \geq B_{1-\alpha_2} \equiv 90.00$$

In the following sections only certain distributions, such as the normal, are assumed if the coefficients a_{ij} or c_j are considered random.

5.6.2 *Input-output Coefficients Random*

Let us consider the case where the input-output coefficients a_{ij} are random, as in the following model

$$\begin{aligned} \text{maximize} \quad & \mathbf{c}'\mathbf{x} \\ \text{subject to} \quad & A\mathbf{x} \leq \mathbf{b} \\ & \mathbf{x} \geq 0 \end{aligned}$$

Assume each a_{ij} is normally and independently distributed with mean $\mu_{a_{ij}}$ and variance $\sigma^2_{a_{ij}}$, i.e., $a_{ij} \sim N(\mu_{a_{ij}}, \sigma^2_{a_{ij}})$. Moreover

$$\sum_j a_{ij}x_j \sim N\left(\sum_j \mu_{a_{ij}}x_j, \sum_j \sigma^2_{a_{ij}}x_j^2\right)$$

$$\sum_j x_j E[a_{ij}] - \sum_j a_{ij}x_j \sim N(0, \mathbf{x}'V\mathbf{x})$$

where

$$V = \begin{bmatrix} \sigma^2_{a_{i1}} & 0 & \cdots & 0 \\ 0 & \sigma^2_{a_{i2}} & & \vdots \\ \vdots & & \ddots & 0 \\ 0 & \cdots & 0 & \sigma^2_{a_{in}} \end{bmatrix}$$

To convert this stochastic model into a deterministic model, let us consider

$$P\{\Sigma a_{ij}x_j \leq b_i\} \geq \alpha_i \qquad i = 1, 2, \ldots, m$$

Then

$$P\left\{-b_i \leq -\sum_j a_{ij}x_j\right\}$$

$$= P\left\{\frac{-b_i + \sum_j x_j E[a_{ij}]}{(\mathbf{x}'V\mathbf{x})^{1/2}} \leq \frac{-\sum_j a_{ij}x_j + \sum_j x_j E[a_{ij}]}{(\mathbf{x}'V\mathbf{x})^{1/2}}\right\} \geq \alpha_i$$

if and only if

$$\frac{-b_i + \Sigma x_j E[a_{ij}]}{(\mathbf{x}'V\mathbf{x})^{1/2}} \leq K_{\alpha_i}$$

or $\Sigma x_j E[a_{ij}] - K_{\alpha_i}(\mathbf{x}'V\mathbf{x})^{1/2} \leq b_i$. Therefore we have the following deterministic model

$$\begin{array}{ll} \text{maximize} & \mathbf{c}'\mathbf{x} \\ \text{subject to} & \sum_j \mu_{a_{ij}}x_j - K_{\alpha_i}(\mathbf{x}'V\mathbf{x})^{1/2} \leq b_i \qquad i = 1, 2, \ldots, m \\ & \qquad\qquad x_j \geq 0 \qquad j = 1, 2, \ldots, n \end{array}$$

Note that this deterministic model is nonlinear. For this particular formulation certain nonlinear programming algorithms can be used to obtain an optimal solution.

5.6.3 *Input-output Coefficients and Constraint Requirements Random*

Now consider the case $b_i \sim N(\mu_{b_i}, \sigma^2_{b_i})$ and $a_{ij} \sim N(\mu_{a_{ij}}, \sigma^2_{a_{ij}})$ (for all i and j) where b_i and a_{ij} are independent. These constraints can be converted into deterministic constraints. Note that

$$b_i - \sum_{j=1}^{n} a_{ij}x_j \sim N(\mu^i_{\mathbf{b}-A\mathbf{x}}, \sigma^{i2}_{\mathbf{b}-A\mathbf{x}})$$

with mean

$$E[b_i] - \sum_{j=1}^{n} x_j E[a_{ij}] = \mu^i_{\mathbf{b}-A\mathbf{x}}$$

and variance

$$\sigma^2_{b_i} + \sum_{j=1}^{n} x_j^2 \sigma^2_{a_{ij}} = \sigma^{i2}_{\mathbf{b}-A\mathbf{x}}$$

Hence

$$P\{0 \le b_i - \sum_j a_{ij}x_j\}$$

$$= P\left\{\frac{-E[b_i] + \sum_j x_j E[a_{ij}]}{\sigma^i_{\mathbf{b}-A\mathbf{x}}} \le \frac{\left(b_i - \sum_j a_{ij}x_j\right) - (\mu^i_{\mathbf{b}-A\mathbf{x}})}{\sigma^i_{\mathbf{b}-A\mathbf{x}}}\right\} \ge \alpha_i$$

if and only if $-E[b_i] + \sum_j x_j E[a_{ij}] \le K_{\alpha_i}\sigma^i_{\mathbf{b}-A\mathbf{x}}$ or

$$\sum_j x_j E[a_{ij}] - K_{\alpha_i}\left[\sigma^2_{b_i} + \sum_j x_j^2 \sigma^2_{a_{ij}}\right]^{1/2} \le E[b_i]$$

5.6.4 *Cost Vector Random*

Linear programming problems where some or all cost coefficients are not known with certainty arise frequently. For instance, the price of producing a certain commodity varies from year to year; the profit of producing this commodity might be c_ℓ this year and c_ℓ^* next year. Such stochastic linear models can be converted to deterministic models but, as in the previous section, such models will be nonlinear.

Assume $c_j \sim N(\mu_{c_j}, \sigma_{c_j}^2)$; then $\mathbf{c} \sim N(\boldsymbol{\mu}_c, V)$ and $\mathbf{c}'\mathbf{x} \sim N(\boldsymbol{\mu}_c'\mathbf{x}, \mathbf{x}'V\mathbf{x})$. Let us assume that we can formulate the net returns $\mathbf{c}'\mathbf{x}$ into probabilistic statements in the following manner

$$\begin{aligned} \text{maximize} \quad & v \\ \text{subject to} \quad & P\{\mathbf{c}'\mathbf{x} \geq v\} \geq \alpha \\ & A\mathbf{x} \leq \mathbf{b} \\ & \mathbf{x} \geq 0 \end{aligned}$$

Then

$$P\{\mathbf{c}'\mathbf{x} \geq v\} = P\{(v - \boldsymbol{\mu}_c'\mathbf{x})/(\mathbf{x}'V\mathbf{x})^{1/2} \leq (\mathbf{c}'\mathbf{x} - \boldsymbol{\mu}_c'\mathbf{x})/(\mathbf{x}'V\mathbf{x})^{1/2}\} \geq \alpha$$

if and only if $(v - \boldsymbol{\mu}_c'\mathbf{x})/(\mathbf{x}'V\mathbf{x})^{1/2} \leq K_\alpha$. Therefore we can convert this stochastic problem into the following nonlinear deterministic problem

$$\begin{aligned} \text{maximize} \quad & \boldsymbol{\mu}_c'\mathbf{x} + K_\alpha(\mathbf{x}'V\mathbf{x})^{1/2} \\ \text{subject to} \quad & A\mathbf{x} \leq \mathbf{b} \\ & \mathbf{x} \geq 0 \end{aligned}$$

5.6.5 *Cost Vector and Constraint Requirements Random*

If $c_j \sim N(\mu_{c_j}, \sigma_{c_j}^2)$ and $b_i \sim N(\mu_{b_i}, \sigma_{b_i}^2)$ where c_j and b_i are independent, then from the previous sections we have the following approximate deterministic model

$$\begin{aligned} \text{maximize} \quad & \boldsymbol{\mu}_c'\mathbf{x} + K_\alpha(\mathbf{x}'V\mathbf{x})^{1/2} \\ \text{subject to} \quad & \textstyle\sum_j a_{ij}x_j \leq E[b_i] + K_{\alpha_i}\sigma_{b_i} \qquad i = 1, 2, \ldots, m \\ & x_j \geq 0 \qquad j = 1, 2, \ldots, n \end{aligned}$$

EXERCISES

1. Consider the linear model $y = \beta_0 + \beta_1 x$ where (x, y) are $(1, 1)$, $(2, 2)$, $(3, 1)$, and $(4, 2)$.
   a) Determine estimates for β_0 and β_1 under the criteria of minimizing the sum of the absolute deviations and minimizing the maximum deviation.
   b) If the above problem is in primal form under L_1, show that it can be solved without introducing artificial variables. Also, discuss ways the data can be centered so the necessary conditions can be used to save needless iterations. (Hint: Consider the mean and/or median.)

2. Assume x is a nonnegative random variable bounded by M and $E[x] = \mu$. Show that, for $a < M$

   $$P\{x \geq a\} \leq \min(\mu/a, 1)$$

   (Hint: Consider $f(x) \leq y_0 + u_1 x$ where $f(x) = 0$ for $x < a$ and $f(x) = 1$ for $x \geq a$. Show that this inequality is sharp when either $a \geq \mu$ or $a \leq \mu$.)
3. Suppose that x takes any real scalar value with $E[x] = 0$ and $E[x^2] = \theta$. Show that

   $$P\{\,|x| > a\} \leq \min(\theta/a^2, 1)$$

   Also show that this inequality is sharp.
4. Suppose $x \in [-M, M]$ and $E[x] = 0$. Calculate a bound for $P\{x \geq a\}$.
5. If a linear programming problem has unrestricted variables, the problem can be reconstructed by introducing two new variables for each unrestricted variable. For instance, if x_j is unrestricted, x_j can be expressed as $x_j^{(0)} - x_j^{(1)}$ where $x_j^{(0)}$ and $x_j^{(1)}$ are nonnegative.

   Another transformation is to introduce to the model only one new variable x_{n+1} such that $x_{n+1} = \max(0, -\min_j x_j)$; then $\bar{x}_j = x_j + x_{n+1}$ $(j = 1, 2, \ldots, n)$.
   a) Discuss the advantages of using the second transformation.
   b) Consider the linear model $y = \beta_0 + \beta_1 x$. Exhibit how the tableau for this problem under the criterion of L_1 is constructed so that $y = \bar{\beta}_0 + \bar{\beta}_1 x - \bar{\beta}_2(1 + x)$ where $\bar{\beta}_j = \beta_j + \bar{\beta}_2$ $(j = 0, 1)$.
6. In Exercise 1 under L_1, set up the initial tableau under alternative *b*) in Exercise 5; also determine the best regression estimates using this smaller tableau.
7. Often some of the components of **y** are negative. Discuss how to determine an immediate initial basis without proceeding to the next tableau. (Hint: Consider the transformation used in the dual simplex algorithm.)
8. Consider the following set of observations (x_i, y_i)

   $$\{(0, -.187), (1, 1.037), (2, 1.907), (3, 3.114), (4, 3.515)\}$$

   a) Under L_1, show that $\beta_0 = .167$ and $\beta_1 = .870$.
   b) Is this estimator unbiased?
   c) Solve this problem via the simplex algorithm and determine the set of alternative optimal solutions.
   d) If the median of n values, n even, is taken to be $n/2$, show that by pivoting on the median of the positive ratios in *c*) that an optimal solution can be determined in one iteration.
9. Consider the following set of observations

   $$\{(.5, .583), (2, 1.746), (3, 2.733), (4, 3.683), (5, 5.494)\}$$

   and the linear model $y = \beta_0 + \beta_1 x$.
   a) Using the median of the positive ratios, show that after three iterations the problem will cycle.

10. Consider the linear model $y = \beta_0 + \beta_1 x$.
    a) Under the criterion of minimizing the sum of absolute deviations, set up this problem as three linear programming problems (without considering the dual problem). (Hint: Transformations involving $1'e_1$ and $1'e_2$ will be needed.)
    b) Show that two dual problems of the problems in *a*) are identical or provide a counterexample.
11. Consider the model

$$\begin{aligned} \text{maximize} \quad & \mathbf{c}'\mathbf{x} \\ \text{subject to} \quad & A\mathbf{x} \leq \mathbf{b} \\ & \mathbf{x} \geq 0 \end{aligned}$$

    Under the following assumptions, convert the stochastic model into a deterministic model via chance-constrained programming.
    a) One input-output coefficient $a_{i^*j^*}$ is normally distributed with mean $\mu_{a_{i^*j^*}}$ and variance $\sigma^2_{a_{i^*j^*}}$.
    b) To *a*), add the assumption that each constraint coefficient $b_i \sim \text{NID}(\mu_{b_i}, \sigma^2_{b_i})$ and b_i and $a_{i^*j^*}$ are independent.
12. Prove that for a given set of observations, if there exist multiple optimal hyperplanes under L_1, then any convex combination of these optimal hyperplanes is also optimal under L_1.
13. Prove that if the number of observations N is odd, then any optimal hyperplane under L_1 passes through at least one observation point; and prove that for N odd, if there exist two optimal hyperplanes under L_1 which pass through at least p observation points, then both hyperplanes have one observation point in common.
14. Show that the L_1 estimator and L_2 estimator are equivalent if $X'(I\mathbf{e}_1 - I\mathbf{e}_2) = 0$.

References

Appa, G., and Smith, C. 1973. On L_1 and Chebyshev estimation. *J. Math. Program.* 5:73–87.

Barrodale, I., and Young, A. 1966. Algorithms for best L_1 and L_∞ linear approximations on discrete set. *Numer. Math.* 8:295–306.

Charnes, A., and Cooper, W. W. 1959. Chance-constrained programming. *Manage. Sci.* 6:73–80.

Draper, N. R., and Smith, H. 1966. *Applied Regression Analysis.* Wiley, New York.

Francis, R. L., and Meeks, H. D. 1972. On saddle point conditions and the generalized Neyman-Pearson lemma. *Aust. J. Stat.* 14:73–78.

Francis, R. L., and Wright, G. 1969. Some duality relationships for the generalized Neyman-Pearson problem. *J. Optim. Theory Appl.* 4:394–412.

Isii, K. 1964. Inequalities of the types of Chebyshev and Cramer-Rao and mathematical programming. *Ann. Inst. Stat. Math.* 16:277–93.

Karst, Otto J. 1958. Linear curve fitting using least deviations. *JASA* 53:118–32.

Kempthorne, O., and Folks, L. 1971. *Probability, Statistics, and Data Analysis.* Iowa State Univ. Press, Ames.

Kingman, J. F. C. 1963. On inequalities of the Tchebychev type. *Proc. Cambridge Phil. Soc.* 59:135–46.

Kiountouzis, E. A. 1973. Linear programming techniques in regression analysis. *Appl. Stat.* 22:69–73.

Lehmann, E. 1959. *Testing Statistical Hypotheses.* Wiley, New York.

Mood, A. M., and Graybill, F. A. 1963. *Introduction to the Theory of Statistics.* McGraw-Hill, New York.

Robers, P. D., and Ben-Israel, A. 1969. An interval programming algorithm for discrete linear L_1 approximation problems. *J. Approx. Theory* 2:323–36.

Sielken, R. L., and Hartley, H. O. 1973. Two linear programming algorithms for unbiased estimation of linear models. *J. Am. Stat. Assoc.* 68:639–41.

Stiefel, E. L. 1960. Note on Jordan elimination, linear programming, and Tchebycheff approximation. *Numer. Math.* 2:1–17.

Wagner, H. M. 1959. Linear programming techniques for regression analysis. *JASA* 54:206–12.

Wegner, P. 1963. Relations between multivariate statistics and mathematical programming. *Appl. Stat.* 12:146–50.

Whittle, P. 1971. *Optimization under Constraints: Theory and Applications of Nonlinear Programming.* Wiley, New York.

CHAPTER SIX

SADDLE POINT OPTIMALITY CRITERIA OF NONLINEAR PROGRAMMING PROBLEMS

In this chapter optimality criteria of the saddle point type for nonlinear programming problems are established. In the discussion the nonlinear programming problems are of the form

$$\begin{aligned} &\text{minimize} \quad F(\mathbf{y}) \\ &\text{subject to} \quad \mathbf{f}(\mathbf{y}) \leq 0 \\ &\qquad\qquad\quad \mathbf{y} \in L \end{aligned}$$

where $F(\mathbf{y})$ and $\mathbf{f}(\mathbf{y})$ are arbitrary functions in E^m and L is an arbitrary set in E^m. A sample problem is

$$\begin{aligned} &\text{minimize} \quad (y-1)^2 - 2y \\ &\text{subject to} \quad -y + 1 \leq 0 \\ &\qquad\qquad\quad y \in Q^+ \end{aligned}$$

where Q^+ is the positive orthant. The optimal solution is $y^0 = 2$. As in Chapter 4, the saddle point optimality criterion for this problem is:

A necessary and sufficient condition that y^0 is an optimal solution is that there exists a real number x^0 such that

$$\phi(y^0, x) \leq \phi(y^0, x^0) \leq \phi(y, x^0) \qquad \forall y \in Q^+ \qquad \forall x \in Q^+$$

where $\phi(y,x) = (y-1)^2 - 2y + x(-y+1)$. Here $x^0 = 0$ and $\phi(y,x)$ has a saddle point solution at $y^0 = 2$, $x^0 = 0$. Thus $\phi(y, x^0)$ has a global minimum or saddle point over the domain $y \geq 0$ at the point $y^0 = 2$. The y^0 part of the saddle point solution represents a solution of the minimization problem.

The saddle point criterion for the above example is both necessary and sufficient for y^0 to be a solution of the minimization problem. This is not always the case. The above saddle point condition is a sufficient condition without any requirements. However, to establish the necessity of the above saddle point condition we need, in addition to the convexity of a certain auxiliary set, a regularity condition or a constraint qualification. The optimality criteria of this chapter are presented without differentiability assumptions on the functions involved. Consider the following two problems

Problem 1

$$\begin{aligned} &\text{minimize} \quad F(\mathbf{y}) \\ &\text{subject to} \quad \mathbf{f}(\mathbf{y}) \leq 0 \\ &\qquad\qquad\quad \mathbf{y} \in L \end{aligned} \tag{6.1}$$

Problem 2 (Saddle Value Problem)

find vectors $\mathbf{y}^0 \in L$ and $\mathbf{x}^0 \geq 0$ such that

$$\phi(\mathbf{y}^0, \mathbf{x}) \leq \phi(\mathbf{y}^0, \mathbf{x}^0) \leq \phi(\mathbf{y}, \mathbf{x}^0) \qquad \forall \mathbf{x} \geq 0 \qquad \forall \mathbf{y} \in L$$

where $\phi(\mathbf{y}, \mathbf{x}) = F(\mathbf{y}) + \mathbf{x}'\mathbf{f}(\mathbf{y})$ and

$$\mathbf{f}(\mathbf{y}) = \begin{bmatrix} f_1(\mathbf{y}) \\ \vdots \\ f_n(\mathbf{y}) \end{bmatrix}$$

and each $f_i : E^m \to E^1 \equiv R$.

Theorems 6.1 and 6.2 establish that Problem 1, under a weak regularity assumption presented later, has an optimal solution $\mathbf{y}^0$ if and only if there exists $\mathbf{x}^0$ such that $(\mathbf{y}^0, \mathbf{x}^0)$ is a solution of Problem 2. Moreover, $\mathbf{x}^0$ is constructed explicitly as indicated in the necessity part of the argument. Furthermore, if $(\mathbf{y}^0, \mathbf{x}^0)$ is a solution of Problem 2, then $\mathbf{y}^0$ is an optimal solution of Problem 1 and the following four conditions hold:

(1) $\mathbf{f}(\mathbf{y}^0) \leq 0$
(2) $\mathbf{y}^0 \in L$
(3) $\mathbf{x}^{0\prime}\mathbf{f}(\mathbf{y}^0) = 0$
(4) $\mathbf{x}^0 \geq 0$

For reference purposes these four conditions will be denoted as *conditions C*. Also, if $\mathbf{y}^0$ is an optimal solution of Problem 1, then there exists $\mathbf{x}^0 \geq 0$ such that $(\mathbf{y}^0, \mathbf{x}^0)$ is a solution of Problem 2 and $\mathbf{y}^0$, $\mathbf{x}^0$ satisfy conditions C. Note that if L is the positive orthant, F is a convex function, and $\mathbf{f}$ is a componentwise convex function, then Problems 1 and 2 are the types considered by the classical Kuhn-Tucker theory discussed here and in Chapter 7.

6.1 Kuhn-Tucker Theory

THEOREM 6.1. *In order for* $\mathbf{y}^0$ *to be an optimal solution of Problem 1, it is sufficient that* $\mathbf{y}^0$ *and some* $\mathbf{x}^0 \geq 0$ *be a saddle point solution of Problem 2.*

Proof. If $(\mathbf{y}^0, \mathbf{x}^0)$ is a saddle point solution, then

$$F(\mathbf{y}^0) + \mathbf{x}'\mathbf{f}(\mathbf{y}^0) \leq F(\mathbf{y}^0) + \mathbf{x}^{0\prime}\mathbf{f}(\mathbf{y}^0) \leq F(\mathbf{y}) + \mathbf{x}^{0\prime}\mathbf{f}(\mathbf{y})$$
$$\forall \mathbf{x} \geq 0 \qquad \forall \mathbf{y} \in L$$

If $\mathbf{f}(\mathbf{y}^0) \nleq 0$, then there exists $\tilde{\mathbf{x}} \geq 0$ such that $\tilde{\mathbf{x}}'\mathbf{f}(\mathbf{y}^0) > 0$. But $\tilde{\mathbf{x}} \geq 0$ implies that $k\tilde{\mathbf{x}} \geq 0$ for all $k \geq 0$; therefore there exists $\bar{\mathbf{x}} = k\tilde{\mathbf{x}} \geq 0$ such that

$$\bar{\mathbf{x}}'\mathbf{f}(\mathbf{y}^0) = k\tilde{\mathbf{x}}'\mathbf{f}(\mathbf{y}^0) > \mathbf{x}^{0\prime}\mathbf{f}(\mathbf{y}^0)$$

and the left side of the above inequality is violated, hence $\mathbf{f}(\mathbf{y}^0) \leq 0$.

Also $\mathbf{x}^{0\prime}\mathbf{f}(\mathbf{y}^0) = 0$, because if $\mathbf{x}^{0\prime}\mathbf{f}(\mathbf{y}^0) \neq 0$, then since $\mathbf{x}^0 \geq 0$ and $\mathbf{f}(\mathbf{y}^0) \leq 0$, we have $\mathbf{x}^{0\prime}\mathbf{f}(\mathbf{y}^0) = k < 0$, implying

$$\mathbf{x}'\mathbf{f}(\mathbf{y}^0) \leq k < 0 \qquad \forall \mathbf{x} \geq 0$$

which is contradicted by $\mathbf{x} = 0$. Therefore

$$F(\mathbf{y}^0) \leq F(\mathbf{y}) + \mathbf{x}^{0\prime}\mathbf{f}(\mathbf{y}) \qquad \forall \mathbf{y} \in L$$

and since $\mathbf{x}^0 \geq 0$, this implies that

$$F(\mathbf{y}^0) \leq F(\mathbf{y}) \qquad \forall \mathbf{y} \in L \quad \text{with } \mathbf{f}(\mathbf{y}) \leq 0$$

Hence, $\mathbf{y}^0$ solves Problem 1, and $(\mathbf{y}^0, \mathbf{x}^0)$ satisfies conditions C.

To prove the converse of Theorem 6.1 we need some preliminary results. If Problem 1 has an optimal solution $\mathbf{y}^0$, the following two point sets K^1 and K^2 in E^{n+1} are defined.

DEFINITION 6.1. *Let* $\bar{\mathbf{z}}$ *be a point in* E^n *and let* K^1 *be the set of all points* $\mathbf{z} = (z_0, \bar{\mathbf{z}}) \in E^{n+1}$ *with the property that there exists at least one* $\mathbf{y} \in L$ *such that* $-z_0 + F(\mathbf{y}) \leq 0, \mathbf{f}(\mathbf{y}) + \bar{\mathbf{z}} \leq 0$, *i.e.*

$$K^1 = \{\mathbf{z} \mid -z_0 + F(\mathbf{y}) \leq 0, \mathbf{f}(\mathbf{y}) + \bar{\mathbf{z}} \leq 0 \quad \textit{for some } \mathbf{y} \in L\}$$

Note that K^1 *is nonempty since* $(z_0, \bar{\mathbf{z}}) = (F(\mathbf{y}^0), 0) \in K^1$.

DEFINITION 6.2. *Let K^2 be the set of all points* $\mathbf{z} = (z_0, \bar{\mathbf{z}}) \in E^{n+1}$ *such that* $-z_0 + F(\mathbf{y}^0) > 0$, $-\bar{\mathbf{z}} \leq 0$, *i.e.*

$$K^2 = \{\mathbf{z} \mid -z_0 + F(\mathbf{y}^0) > 0, -\bar{\mathbf{z}} \leq 0\}$$

where $\mathbf{y}^0$ *is an optimal solution of the minimization problem (6.1). For any* $\delta > 0$, *note that* $(F(\mathbf{y}^0) - \delta, 0) \in K^2$.

LEMMA 6.1. *K^2 is convex.*

Proof. Let $\mathbf{z}^1$ and $\mathbf{z}^2$ be two arbitrary points in K^2, and let $\mathbf{y}^0$ be an optimal solution of Problem 1; then

$$\begin{aligned} -z_0^1 + F(\mathbf{y}^0) &> 0 \\ -\bar{\mathbf{z}}^1 &\leq 0 \\ -z_0^2 + F(\mathbf{y}^0) &> 0 \\ -\bar{\mathbf{z}}^2 &\leq 0 \end{aligned}$$

Consider

$$\alpha \mathbf{z}^1 + (1 - \alpha)\mathbf{z}^2 = \tilde{\mathbf{z}} \quad \text{for any } \alpha \in [0, 1] \tag{6.2}$$

then $\alpha(-z_0^1 + F(\mathbf{y}^0)) + (1 - \alpha)(-z_0^2 + F(\mathbf{y}^0)) > 0$, so that from (6.2) it follows that

$$\begin{aligned} &-\tilde{z}_0 + F(\mathbf{y}^0) > 0 \\ &\alpha(-\bar{\mathbf{z}}^1) + (1 - \alpha)(-\bar{\mathbf{z}}^2) = -\tilde{\bar{\mathbf{z}}} \leq 0 \end{aligned}$$

Therefore K^2 is convex.

LEMMA 6.2. $K^1 \cap K^2 = \phi$.

Proof. Suppose $K^1 \cap K^2 \neq \phi$; then there exists $\mathbf{z}^*$ such that $\mathbf{z}^* \in K^1$ and $\mathbf{z}^* \in K^2$. If $\mathbf{z}^* \in K^1$, there exists $\mathbf{y}^* \in L$ such that $F(\mathbf{y}^*) - z_0^* \leq 0$ and $\mathbf{f}(\mathbf{y}^*) + \bar{\mathbf{z}}^* \leq 0$.

If $\mathbf{z}^* \in K^2$, then $F(\mathbf{y}^0) - z_0^* > 0$, $-\bar{\mathbf{z}}^* \leq 0$. It follows that $\mathbf{f}(\mathbf{y}^*) \leq 0$ and $F(\mathbf{y}^0) > z_0^* \geq F(\mathbf{y}^*)$. But $\mathbf{y}^* \in L$, which contradicts the assumption that $\mathbf{y}^0$ is an optimal solution of the defined minimization problem.

LEMMA 6.3. *If K^1 and K^2 are two convex sets, then there exists a hyperplane* $\mathbf{v}'\mathbf{z} = \beta$ $(\mathbf{v} \neq 0)$ *which separates K^1 and K^2, such that*

$$\mathbf{v}'\mathbf{z}^1 \geq \mathbf{v}'\mathbf{z}^2 \qquad \forall \mathbf{z}^1 \in K^1 \qquad \forall \mathbf{z}^2 \in K^2$$

Proof. Consider the set $R = \{\mathbf{z}^1 - \mathbf{z}^2 \mid \mathbf{z}^1 \in K^1, \mathbf{z}^2 \in K^2\}$. The set R is convex, since for any two vectors $\hat{\mathbf{z}}_3, \bar{\mathbf{z}}_3 \in R$ and any $\alpha \in [0, 1]$

$$\begin{aligned} \alpha\hat{\mathbf{z}}_3 + (1 - \alpha)\bar{\mathbf{z}}_3 &= \alpha(\hat{\mathbf{z}}_3^1 - \hat{\mathbf{z}}_3^2) + (1 - \alpha)(\bar{\mathbf{z}}_3^1 - \bar{\mathbf{z}}_3^2) \\ &= [\alpha\hat{\mathbf{z}}_3^1 + (1 - \alpha)\bar{\mathbf{z}}_3^1] - [\alpha\hat{\mathbf{z}}_3^2 + (1 - \alpha)\bar{\mathbf{z}}_3^2] \\ &= \tilde{\mathbf{z}}_3^1 - \tilde{\mathbf{z}}_3^2 \end{aligned}$$

where $\tilde{\mathbf{z}}_3^1 \in K^1, \tilde{\mathbf{z}}_3^2 \in K^2$.

Also the vector 0 is not an interior point of R; otherwise for some $\alpha \in (0, 1)$ and $\hat{\mathbf{z}}_3, \bar{\mathbf{z}}_3 \in R$

$$0 = \alpha\hat{\mathbf{z}}_3 + (1 - \alpha)\bar{\mathbf{z}}_3 = \tilde{\mathbf{z}}_3^1 - \tilde{\mathbf{z}}_3^2$$

which implies that $\tilde{\mathbf{z}}_3^1 = \tilde{\mathbf{z}}_3^2$ or $K^1 \cap K^2 \neq \phi$, which contradicts Lemma 6.2.

Therefore there exists a hyperplane containing 0 such that all of R lies in one closed half-space produced by the hyperplane. Hence there exists $\mathbf{v} \in E^{n+1}$ such that

$$\mathbf{v}'(\mathbf{z}^1 - \mathbf{z}^2) \geq 0 \quad \text{for all } \mathbf{z}^1 - \mathbf{z}^2 \in R$$

or $\mathbf{v}'\mathbf{z}^1 \geq \mathbf{v}'\mathbf{z}^2$ for all $\mathbf{z}^1 \in K^1$ and all $\mathbf{z}^2 \in K^2$.

LEMMA 6.4. *Let L be an arbitrary convex cone such that if* $\mathbf{x} \in L$ $(\mathbf{x} \neq 0)$ *and* $\mathbf{y} \in \tilde{L}^*$, *the interior of the polar cone of L, then* $\mathbf{x}'\mathbf{y} \neq 0$.

Proof. Suppose $\mathbf{x}'\mathbf{y} = 0$. Since $\mathbf{y} \in \tilde{L}^*$ and for any δ sufficiently small such that $\mathbf{y} + \boldsymbol{\delta} \in L^*$

$$0 \leq \mathbf{x}'(\mathbf{y} + \delta) = \mathbf{x}'\mathbf{y} + \mathbf{x}'\delta = \mathbf{x}'\delta$$

However, for sufficiently large $k > 0$, let $\boldsymbol{\delta} = -\mathbf{x}/k$; then

$$\mathbf{x}'\delta = -\mathbf{x}'\mathbf{x}/k = -|\mathbf{x}|^2/k < 0$$

since $\mathbf{x} \neq 0$. But this contradicts the above inequality, i.e., $0 \leq \mathbf{x}'\boldsymbol{\delta}$, therefore $\mathbf{x}'\mathbf{y} \neq 0$. (See Exercise 9.)

Before proving Theorem 6.2, the converse of Theorem 6.1 when $\mathbf{y}^0$ is an optimal solution of (6.1), we need a weak regularity assumption.

Regularity Assumption. There exists at least one $\mathbf{y}^* \in L$ such that $\mathbf{f}(\mathbf{y}^*) < 0$. This constraint qualification is often referred to as Slater's Condition.

THEOREM 6.2. *Assume K^1 is convex and the regularity assumption holds. Then for $\mathbf{y}^0$ to be an optimal solution of Problem 1, it is necessary that $\mathbf{y}^0$ and some $\mathbf{x}^0 \geq 0$ be a saddle point solution of Problem 2. Moreover, $(\mathbf{y}^0, \mathbf{x}^0)$ satisfies conditions C.*

Proof. (i) Consider the two point sets K^1 and K^2 defined above; then it follows from Lemmas 6.1 and 6.2 that K^1 and K^2 are convex disjoint point sets in E^{n+1}. Consequently, by Lemma 6.3 there is a separating hyperplane $\mathbf{v}'\mathbf{z} = \beta$ $(\mathbf{v} \neq 0)$ such that

$$\mathbf{v}'\mathbf{z}^1 \geq \mathbf{v}'\mathbf{z}^2 \qquad \forall \mathbf{z}^1 \in K^1 \qquad \forall \mathbf{z}^2 \in K^2 \tag{6.3}$$

This inequality is also true for all $\mathbf{z}^2 \in \bar{K}^2$, the closure of K^2 relative to E^{n+1}.

(ii) Our first aim is to show by a standard argument that

$$\mathbf{v} \in \mathfrak{L}^* = \{\mathbf{v} \mid v_0 \geq 0, -\bar{\mathbf{v}} \geq 0\}$$

i.e., the negative of the polar cone of $-\mathfrak{L} = \{\boldsymbol{\xi} \mid \xi_0 \leq 0, -\bar{\boldsymbol{\xi}} \leq 0\}$. Hence, define v_0 to be the first component of $\mathbf{v}$ and $\bar{\mathbf{v}}$ to be the vector composed of the other components of $\mathbf{v}$; similarly for $\boldsymbol{\xi}$.

Then (6.3) implies that for any $\mathbf{z}^1 \in K^1$, $\mathbf{z}^2 \in \bar{K}^2$

$$-v_0 F(\mathbf{y}^0) + \mathbf{v}'\mathbf{z}^1 \geq v_0(z_0^2 - F(\mathbf{y}^0)) + \bar{\mathbf{v}}'\bar{\mathbf{z}}^2$$

or, expressed equivalently

$$-v_0 F(\mathbf{y}^0) + \mathbf{v}'\mathbf{z}^1 \geq v_0 \xi_0 + \bar{\mathbf{v}}'\bar{\xi} \quad \forall \xi \in -\mathfrak{L} \tag{6.4}$$

Now suppose that $(v_0, \bar{\mathbf{v}}) \notin \mathfrak{L}^*$; then there exists a vector $-\bar{\mathbf{u}} \geq 0$ and a scalar $u_0 \geq 0$ such that $u_0 v_0 + \bar{\mathbf{u}}'\bar{\mathbf{v}} < 0$, that is, a $\boldsymbol{\xi} = -\mathbf{u}$ such that $\xi_0 v_0 + \bar{\xi}'\bar{\mathbf{v}} > 0$. Therefore, it is possible to choose $k^* > 0$ such that the inequality (6.4) is violated, and therefore $(v_0, \bar{\mathbf{v}}) \in \mathfrak{L}^* = \{\mathbf{v} \mid v_0 \geq 0, \bar{\mathbf{v}} \leq 0\}$.

(iii) We next establish that $v_0 > 0$. Consider any $\mathbf{y} \in L$. Then

$$(F(\mathbf{y}), -\mathbf{f}(\mathbf{y})) \in K^1 \qquad (F(\mathbf{y}^0), 0) \in \bar{K}^2$$

Then since $(F(\mathbf{y}), -\mathbf{f}(\mathbf{y})) \in K^1$ and $(F(\mathbf{y}^0), 0)$ is on the boundary of $\bar{K}^2$,

(6.3) yields

$$v_0 F(\mathbf{y}) - \bar{\mathbf{v}}'\mathbf{f}(\mathbf{y}) \geq v_0 F(\mathbf{y}^0) \tag{6.5}$$

Now suppose that $v_0 = 0$; then, considering $\mathbf{y}^*$ which satisfies the regularity assumption

$$-\bar{\mathbf{v}}'\mathbf{f}(\mathbf{y}^*) \geq 0 \quad \text{or} \quad \bar{\mathbf{v}}'\mathbf{f}(\mathbf{y}^*) \leq 0$$

But $\bar{\mathbf{v}} \leq 0$ and $\mathbf{f}(\mathbf{y}^*) \leq 0$ imply that $\bar{\mathbf{v}}'\mathbf{f}(\mathbf{y}^*) \geq 0$; hence

$$\bar{\mathbf{v}}'\mathbf{f}(\mathbf{y}^*) = 0 \tag{6.6}$$

But in view of the regularity assumption, (6.6) is true only if $\bar{\mathbf{v}} = 0$, contradicting the existence of the separating hyperplane in (i). Hence, $v_0 > 0$ and from (6.5) we have

$$F(\mathbf{y}) - (1/v_0)\bar{\mathbf{v}}'\mathbf{f}(\mathbf{y}) \geq F(\mathbf{y}^0) \qquad \forall \mathbf{y} \in L$$

(iv) In view of (iii) we can define the n-dimensional vector

$$\mathbf{x}^0 = -(1/v_0)\bar{\mathbf{v}} \tag{6.7}$$

Since $\bar{\mathbf{v}} \leq 0$ and $v_0 > 0$, then $\mathbf{x}^0 \geq 0$ and

$$F(\mathbf{y}) + \mathbf{x}^{0\prime}\mathbf{f}(\mathbf{y}) \geq F(\mathbf{y}^0) \qquad \forall \mathbf{y} \in L \tag{6.8}$$

so that $\phi(\mathbf{y}, \mathbf{x}^0) \geq F(\mathbf{y}^0)$ for all $\mathbf{y} \in L$. Now it remains to show that $\mathbf{x}^{0\prime}\mathbf{f}(\mathbf{y}^0) = 0$. In particular, for $\mathbf{y} = \mathbf{y}^0$ in (6.8) $\mathbf{x}^{0\prime}\mathbf{f}(\mathbf{y}^0) \geq 0$. However, $\mathbf{x}^0 \geq 0$ and $\mathbf{f}(\mathbf{y}^0) \leq 0$ imply that

$$\mathbf{x}^{0\prime}\mathbf{f}(\mathbf{y}^0) \leq 0 \tag{6.9}$$

so it must be that $\mathbf{x}^{0\prime}\mathbf{f}(\mathbf{y}^0) = 0$. Hence $\phi(\mathbf{y}^0, \mathbf{x}^0) = F(\mathbf{y}^0)$. Now taking into consideration that (6.9) also holds for any $\mathbf{x} \geq 0$, then

$$F(\mathbf{y}^0) \geq F(\mathbf{y}^0) + \mathbf{x}'\mathbf{f}(\mathbf{y}^0) = \phi(\mathbf{y}^0, \mathbf{x}) \qquad \forall \mathbf{x} \geq 0$$

We therefore have

$$\phi(\mathbf{y}^0, \mathbf{x}) \leq \phi(\mathbf{y}^0, \mathbf{x}^0) \leq \phi(\mathbf{y}, \mathbf{x}^0) \qquad \forall \mathbf{x} \geq 0 \qquad \forall \mathbf{y} \in L$$

Thus $(\mathbf{y}^0, \mathbf{x}^0)$ is a saddle point solution of $\phi(\mathbf{y}, \mathbf{x})$ and moreover, $(\mathbf{y}^0, \mathbf{x}^0)$ satisfies conditions C.

Theorem 6.2 is an equivalence theorem between Problems 1 and 2 that differs from Theorem 4.1 in Chapter 4 since $F(\mathbf{y})$ and $\mathbf{f}(\mathbf{y})$ are not necessarily linear and L is an arbitrary set in E^m. Figure 6.1 summarizes the results given in Theorems 6.1 and 6.2.

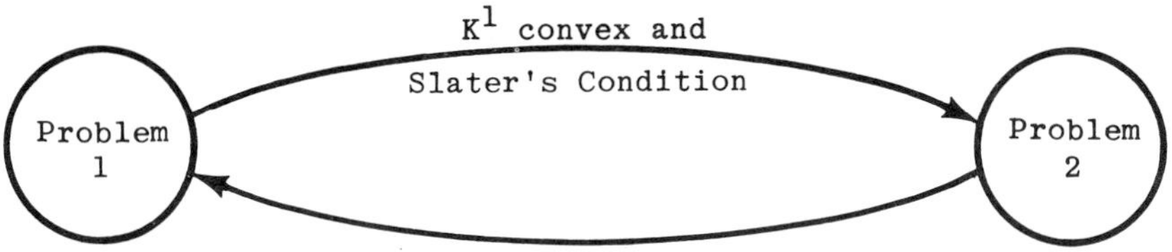

Figure 6.1. Equivalence relationship of nonlinear programming problems.

As stated in the introduction, some conditions are needed to prove the necessity of the saddle point criterion. The following two examples illustrate nonlinear problems whose Lagrangian function does not have a saddle point solution.

Example 6.1

Consider the problem

$$\begin{aligned} &\text{minimize} && -y \\ &\text{subject to} && y^2 \leq 0 \\ &&& y \geq 0 \end{aligned}$$

Since $y^0 = 0$ is the only feasible point, then y^0 must necessarily solve the problem. Let us now determine if there exists $x^0 \geq 0$ such that (y^0, x^0) is a saddle point solution of $\phi(y, x) = -y + xy^2$. If the saddle value problem possesses a nonnegative Lagrange multiplier, then either $x^0 = 0$ or $x^0 > 0$.

Assume $x^0 = 0$. Then there exists $y > 0$ which violates the inequality $\phi(y^0, x^0) \leq \phi(y, x^0)$ which must hold for all $y \geq 0$. Specifically, since $\phi(y^0, x^0) = 0$ and $\phi(y, x^0) = -y$, then $0 \nleq -\bar{y}$ for any $\bar{y} > 0$.

Assume $x^0 > 0$. Again $\phi(y^0, x^0) \leq \phi(y, x^0)$ must hold for all $y \geq 0$; in particular, $0 \leq -1 + x^0 y$ must hold for all $y \geq 0$. However, for $0 < y < 1/x^0$ this inequality is violated.

Therefore there does not exist a nonnegative Lagrange multiplier x^0 such that (y^0, x^0) solves the saddle value problem. Note also that the regularity assumption is not satisfied; i.e., there does not exist $y^* \geq 0$ such that $y^{*2} < 0$, but K^1 is convex. In particular

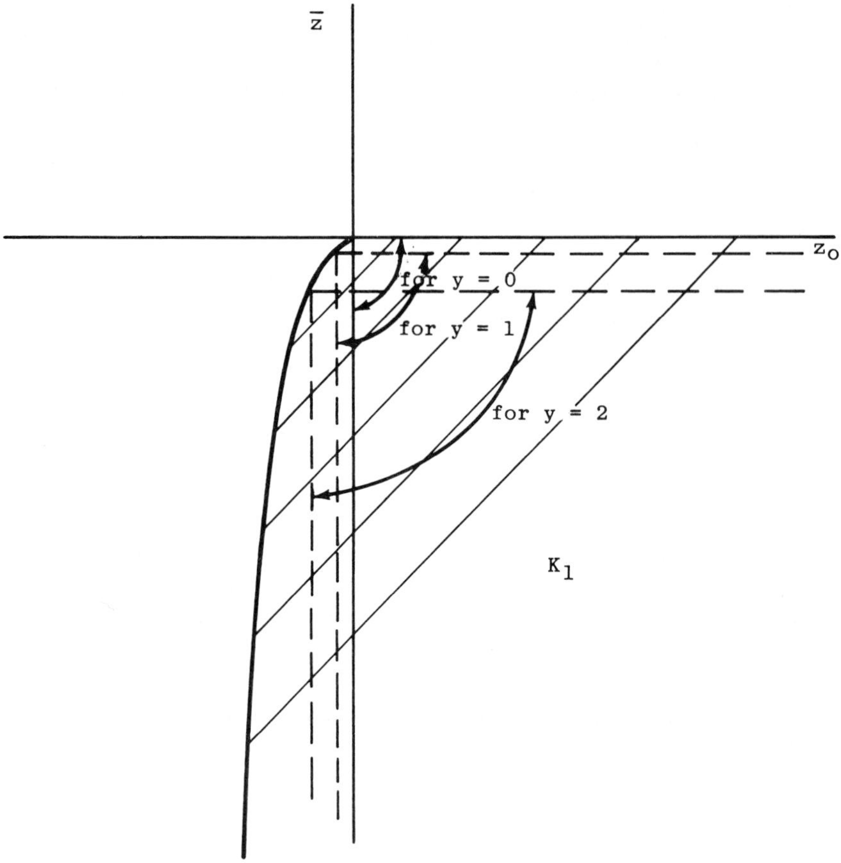

Figure 6.2. K^1 **associated with Example 6.1.**

$$K^1 = \{\mathbf{z} \mid z_0 + y \geq 0, y^2 + \bar{z} \leq 0 \quad \text{for some } y \geq 0\}$$

is convex as is shown in Fig. 6.2.

EXAMPLE 6.2

Consider the problem

$$\begin{array}{ll} \text{minimize} & y_1 + y_2 \\ \text{subject to} & y_2 - y_1^3 \leq 0 \\ & y_2 \geq 0 \\ & y_1 \quad \text{unrestricted} \end{array}$$

Here $\mathbf{y}^0 = \begin{bmatrix} 0 \\ 0 \end{bmatrix}$ solves the above problem; and moreover, the regularity assumption is satisfied by $\bar{\mathbf{y}} = \begin{bmatrix} 1 \\ 0 \end{bmatrix}$. Let us try to determine if there exists $x^0 \geq 0$ such that

$$\phi(\mathbf{y}^0, x) \leq \phi(\mathbf{y}^0, x^0) \leq \phi(\mathbf{y}, x^0) \qquad \forall y_2 \geq 0, \qquad \forall y_1$$

where $\phi(\mathbf{y}, x) = y_1 + y_2 + x(y_2 - y_1^3)$.

Assume $x^0 = 0$; then $0 \leq y_1 + y_2$ must hold for all $y_2 \geq 0$ and y_1. But for $y_1 = -1, y_2 = 0$, this inequality is violated.

Assume $x^0 > 0$; then $0 \leq y_1 + y_2 + x^0(y_2 - y_1^3)$ for all nonnegative y_2 and all $y_1 \in E^1$. However, for $y_2 = 0$ and any y_1 such that $y_1^2 > 1/x^0$, the above inequality is violated.

Therefore the above example does not have a saddle point solution even though the regularity assumption is satisfied. It is left as an exercise to show that in this example the set K^1 is not convex.

We should remark that the conditions of Theorem 6.2 are sufficient but not necessary to establish the existence of a saddle point solution of Problem 2. In particular, one or both conditions in Theorem 6.2 can be violated and a solution to Problem 2 still can exist.

The following example illustrates that the saddle value problem can have an optimal solution even though the regularity assumption does not hold. Consider the problem

$$\begin{aligned} &\text{minimize} \quad y \\ &\text{subject to} \quad y^3 \leq 0 \\ &\qquad\qquad\quad\; y \geq 0 \end{aligned}$$

Clearly, $y^0 = 0$ solves this problem and any $x^0 \geq 0$ solves the saddle value problem; hence this example also illustrates that the saddle value problem can have alternative solutions.

The following illustrates that with both conditions of Theorem 6.2 violated, the associated saddle value problem still has a saddle value solution. In particular, consider the problem

$$\begin{aligned} &\text{minimize} \quad -y \\ &\text{subject to} \quad y \leq 0 \\ &\qquad\qquad\quad\; y \in L = \{y \in E^1 \mid y = 0 \text{ or } 1\} \end{aligned}$$

Clearly, $y^0 = 0$ solves this problem and the reader can verify that

$(y^0, x^0) = (0, 1)$ is one saddle value solution. Also, note that Slater's Condition and the K^1 assumption are violated.

6.2 Sufficient Conditions on K^1

The following lemmas establish some sufficient conditions which ensure that the set K^1 is convex.

LEMMA 6.5. *If L is convex and F and* $\mathbf{f}$ *are linear functions of* $\mathbf{y}$, *then the set K^1 is convex.*

Proof. Let $\mathbf{z}^1$ and $\mathbf{z}^2$ be two arbitrary points in K^1. Then there exists $\mathbf{y}^1 \in L$ such that

$$-z_0^1 + F(\mathbf{y}^1) \leq 0 \qquad \mathbf{f}(\mathbf{y}^1) + \bar{\mathbf{z}}^1 \leq 0 \tag{6.10}$$

and there exists $\mathbf{y}^2 \in L$ such that

$$-z_0^2 + F(\mathbf{y}^2) \leq 0 \qquad \mathbf{f}(\mathbf{y}^2) + \bar{\mathbf{z}}^2 \leq 0 \tag{6.11}$$

Inasmuch as F and $\mathbf{f}$ are linear and L is convex, then for $\mathbf{y}^1, \mathbf{y}^2 \in L$ and any $\alpha \in [0, 1]$

$$\alpha\mathbf{y}^1 + (1 - \alpha)\mathbf{y}^2 \in L \tag{6.12}$$

$$\alpha F(\mathbf{y}^1) + (1 - \alpha)F(\mathbf{y}^2) = F(\alpha\mathbf{y}^1 + (1 - \alpha)\mathbf{y}^2) \tag{6.13}$$

$$\alpha\mathbf{f}(\mathbf{y}^1) + (1 - \alpha)\mathbf{f}(\mathbf{y}^2) = \mathbf{f}(\alpha\mathbf{y}^1 + (1 - \alpha)\mathbf{y}^2) \tag{6.14}$$

Thus consider $\alpha\mathbf{z}^1 + (1 - \alpha)\mathbf{z}^2 = \tilde{\mathbf{z}}$ for any $\alpha \in [0, 1]$. Therefore it follows from equations (6.10) through (6.14) that

$$-\tilde{z}_0 + F(\alpha\mathbf{y}^1 + (1 - \alpha)\mathbf{y}^2) \leq 0$$
$$\mathbf{f}(\alpha\mathbf{y}^1 + (1 - \alpha)\mathbf{y}^2) + \bar{\tilde{\mathbf{z}}} \leq 0$$

where $\alpha\mathbf{y}^1 + (1 - \alpha)\mathbf{y}^2 \in L$. Therefore K^1 is convex.

EXAMPLE 6.3

Appealing to Lemma 6.5, K^1 is convex in the following problem

$$\begin{aligned} &\text{minimize} \quad \mathbf{b}'\mathbf{y} \\ &\text{subject to} \quad \mathbf{c} - A'\mathbf{y} \leq 0 \\ &\qquad\qquad\qquad \mathbf{y} \in L \end{aligned}$$

where L is a closed convex set in E^m.

Moreover if the feasible region is polyhedral then, in view of Chapter 4, the regularity assumption is not needed. If the feasible region is not necessarily polyhedral the regularity assumption in Theorem 6.2 is needed.

DEFINITION 6.3. *A function $f(\mathbf{y})$ is said to be convex over a convex set Y in E^m if for any two vectors $\mathbf{y}^1$ and $\mathbf{y}^2$ in Y and for all $\alpha \in [0, 1]$*

$$f(\alpha\mathbf{y}^1 + (1 - \alpha)\mathbf{y}^2) \leq \alpha f(\mathbf{y}^1) + (1 - \alpha)f(\mathbf{y}^2) \tag{6.15}$$

LEMMA 6.6. *If $\{(F(\mathbf{y}), \mathbf{f}(\mathbf{y})) \mid \mathbf{y} \in L\}$ is convex for an arbitrary set L, then K^1 is convex.*

Proof. Let $\mathbf{z}^1, \mathbf{z}^2 \in K^1$; then there exist $\mathbf{y}^1, \mathbf{y}^2 \in L$ such that (6.10) and (6.11) hold. Now since $\{(F(\mathbf{y}), \mathbf{f}(\mathbf{y})) \mid \mathbf{y} \in L\}$ is convex, then there exists $\mathbf{y}^* \in L$ such that for any $\alpha \in [0, 1]$

$$\begin{aligned} \alpha F(\mathbf{y}^1) + (1 - \alpha)F(\mathbf{y}^2) &= F(\mathbf{y}^*) \\ \alpha \mathbf{f}(\mathbf{y}^1) + (1 - \alpha)\mathbf{f}(\mathbf{y}^2) &= \mathbf{f}(\mathbf{y}^*) \end{aligned}$$

Consider $\alpha\mathbf{z}^1 + (1 - \alpha)\mathbf{z}^2 = \tilde{\mathbf{z}}$. It follows that

$$\begin{aligned} -\tilde{z}_0 + F(\mathbf{y}^*) &\leq 0 \\ \mathbf{f}(\mathbf{y}^*) + \tilde{\bar{\mathbf{z}}} &\leq 0 \end{aligned}$$

for some $\mathbf{y}^* \in L$ and K^1 is convex.

EXAMPLE 6.4

Let $F(y) = -|y|, \mathbf{f}(y) = y - 3$, and $L = \{y \mid y \in [0, 1]\}$. Then we have the following problem

$$\begin{aligned} &\text{minimize} \quad -|y| \\ &\text{subject to} \quad y \leq 3 \\ &\qquad\qquad\quad\; y \in [0, 1] \end{aligned}$$

Clearly, $\{(F(y), \mathbf{f}(y)) \mid y \in [0, 1]\}$ is convex; hence from Lemma 6.6, K^1 is convex. Note in view of Definition 6.3 that $F(y)$ is not a convex function.

LEMMA 6.7. *If F is a convex function and $\mathbf{f}$ is a componentwise convex function of $\mathbf{y}$ and L is an arbitrary convex set, then K^1 is convex.*

Proof. Let $\mathbf{z}^1, \mathbf{z}^2 \in K^1$; then there exist $\mathbf{y}^1, \mathbf{y}^2 \in L$ such that (6.10) and (6.11) hold. Thus

$$-z_0^1 + F(\mathbf{y}^1) \le 0 \qquad -z_0^2 + F(\mathbf{y}^2) \le 0$$

so that

$$-\tilde{z}_0 + \alpha F(\mathbf{y}^1) + (1 - \alpha)F(\mathbf{y}^2) \le 0 \quad \text{for any} \quad \alpha \in [0, 1] \tag{6.16}$$

But F is convex, therefore

$$F(\alpha \mathbf{y}^1 + (1 - \alpha)\mathbf{y}^2) \le \alpha F(\mathbf{y}^1) + (1 - \alpha)F(\mathbf{y}^2)$$

Hence from (6.16)

$$-\tilde{z}_0 + F(\alpha \mathbf{y}^1 + (1 - \alpha)\mathbf{y}^2) \le 0$$

where $\alpha \mathbf{y}^1 + (1 - \alpha)\mathbf{y}^2 \in L$. Also

$$f_i(\mathbf{y}^1) + \bar{z}_i^1 \le 0 \qquad f_i(\mathbf{y}^2) + \bar{z}_i^2 \le 0 \qquad i = 1, 2, \ldots, n$$

But each f_i is convex, hence

$$f_i(\alpha \mathbf{y}^1 + (1 - \alpha)\mathbf{y}^2) + \tilde{\bar{z}}_i \le 0 \qquad i = 1, 2, \ldots, n$$
$$\alpha \mathbf{y}^1 + (1 - \alpha)\mathbf{y}^2 \in L$$

Therefore K^1 is convex.

EXAMPLE 6.5

Lemma 6.7 with L defined as the closed positive orthant yields problems considered in the classical Kuhn-Tucker theory; namely, consider the problem

$$\begin{aligned} &\text{minimize} && F(\mathbf{y}) \\ &\text{subject to} && \mathbf{f}(\mathbf{y}) \le 0 \\ &&& \mathbf{y} \ge 0 \end{aligned}$$

and its associated saddle value problem

find vectors $\mathbf{y}^0 \ge 0$ and $\mathbf{x}^0 \ge 0$
such that $\phi(\mathbf{y}^0, \mathbf{x}) \le \phi(\mathbf{y}^0, \mathbf{x}^0) \le \phi(\mathbf{y}, \mathbf{x}^0) \qquad \forall \mathbf{x} \ge 0 \qquad \forall \mathbf{y} \ge 0$

In this situation, the equivalence theorem holds under the regularity assumption (Slater's Condition) and the convexity of F and each f_i. Note that if each f_i is linear and

$$\{(\xi, \eta) \in E^{m+1} \mid \mathbf{f}(\mathbf{y}) + \xi \leq 0, \eta + F(\mathbf{y}) - F(\mathbf{y}^0) \leq 0 \quad \text{for } \mathbf{y} \in L\} \quad (6.17)$$

is a closed convex set, then the constraint qualification is not needed. This was shown in Chapter 4 and was denoted Condition A.

Note that (6.17) is constructed by changing the problem

$$\begin{aligned} &\text{minimize} && F(\mathbf{y}) \\ &\text{subject to} && \mathbf{f}(\mathbf{y}) \leq 0 \\ &&& \mathbf{y} \in L \end{aligned}$$

to

$$\begin{aligned} &-\text{maximize} && -F(\mathbf{y}) \\ &\text{subject to} && -\mathbf{f}(\mathbf{y}) \geq 0 \\ &&& \mathbf{y} \in L \end{aligned}$$

These results are quite similar to those presented in Chapter 4 which hold without constraint qualifications when the orthant domains are polyhedral.

Let us consider the Lagrangian of Problem II in Chapter 4, i.e., $\phi(\mathbf{y}, \mathbf{x}) = \mathbf{b}'\mathbf{y} + \mathbf{x}'(\mathbf{c} - A'\mathbf{y})$. Then we have the following lemma where L is the positive orthant.

LEMMA 6.8. *If* $\mathbf{y}^0 \in L$ *and* $\mathbf{x}^0 \geq 0$ *are such that* $(\mathbf{y}^0, \mathbf{x}^0)$ *is a saddle point solution of the Lagrangian* $\phi(\mathbf{y}, \mathbf{x}) = \mathbf{b}'\mathbf{y} + \mathbf{c}'\mathbf{x} - \mathbf{y}'A\mathbf{x}$, *then*

(*a*) $A'\mathbf{y}^0 - \mathbf{c} \geq 0$
(*b*) $\mathbf{y}^0 \geq 0$
(*c*) $\mathbf{x}^{0\prime}(\mathbf{c} - A'\mathbf{y}^0) = 0$
(*d*) $\mathbf{x}^0 \geq 0$
(*e*) $\mathbf{b} - A\mathbf{x}^0 \geq 0$
(*f*) $\mathbf{y}^{0\prime}(\mathbf{b} - A\mathbf{x}^0) = 0$

Moreover, if there exists a vector $(\mathbf{y}^0, \mathbf{x}^0)$ *that satisfies conditions* (*a*)–(*f*), *then* $(\mathbf{y}^0, \mathbf{x}^0)$ *is a saddle point solution of the saddle value problem.*

Proof. If $(\mathbf{y}^0, \mathbf{x}^0)$ is a solution of Problem 2, then conditions (*a*)–(*d*) are conditions C. Conditions (*e*) and (*f*) follow from (*i*) and (*ii*) of the proof of Lemma 4.5.

Assume that $(\mathbf{y}^0, \mathbf{x}^0)$ satisfies conditions (*a*)–(*f*). Then

$$\begin{aligned}\phi(\mathbf{y}^0, \mathbf{x}) &= \mathbf{b}'\mathbf{y}^0 + \mathbf{x}'(\mathbf{c} - A'\mathbf{y}^0)\\ &\leq \mathbf{b}'\mathbf{y}^0 \quad \text{for all } \mathbf{x} \geq 0\\ &= \phi(\mathbf{y}^0, \mathbf{x}^0) \quad \text{from } (c)\\ &= \mathbf{b}'\mathbf{y}^0 + \mathbf{x}^{0\prime}(\mathbf{c} - A'\mathbf{y}^0)\\ &= \mathbf{c}'\mathbf{x}^0 \quad \text{from } (f)\\ &\leq \mathbf{c}'\mathbf{x}^0 + \mathbf{y}'(\mathbf{b} - A\mathbf{x}^0) \quad \text{for all } \mathbf{y} \geq 0\\ &= \phi(\mathbf{y}, \mathbf{x}^0)\end{aligned}$$

Hence $(\mathbf{y}^0, \mathbf{x}^0)$ is a saddle point solution in view of the above.

6.3 Equivalent Dual Forms

It may appear that there is only one dual problem for any linear programming problem, in particular, a linear dual problem that maps the technological coefficients in some specific fashion. This section displays linear problems LP $(A, \mathbf{b}, \mathbf{c})$ that have a certain set of saddle value solutions; and moreover generate nonclassical dual forms which have the same optimal solution as the classical dual problem. For simplicity the domains are restricted to orthant domains.

Let Δ denote a saddle point solution $(\mathbf{y}^0, \mathbf{x}^0)$ of $\phi(\mathbf{y}, \mathbf{x})$ and $\mathcal{P}(\Delta)$ denote the set of all linear programming problems with Lagrangian functions having at least the saddle point solution Δ. Likewise $\bigcap_{i=1}^{n} \mathcal{P}(\Delta_i)$ will denote the set of LP$(A, \mathbf{b}, \mathbf{c})$ with Lagrangian functions having at least the saddle point solutions $\Delta_1, \Delta_2, \cdots, \Delta_n$.

By Lemma 6.8, the following conditions are necessary and sufficient that $(\mathbf{y}^0, \mathbf{x}^0)$ is a saddle point solution of $\phi(\mathbf{y}, \mathbf{x}) = \mathbf{b}'\mathbf{y} + \mathbf{x}'(\mathbf{c} - A'\mathbf{y})$

$$\mathbf{b} - A\mathbf{x}^0 \geq 0 \qquad \mathbf{y}^{0\prime}(\mathbf{b} - A\mathbf{x}^0) = 0 \qquad \mathbf{x}^0 \geq 0 \tag{6.18}$$

$$\mathbf{c} - A'\mathbf{y}^0 \leq 0 \qquad \mathbf{x}^{0\prime}(\mathbf{c} - A'\mathbf{y}^0) = 0 \qquad \mathbf{y}^0 \geq 0 \tag{6.19}$$

THEOREM 6.3. *Let* $(\mathbf{y}^0, \mathbf{x}^0)$ *be a saddle point solution of the Lagrangian of* $LP(A, \mathbf{b}, \mathbf{c})$ *with all nonzero components. Let* T_1 *and* T_2 *be two nonsingular matrices such that* $T_1 A = A T_2 = A^*$. *Then* $LP(A^*, \mathbf{b}^*, \mathbf{c}^*) \in \mathcal{P}((\mathbf{y}^0, \mathbf{x}^0))$ where $T_1 \mathbf{b} = \mathbf{b}^*$ *and* $T_2' \mathbf{c} = \mathbf{c}^*$.

Proof. Now if $(\mathbf{y}^0, \mathbf{x}^0)$ is a saddle point solution with nonzero components of $\phi(\mathbf{y}, \mathbf{x})$, then conditions (6.18) and (6.19) must hold. Therefore $\mathbf{b} - A\mathbf{x}^0 = 0$ and $A'\mathbf{y}^0 - \mathbf{c} = 0$. Now choose any two nonsingular matrices such that $A^* = T_1 A = A T_2 (A^* \neq A)$. Then from the above we have

$$T_1(\mathbf{b} - A\mathbf{x}^0) = 0 \tag{6.20}$$

$$T_2'(A'\mathbf{y}^0 - \mathbf{c}) = 0 \tag{6.21}$$

Now (6.20) and (6.21) can be written as

$$T_1\mathbf{b} - T_1 A\mathbf{x}^0 = \mathbf{b}^* - A^*\mathbf{x}^0 = 0$$
$$T_2'A'\mathbf{y}^0 - T_2'\mathbf{c} = A^{*\prime}\mathbf{y}^0 - \mathbf{c}^* = 0$$

Hence $\mathrm{LP}(A^*, \mathbf{b}^*, \mathbf{c}^*) \in \mathcal{P}((\mathbf{y}^0, \mathbf{x}^0))$.

THEOREM 6.4. *Assume there exist m^* and n^* zero components of $\mathbf{y}^0$ and $\mathbf{x}^0$, respectively, where $(\mathbf{y}^0, \mathbf{x}^0)$ is a saddle point solution of the Lagrangian function of $LP(A, \mathbf{b}, \mathbf{c})$. Then there exist two matrices T_1 and T_2 such that $T_1 A = A T_2 = A^*$, and $LP(A^*, \mathbf{b}^*, \mathbf{c}^*) \in \mathcal{P}((\mathbf{y}^0, \mathbf{x}^0))$ where $T_1 \mathbf{b} = \mathbf{b}^*$ and $T_2'\mathbf{c} = \mathbf{c}^*$.*

Proof. Assume the first m^* and n^* components of $\mathbf{y}^0$ and $\mathbf{x}^0$ are zero. Then from (6.18) and (6.19) we have that if $\mathrm{LP}(A, \mathbf{b}, \mathbf{c}) \in \mathcal{P}((\mathbf{y}^0, \mathbf{x}^0))$, then

$$(\mathbf{b} - A\mathbf{x}^0)_i \geq 0 \qquad i = 1, 2, \cdots, m^* \tag{6.22}$$

$$(\mathbf{b} - A\mathbf{x}^0)_i = 0 \qquad i = m^* + 1, m^* + 2, \cdots, m \tag{6.23}$$

$$(A'\mathbf{y}^0 - \mathbf{c})_j \geq 0 \qquad j = 1, 2, \cdots, n^* \tag{6.24}$$

$$(A'\mathbf{y}^0 - \mathbf{c})_j = 0 \qquad j = n^* + 1, n^* + 2, \cdots, n \tag{6.25}$$

Let $A = [\bar{A}_{m \times n^*} \mid \bar{\bar{A}}_{m \times (n - n^*)}]$. Choose T_1 and T_2 so

$$T_1 = \begin{bmatrix} I_{m^* \times m^*} & 0 \\ 0 & \bar{T}_1 \end{bmatrix} \qquad T_2 = \begin{bmatrix} I_{n^* \times n^*} & 0 \\ 0 & \bar{T}_2 \end{bmatrix}$$

are nonsingular and

$$T_1\bar{A} = \bar{A} \qquad T_1 A = A T_2 \tag{6.26}$$

Hence applying (6.22) and (6.23) jointly we have

$$(\mathbf{b}^* - A^*\mathbf{x}^0)_i \geq 0 \qquad i = 1, 2, \cdots, m^*$$
$$(\mathbf{b}^* - A^*\mathbf{x}^0)_i = 0 \qquad i = m^* + 1, m^* + 2, \cdots, m$$

Likewise, applying (6.24) and (6.25) jointly to $T_2'(A'\mathbf{y}^0 - \mathbf{c})$ we have

$$(A^{*\prime}\mathbf{y}^0 - \mathbf{c}^*)_j \geq 0 \qquad j = 1, 2, \cdots, n^*$$
$$(A^{*\prime}\mathbf{y}^0 - \mathbf{c}^*)_j = 0 \qquad j = n^* + 1, n^* + 2, \cdots, n$$

Therefore $\mathrm{LP}(A^*, \mathbf{b}^*, \mathbf{c}^*) \in \mathcal{P}((\mathbf{y}^0, \mathbf{x}^0))$.

This procedure can be applied to any $\mathrm{LP}(A, \mathbf{b}, \mathbf{c}) \in \bigcap_{i=1}^{n} \mathcal{P}(\Delta_i)$ by considering (6.18) and (6.19) for each coordinate of Δ_i where each $\Delta_i = (\mathbf{y}_i^0, \mathbf{x}_i^0)$; for example:

$$(\mathbf{b} - A\mathbf{x}_i^0)_{i'} \geq 0 \quad \text{if the } i' \text{ component of at least one } \mathbf{y}_i^0 \text{ is } 0$$
$$(\mathbf{b} - A\mathbf{x}_i^0)_{i'} = 0 \quad \text{otherwise}$$
$$(A'\mathbf{y}_i^0 - \mathbf{c})_{j'} \geq 0 \quad \text{if the } j' \text{ component of at least one } \mathbf{x}_i^0 \text{ is } 0$$
$$(A'\mathbf{y}_i^0 - \mathbf{c})_{j'} = 0 \quad \text{otherwise}$$

Thus, by Theorem 6.4, we can construct an $\mathrm{LP}(A^*, \mathbf{b}^*, \mathbf{c}^*) \in \bigcap_{i=1}^{n} \mathcal{P}(\Delta_i)$.

Hence, given any $\mathrm{LP}(A, \mathbf{b}, \mathbf{c})$ with a certain set of saddle point solutions, we can find another problem with the same set of saddle point solutions as $\mathrm{LP}(A, \mathbf{b}, \mathbf{c})$, i.e., $\mathrm{LP}(A^*, \mathbf{b}^*, \mathbf{c}^*)$ under the mapping of T_1 and T_2, and moreover, construct a nonclassical dual, i.e., maximize $\mathbf{c}^{*\prime}\mathbf{x}$ subject to $A^*\mathbf{x} \leq \mathbf{b}^*$, $(\mathbf{x} \geq 0)$.

6.3.1 Numerical Illustrations

The following example illustrates an equivalent dual form with the same solution as the classical dual problem. Consider the following linear programming problem and its classical dual

(I)

$$\begin{aligned} \text{minimize} \quad & 2y_1 + 3y_2 \\ \text{subject to} \quad & y_1 + 2y_2 \geq 2 \\ & y_2 \geq 1 \\ & 2y_1 - y_2 \geq 3 \\ & y_1, \quad y_2 \geq 0 \end{aligned}$$

(*II*)

$$\begin{array}{ll} \text{maximize} & 2x_1 + x_2 + 3x_3 \\ \text{subject to} & x_1 + 2x_3 \leq 2 \\ & 2x_1 + x_2 - x_3 \leq 3 \\ & x_1, \quad x_2, \quad x_3 \geq 0 \end{array}$$

Now

$$(\mathbf{y}^0, \mathbf{x}^0) = \left(\begin{bmatrix} 2 \\ \\ 1 \end{bmatrix}, \begin{bmatrix} 0 \\ 4 \\ 1 \end{bmatrix} \right)$$

is a saddle point solution of $\phi(\mathbf{y}, \mathbf{x})$; i.e., $\mathbf{y}^0$ is an optimal solution of (I) and $\mathbf{x}^0$ is an optimal solution of (II). Therefore let

$$T_1 = \begin{bmatrix} -1 & 1 \\ \\ 2 & 0 \end{bmatrix} \qquad T_2 = \begin{bmatrix} 1 & 0 & 0 \\ 0 & 1/2 & 5/2 \\ 0 & 1/2 & -3/2 \end{bmatrix}$$

Hence

$$T_1 A = A T_2 = \begin{bmatrix} 1 & 1 & -3 \\ 2 & 0 & 4 \end{bmatrix} = A^*$$

$$T_1 \mathbf{b} = \begin{bmatrix} 1 \\ \\ 4 \end{bmatrix} = \mathbf{b}^* \qquad T_2' \mathbf{c} = \begin{bmatrix} 2 \\ 2 \\ -2 \end{bmatrix} = \mathbf{c}^*$$

Therefore an equivalent dual form is

(*III*)

$$\begin{array}{ll} \text{maximize} & 2x_1 + 2x_2 - 2x_3 \\ \text{subject to} & x_1 + x_2 - 3x_3 \leq 1 \\ & 2x_1 + 4x_3 \leq 4 \\ & x_1, \quad x_2, \quad x_3 \geq 0 \end{array}$$

which also has an optimal solution $x^{0\prime} = (0, 4, 1)$.

Note that (I) and (III) are mutually dual in that they have the same saddle point vector, but it is not necessarily true that they have the same optimal objective value, i.e., $\mathbf{c}^{*\prime}\mathbf{x}^0 = 6 \neq 7 = \mathbf{b}'\mathbf{y}^0$. If we introduce the notion of mutually dual problems so that the two problems have the same optimum, then we must also require that $(T_1\mathbf{b})'\mathbf{y}^0 = \mathbf{b}'\mathbf{y}^0$. Therefore, again considering (I), it is necessary first to find a T_1 such that $(T_1\mathbf{b})'\mathbf{y}^0 = \mathbf{b}'\mathbf{y}^0 = 7$ and, from (6.26), $T_1\overline{A} = \begin{bmatrix} 1 \\ 2 \end{bmatrix}$. One such T_1 and an appropriate T_2 are

$$T_1 = \begin{bmatrix} -1 & 1 \\ 4 & -1 \end{bmatrix} \qquad T_2 = \begin{bmatrix} 1 & 0 & 0 \\ 0 & -1/2 & 15/2 \\ 0 & 1/2 & -3/2 \end{bmatrix}$$

Thus

$$T_1 A = A T_2 = \begin{bmatrix} 1 & 1 & -3 \\ 2 & -1 & 9 \end{bmatrix} = A^*$$

$$T_1\mathbf{b} = \begin{bmatrix} 1 \\ 5 \end{bmatrix} = \mathbf{b}^* \qquad T_2'\mathbf{c} = \begin{bmatrix} 2 \\ 1 \\ 3 \end{bmatrix} = \mathbf{c}^*$$

Hence an equivalent dual form of (I) with the same optimal solution and optimum as (II) is

$$\begin{array}{ll} \text{maximize} & 2x_1 + x_2 + 3x_3 \\ \text{subject to} & x_1 + x_2 - 3x_3 \leq 1 \\ & 2x_1 - x_2 + 9x_3 \leq 5 \\ & x_1, \quad x_2, \quad x_3 \geq 0 \end{array}$$

6.4 Duality with Degenerate and Nondegenerate Domains

In Section 6.1, the Kuhn-Tucker theory was developed over nondegenerate domains. This section discusses the equivalence theorem when some constraints are nondegenerate or expressed as equalities, in particular, constraints of the form

$$\begin{aligned} \mathbf{f}(\mathbf{y}) &\leq 0 \\ \mathbf{h}(\mathbf{y}) &= 0 \\ \mathbf{y} &\in E^m \end{aligned} \tag{6.27}$$

where $\mathbf{h}(\mathbf{y}) = \mathbf{b} - A\mathbf{y}$ and the rows of A are linearly independent.

To establish the equivalence theorem with constraints of the form (6.27) it is required, as in Definitions 6.1 and 6.2, to define K^1 and K^2 appropriately, i.e., so that Lemmas 6.1 and 6.2 remain valid. In this light consider the following two problems

Problem 1-A

$$\begin{aligned} \text{minimize} \quad & F(\mathbf{y}) \\ \text{subject to} \quad & \mathbf{f}(\mathbf{y}) \leq 0 \\ & \mathbf{h}(\mathbf{y}) = 0 \end{aligned}$$

Problem 2-A

$$\begin{aligned} &\text{find } \mathbf{y}^0 \in E^m,\ \mathbf{x}^0 \geq 0, \text{ and } \mathbf{x}_0^0 \in E^k \\ &\text{such that } \phi(\mathbf{y}^0, \mathbf{x}, \mathbf{x}_0) \leq \phi(\mathbf{y}^0, \mathbf{x}^0, \mathbf{x}_0^0) \leq \phi(\mathbf{y}, \mathbf{x}^0, \mathbf{x}_0^0) \\ &\qquad\qquad \forall \mathbf{y} \in E^n \quad \forall \mathbf{x} \geq 0 \quad \forall \mathbf{x}^0 \in E^k \end{aligned}$$

where $\phi(\mathbf{y}, \mathbf{x}, \mathbf{x}_0) = F(\mathbf{y}) + \mathbf{x}'\mathbf{f}(\mathbf{y}) + \mathbf{x}_0'\mathbf{h}(\mathbf{y})$.

Consider also the generalized regularity assumption:

Condition I. There exists at least one $\mathbf{y}^* \in E^m$ such that $\mathbf{h}(\mathbf{y}^*) = 0$ and $\mathbf{f}(\mathbf{y}^*) < 0$.

DEFINITION 6.4. *Let $\bar{K}^1$ be the set of points $(z_0, \bar{\mathbf{z}}, \mathbf{z}_1) \in E^{n+k+1}$, where $z_0 \in E^1$ and $\bar{\mathbf{z}} \in E^n$ and $\mathbf{z}_1 \in E^k$, such that there exists at least one $\mathbf{y} \in E^m$ where $-\mathbf{z}_0 + F(\mathbf{y}) \leq 0$, $\mathbf{f}(\mathbf{y}) + \bar{\mathbf{z}} \leq 0$, and $\mathbf{h}(\mathbf{y}) - \mathbf{z}_1 = 0$, i.e.,*

$$\bar{K}^1 = \left\{ \mathbf{z} \,\middle|\, \begin{aligned} -z_0 + F(\mathbf{y}) &\leq 0 \\ \mathbf{f}(\mathbf{y}) + \bar{\mathbf{z}} &\leq 0 \quad \textit{for some } \mathbf{y} \in E^m \\ \mathbf{h}(\mathbf{y}) - \mathbf{z}_1 &= 0 \end{aligned} \right\}$$

DEFINITION 6.5. *Let $\bar{K}^2$ be the set of points in E^{n+k+1} such that $-z_0 + F(\mathbf{y}^0) > 0$, $-\bar{\mathbf{z}} \leq 0$, and $\mathbf{z}_1 = 0 \in E^k$, i.e.*

$$\bar{K}^2 = \{\mathbf{z} \mid -z_0 + F(\mathbf{y}^0) > 0,\ -\bar{\mathbf{z}} \leq 0,\ \mathbf{z}_1 = 0 \in E^k\}$$

where $\mathbf{y}^0$ is an optimal solution of Problem 1-A.

With the above definitions, Lemmas 6.9 and 6.10 follow from Lemmas 6.1 and 6.2, respectively.

LEMMA 6.9. *The set* $\bar{K}^2$ *is convex.*

LEMMA 6.10. *The set* $\bar{K}^1 \cap \bar{K}^2$ *is nonempty.*

It is shown below, in the spirit of Section 6.1 under Condition I and under the assumption that $\bar{K}^1$ is convex, that if $\mathbf{y}^0$ solves Problem 1-A then there exists a vector $(\mathbf{x}^0, \mathbf{x}_0^0)$ such that $(\mathbf{y}^0, \mathbf{x}^0, \mathbf{x}_0^0)$ solves Problem 2-A and $\mathbf{x}^{0\prime}\mathbf{f}(\mathbf{y}^0) = 0$. As before, no conditions are required for the converse.

LEMMA 6.11. *If* $\mathbf{s}'(\mathbf{b} - A\mathbf{y}) \geq 0$ *for some* $\mathbf{s} \neq 0$ *and all* $\mathbf{y} \in E^m$, *then* $A'\mathbf{s} = 0$.

Proof. Assume that $A'\mathbf{s} \neq 0$. If $\mathbf{s}'\mathbf{b} \leq 0$, then pick $\mathbf{y}^* = \delta A'\mathbf{s}/\mathbf{s}'AA'\mathbf{s}$ where $\delta = \epsilon + \mathbf{s}'\mathbf{b}$ for any $\epsilon > 0$. This implies that $\mathbf{s}'(\mathbf{b} - A'\mathbf{y}^*) < 0$. If $\mathbf{s}'\mathbf{b} > 0$, then for $\mathbf{y}^* = 2A'\mathbf{s}\mathbf{b}'\mathbf{s}/\mathbf{s}'AA'\mathbf{s}$ we have $\mathbf{s}'(\mathbf{b} - A\mathbf{y}^*) < 0$. Hence $A'\mathbf{s} = 0$.

This lemma is used in the proof of the following theorem.

THEOREM 6.5. *Assume* $\bar{K}^1$ *is convex and Condition I is satisfied. If* $\mathbf{y}^0$ *solves Problem 1-A, then there exists* $\mathbf{x}^0 \geq 0$ $(\mathbf{x}_0^0 \in E^k)$ *such that*

(*i*) $(\mathbf{y}^0, \mathbf{x}^0, \mathbf{x}_0^0)$ *solves Problem 2-A*
(*ii*) $\mathbf{x}^{0\prime}\mathbf{f}(\mathbf{y}^0) = 0$ *and* $\phi(\mathbf{y}^0, \mathbf{x}^0, \mathbf{x}_0^0) = F(\mathbf{y}^0)$

Proof. In view of (i) and (ii) in the proof of Theorem 6.2, there is a separating hyperplane $\mathbf{v}'\mathbf{z} = \beta$ $(\mathbf{v} \neq 0)$ such that

$$\mathbf{v}'\hat{\mathbf{z}} \geq \mathbf{v}'\tilde{\mathbf{z}} \qquad \hat{\mathbf{z}} \in \bar{K}^1 \qquad \tilde{\mathbf{z}} \in \bar{\bar{K}}^2 \tag{6.28}$$

where $\bar{\bar{K}}^2$ is the closure of $\bar{K}^2$ relative to E^{n+k+1} and

$$\mathbf{v} = (v_1, \mathbf{v}_2, \mathbf{v}_3) \in \{\mathbf{v} \mid v_1 \geq 0, \mathbf{v}_2 \leq 0, \mathbf{v}_3 \in E^k\} \tag{6.29}$$

To next establish that $v_1 > 0$ or, in view of (6.29), that $v_1 \neq 0$. Consider any $\mathbf{y} \in E^m$; then $(F(\mathbf{y}), -\mathbf{f}(\mathbf{y}), \mathbf{h}(\mathbf{y})) \in \bar{K}^1$. With $(F(\mathbf{y}^0), 0, 0) \in \bar{\bar{K}}^2$, then (6.28) implies that

$$v_1 F(\mathbf{y}) - \mathbf{v}_2'\mathbf{f}(\mathbf{y}) + \mathbf{v}_3'\mathbf{h}(\mathbf{y}) \geq v_1 F(\mathbf{y}^0) \qquad \forall \mathbf{y} \in E^m \tag{6.30}$$

where the components of $\mathbf{v}_3$ are the last k components of $\mathbf{v}$. Assume that $v_1 = 0$; then from (6.30)

$$-\mathbf{v}_2'\mathbf{f}(\mathbf{y}) + \mathbf{v}_3'\mathbf{h}(\mathbf{y}) \geq 0 \qquad \forall \mathbf{y} \in E^m \tag{6.31}$$

However, considering $\mathbf{y}^*$ of Condition I, if $\mathbf{h}(\mathbf{y}^*) = 0$ and $\mathbf{f}(\mathbf{y}^*) < 0$, then for any $\mathbf{r} > 0$ and any $\mathbf{s} \in E^k$, we have

$$\mathbf{r}'\mathbf{f}(\mathbf{y}^*) + \mathbf{s}'\mathbf{h}(\mathbf{y}^*) < 0 \tag{6.32}$$

which contradicts (6.31) if $\mathbf{r} \neq 0$. However, if $\mathbf{r} = 0$, then in view of (6.32) and Lemma 6.11, $A'\mathbf{s} = 0$ (for some $\mathbf{s} \neq 0$), contradicting the assumption that the rows of A are linearly independent. Thus $v_1 \neq 0$.

Let

$$\mathbf{x}^0 = -(1/v_1)\mathbf{v}_2 > 0 \tag{6.33}$$

$$\mathbf{x}_0^0 = (1/v_1)\mathbf{v}_3 \in E^k \tag{6.34}$$

then (6.30) can be expressed as

$$F(\mathbf{y}) + \mathbf{x}^{0\prime}\mathbf{f}(\mathbf{y}) + \mathbf{x}_0^{0\prime}\mathbf{h}(\mathbf{y}) \geq F(\mathbf{y}^0) \qquad \forall \mathbf{y} \in E^m \tag{6.35}$$

If $\mathbf{y} = \mathbf{y}^0$ in (6.35), then $\mathbf{x}^{0\prime}\mathbf{f}(\mathbf{y}^0) \geq 0$. However $\mathbf{x}^0 \geq 0$ and $\mathbf{f}(\mathbf{y}^0) \leq 0$ imply that $\mathbf{x}^{0\prime}\mathbf{f}(\mathbf{y}^0) \leq 0$. Thus $\mathbf{x}^{0\prime}\mathbf{f}(\mathbf{y}^0) = 0$ and

$$\phi(\mathbf{y}^0, \mathbf{x}, \mathbf{x}_0) = F(\mathbf{y}^0) + \mathbf{x}'\mathbf{f}(\mathbf{y}^0) + \mathbf{x}_0'\mathbf{h}(\mathbf{y}^0) \leq F(\mathbf{y}^0) = \phi(\mathbf{y}^0, \mathbf{x}^0, \mathbf{x}_0^0)$$
$$\forall \mathbf{x} \geq 0 \qquad \forall \mathbf{x}_0 \in E^k$$

And in view of (6.35)

$$\phi(\mathbf{y}, \mathbf{x}^0, \mathbf{x}_0^0) = F(\mathbf{y}) + \mathbf{x}^{0\prime}\mathbf{f}(\mathbf{y}) + \mathbf{x}_0^{0\prime}\mathbf{h}(\mathbf{y}) \geq F(\mathbf{y}^0) = \phi(\mathbf{y}^0, \mathbf{x}^0, \mathbf{x}_0^0)$$
$$\forall \mathbf{y} \in E^m$$

Hence, conclusions (*i*) and (*ii*) hold.

Note that if the rows of A are not linearly independent, then the linearly dependent rows can be deleted without changing the solution of Problem 1-A, and in this case, the Lagrange multipliers corresponding to the linearly dependent rows of the associated Lagrangian function are zero in the saddle point solution in view of the preceding theorem.

THEOREM 6.6. *If* $(\mathbf{y}^0, \mathbf{x}^0, \mathbf{x}_0^0)$ *solves Problem 2-A, then* $\mathbf{y}^0$ *solves Problem 1-A and* $\mathbf{x}^{0\prime}\mathbf{f}(\mathbf{y}^0) = 0$.

Proof. The proof follows from Theorem 6.1 with the Lagrangian defined as $F(\mathbf{y}) + \mathbf{x}'\mathbf{f}(\mathbf{y}) + \mathbf{x}_0'\mathbf{h}(\mathbf{y})$.

Exercises

1. Let the domains in Problems 1 and 2 be orthant domains. Show that if $(\mathbf{y}^0, \mathbf{x}^0)$ solves Problem 2, then $\mathbf{y}^0$ solves Problem 1.
2. Let $\phi(\mathbf{y}, \mathbf{x}) = F(\mathbf{y}) + \mathbf{x}'\mathbf{f}(\mathbf{y})$ where F and $\mathbf{f}$ are convex differentiable functions of $\mathbf{y}$ ($\mathbf{y} \geq 0$).
   a) Assume F and $\mathbf{f}$ are linear; express the six necessary and sufficient conditions in Lemma 6.8 in terms of $\phi_\mathbf{x}$ and $\phi_\mathbf{y}$ where

$$\phi_\mathbf{x} = \left.\frac{\partial \phi}{\partial \mathbf{x}}\right|_{(\mathbf{y}^0, \mathbf{x}^0)} \qquad \phi_\mathbf{y} = \left.\frac{\partial \phi}{\partial \mathbf{y}}\right|_{(\mathbf{y}^0, \mathbf{x}^0)}$$

   b) If F and $\mathbf{f}$ are convex, then the six conditions expressed in terms of $\phi_\mathbf{x}$ and $\phi_\mathbf{y}$ in *a*) are known as the Kuhn-Tucker conditions. Show that if F and $\mathbf{f}$ are convex, then the Kuhn-Tucker conditions are sufficient to ensure that $\mathbf{y}^0$ solves Problem 1.
3. Let

$$L_1^* = \{\mathbf{y}^* \in E^3 \mid y_3^* \geq 0, y_1^* = 0 = y_2^*\}$$
$$L_2^* = \{\mathbf{y}^* \in E^3 \mid y_1^{*2} + y_2^{*2} - y_3^{*2} \leq 0, y_3^* \geq 0\}$$

   a) Express the following in terms of inequality restrictions

   (1) $A'\mathbf{y} - \mathbf{c} \in L_2^*$
   (2) $\mathbf{b} - A\mathbf{x} \in L_1$

   b) Express the following problem and its dual in terms of inequality restrictions

$$\begin{aligned} \text{minimize} \quad & \mathbf{b}'\mathbf{y} \\ \text{subject to} \quad & A'\mathbf{y} - \mathbf{c} \in L_2^* \\ & \mathbf{y} \in L_1^* \end{aligned}$$

4. Prove that if
   a) $\begin{pmatrix} F \\ \mathbf{f} \end{pmatrix}$ is separable, i.e., $F(\mathbf{y}) = \sum_{i=1}^m F_i(y_i)$,
   b) L is a Cartesian product (of the form $L = \Pi_{i=1}^m Y_i$), and
   c) $\{(F_i(y_i), \mathbf{f}_i(y_i)) \mid y_i \in L\}$ is convex for all i
   then $\{(F(\mathbf{y}), \mathbf{f}(\mathbf{y})) \mid \mathbf{y} \in L\}$ is convex.
5. Appealing to Exercise 4 and Lemma 6.6, show that in the following problem K^1 is convex and the regularity assumption is not needed.

$$\begin{aligned} \text{minimize} \quad & y_1^2 + y_2 \\ \text{subject to} \quad & y_1^2 \in Q_1^+ \\ & \mathbf{y} \in L = \{(y_1, y_2) \mid y_2 \geq 0, y_1 = \pm 1\} \end{aligned}$$

6. Show that if $F(\mathbf{y})$ is a monotonically decreasing function over L, then Condi-

tion A in Chapter 4 is satisfied; i.e., the set T is closed and convex

$$T = \{(\xi, \eta) \in E^{m+1} \mid \mathbf{f}(y) + \xi \leq 0, \eta + F(\mathbf{y}) - F(\mathbf{y}^0) \leq 0 \quad \text{for } \mathbf{y} \in L\}$$

Here assume $\mathbf{f}(\mathbf{y})$ is linear and L is the nonnegative orthant in E^m.

7. Give some examples appealing to the results in Exercise 6.
8. Establish at least two nonclassical mutually dual problems for the following linear programming models

a) maximize $x_1 + x_2$
subject to $x_1 + x_2 \leq 1$
$x_1, \quad x_2 \geq 0$

b) maximize $x_1 + x_2$
subject to $x_1 \leq 1$
$x_2 \leq 1$
$x_1, x_2 \leq 1$

c) maximize $x_1 + 2x_2$
subject to $1 \leq x_1 + x_2 \leq 2$
$x_1, \quad x_2 \geq 0$

9. Let L and C be two arbitrary convex cones. Show that if $\mathbf{f}(\mathbf{y}) \in C$, then the statement and proof of Theorem 6.2 can be easily extended to cone domains. (Hint: $\mathbf{x} \in C^*$, and appeal to Lemma 6.4 and the regularity condition at (6.6).)

References

Ben-Israel, Adi. 1969. Linear equations and inequalities on finite dimensional, real or complex, vector spaces: a unified theory. *J. Math. Anal. Appl.* 27:367–89.

Berman, A., and Ben-Israel, Adi. 1971. Linear inequalities, mathematical programming and matrix theory. *J. Math. Program.* 1:291–300.

Karlin, S. 1959. *Mathematical Methods and Theory of Games, Programming, and Economics.* Addison-Wesley, Reading, Mass.

Kuhn, H. W., and Tucker, A. W. 1950. Nonlinear programming. 2nd Berkeley Symp. Proc. Math. Stat. Probab., pp. 481–92. Univ. Calif. Press, Berkeley.

Kunzi, Hans, and Krelle, Wilhelm. 1966. *Nonlinear Programming.* Blaisdell, Waltham, Mass.

Mangasarian, O. L. 1969. *Nonlinear Programming.* McGraw-Hill, New York.

Slater, M. 1951. Lagrange multipliers revisited: a contribution to nonlinear programming. Rand Corp. Rep. RM-676. Santa Monica, Calif.

Sposito, V. A. 1974. Modified regularity conditions for nonlinear programming problems over mixed cone domains. *J. Math. Program.* 6:167–79.

Sposito, V. A., and David, H. T. 1971. Saddle point optimality criteria of nonlinear programming problems over cones without differentiability. *SIAM J. Appl. Math.* 20:698–702.

Van Slyke, R. M., and Wets, R. J. 1968. A duality theory for abstract mathematical programs with applications to optimal control theory. *J. Math. Anal. Appl.* 22:679–706.

CHAPTER SEVEN

SADDLE POINT CHARACTERIZATION AND QUADRATIC PROGRAMMING

In Chapter 6 we developed saddle point optimality criteria without differentiability assumptions. However, many types of programming problems involve differentiable functions such as linear and quadratic functions. For some problems, efficient computational techniques and quadratic duality are highly dependent on differentiability assumptions.

In this chapter we establish necessary and sufficient saddle point optimality criteria. For the necessary saddle point optimality criteria we need only differentiability; the necessary conditions that characterize a saddle point solution when the domains are restricted to orthant domains are known as the Kuhn-Tucker conditions (1950). For sufficiency we need convexity and concavity conditions as well as differentiability; under convexity, the Kuhn-Tucker conditions are sufficient to characterize an optimal solution of the nonlinear programming problem (Problem 1).

As defined in Chapter 6, Problem 1 is

$$\begin{aligned}&\text{minimize} && F(\mathbf{y})\\ &\text{subject to} && \mathbf{f}(\mathbf{y}) \le 0\\ & && \mathbf{y} \in L\end{aligned}$$

and its associated saddle value problem is

$$\begin{aligned}&\text{find } (\mathbf{y}^0, \mathbf{x}^0) \in L \times Q_n^+\\ &\text{such that } \phi(\mathbf{y}^0, \mathbf{x}) \le \phi(\mathbf{y}^0, \mathbf{x}^0) \le \phi(\mathbf{y}, \mathbf{x}^0) \qquad \forall(\mathbf{y}, \mathbf{x}) \in L \times Q_n^+\end{aligned}$$

where $\phi(\mathbf{y}, \mathbf{x}) = F(\mathbf{y}) + \mathbf{x}'\mathbf{f}(\mathbf{y})$.

The Kuhn-Tucker conditions over orthant domains have been key ingredients in the development of quadratic programming algorithms in the past decade. Some quadratic programming algorithms based on the Kuhn-Tucker conditions are described in this chapter.

7.1 Saddle Point Characterization

In this section we develop necessary and sufficient conditions characterizing a saddle point solution of the saddle value problem. Let L be the positive orthant; $f(\mathbf{z})$ is defined as a function from L into the reals possessing a gradient vector at $\mathbf{z}_0$ denoted by $f_{\mathbf{z}}(\mathbf{z}_0)$.

In the main results of this section $f(\mathbf{z})$ denotes either $\phi(\mathbf{y}, \mathbf{x}^0)$ or $\phi(\mathbf{y}^0, \mathbf{x})$ as in Theorems 7.1 and 7.2 that establish necessary and sufficient conditions for $(\mathbf{y}^0, \mathbf{x}^0)$ to be a saddle point of $\phi(\mathbf{y}, \mathbf{x})$.

LEMMA 7.1. *If $f(\mathbf{z}) - f(\mathbf{z}_0) \geq$ (resp., $\leq$) 0 on L, then*

(i) $f'_{\mathbf{z}}(\mathbf{z}_0)(\mathbf{z} - \mathbf{z}_0) \geq$ *(resp.,* $\leq$*)* 0
(ii) $f'_{\mathbf{z}}(\mathbf{z}_0)\mathbf{z}_0 = 0$
(iii) $f_{\mathbf{z}}(\mathbf{z}_0) \geq$ *(resp.,* $\leq$*)* 0

Proof. It suffices to consider

$$f(\mathbf{z}) - f(\mathbf{z}_0) \geq 0 \quad \text{on } L \tag{7.1}$$

Expanding $f(\mathbf{z}_0 + k(\bar{\mathbf{z}} - \mathbf{z}_0))$ about $\mathbf{z}_0$ for any $k > 0$ and $\bar{\mathbf{z}} \in L$, we have

$$\begin{aligned} f(\mathbf{z}_0 + k(\bar{\mathbf{z}} - \mathbf{z}_0)) &= f(\mathbf{z}_0) + f'_{\mathbf{z}}(\mathbf{z}_0)k(\bar{\mathbf{z}} - \mathbf{z}_0) + o(k \mid \bar{\mathbf{z}} - \mathbf{z}_0 \mid) \\ &= f(\mathbf{z}_0) + f'_{\mathbf{z}}(\mathbf{z}_0)k(\bar{\mathbf{z}} - \mathbf{z}_0) + o(k) \end{aligned} \tag{7.2}$$

Thus for k small enough, $\mathbf{z}_0 + k(\bar{\mathbf{z}} - \mathbf{z}_0) \in L$, and in view of (7.1), it follows that $0 \leq f'_{\mathbf{z}}(\mathbf{z}_0)(\mathbf{z} - \mathbf{z}_0) + o(k)/k$ which implies that

$$f'_{\mathbf{z}}(\mathbf{z}_0)(\mathbf{z} - \mathbf{z}_0) \geq 0 \quad \text{on } L \tag{7.3}$$

and conclusion (*i*) holds. Moreover, (*ii*) holds by letting $\mathbf{z} = 0$ and $\mathbf{z} = 2\mathbf{z}_0$ in (7.3).

Show next that $f_{\mathbf{z}}(\mathbf{z}_0) \geq 0$. Considering (*i*) and (*ii*), we have that $f'_{\mathbf{z}}(\mathbf{z}_0)\mathbf{z} \geq 0$ on L, which implies that $f'_{\mathbf{z}}(\mathbf{z}_0) \geq 0$. Thus conclusion (*iii*) holds.

LEMMA 7.2. *Suppose $f_{\mathbf{z}}(\mathbf{z}_0) \geq$ (resp., $\leq$) 0 on L and $f(\mathbf{z}) \geq$ (resp., $\leq$) $f(\mathbf{z}_0) + f'_{\mathbf{z}}(\mathbf{z}_0)\,\mathbf{z}$ on L, then $f(\mathbf{z}) - f(\mathbf{z}_0) \geq$ (resp., $\leq$) 0 on L.*

Proof. The product $f'_{\mathbf{z}}(\mathbf{z}_0)\,\mathbf{z}$ is nonnegative (nonpositive) for all $\mathbf{z}$ in L. Thus, $f(\mathbf{z}) - f(\mathbf{z}_0) \geq$ (resp., $\leq$) 0 on L.

THEOREM 7.1. *Let $\phi(\mathbf{y}, \mathbf{x})$ be a function from $Q_m^+ \times Q_n^+$ into the reals. Suppose that $\phi(\mathbf{y}, \mathbf{x})$ has continuous first derivatives $\phi_{\mathbf{x}}$ and $\phi_{\mathbf{y}}$ at $(\mathbf{y}^0, \mathbf{x}^0)$. Then the conditions*

(a) $\phi'_{\mathbf{x}}(\mathbf{y}^0, \mathbf{x}^0)(\mathbf{x}^0) = 0$
(b) $\phi_{\mathbf{x}}(\mathbf{y}^0, \mathbf{x}^0) \leq 0$
(c) $\mathbf{x}^0 \geq 0$
(d) $\phi'_{\mathbf{y}}(\mathbf{y}^0, \mathbf{x}^0)\,\mathbf{y}^0 = 0$
(e) $\phi_{\mathbf{y}}(\mathbf{y}^0, \mathbf{x}^0) \geq 0$
(f) $\mathbf{y}^0 \geq 0$

are necessary for $(\mathbf{y}^0, \mathbf{x}^0)$ *to be a saddle point of* $\phi(\mathbf{y}^0, \mathbf{x}^0)$ *where* $(\mathbf{y}^0, \mathbf{x}^0)$ *is in* $Q_m^+ \times Q_n^+$. *Conditions (a)–(f) are known as the "Kuhn-Tucker conditions."*

Proof (necessity). The point $(\mathbf{y}^0, \mathbf{x}^0)$ is a saddle point of $\phi(\mathbf{y}, \mathbf{x})$ on $Q_m^+ \times Q_n^+$. Therefore $\phi(\mathbf{y}^0, \mathbf{x}) \leq \phi(\mathbf{y}^0, \mathbf{x}^0)$ on Q_n^+ with $\mathbf{x}^0$ satisfying (*c*). However, (*ii*) and (*iii*) in Lemma 7.1 yield (*a*) and (*b*) with $f(\mathbf{z}) = \phi(\mathbf{y}^0, \mathbf{x})$, $f(\mathbf{z}_0) = \phi(\mathbf{y}^0, \mathbf{x}^0)$, and $L = Q_n^+$ (with $\leq$ on L). Conditions (*d*), (*e*), and (*f*) hold by a similar argument with $f(\mathbf{z}) = \phi(\mathbf{y}, \mathbf{x}^0)$ and $f(\mathbf{z}_0) = \phi(\mathbf{y}^0, \mathbf{x}^0)$.

EXAMPLE 7.1

In view of the above discussion, a set of necessary conditions are established to characterize the optimal solution of the saddle value problem. If we have the problem

$$\begin{aligned} \text{minimize} \quad & (y - 1)^2 \\ \text{subject to} \quad & y \leq 1 \\ & y \geq 0 \end{aligned}$$

then the Lagrangian function of the saddle value problem is

$$\phi(y, x) = (y - 1)^2 + x(y - 1)$$

Clearly, the vector $(y^*, x^*) = (1, 0)$ satisfies the Kuhn-Tucker conditions; i.e.,

(a) $\phi_{\mathbf{x}}(y^*, x^*)x^* = 0$ or $(y^* - 1)x^* = 0$
(b) $\phi_{\mathbf{x}}(y^*, x^*) \leq 0$ or $y^* - 1 \leq 0$
(c) $x^* \geq 0$
(d) $\phi_{\mathbf{y}}(y^*, x^*)\,y^* = 0$ or $2(y^* - 1)\,y^* + x^* y^* = 0$
(e) $\phi_{\mathbf{y}}(y^*, x^*) \geq 0$ or $2(y^* - 1) + x^* \geq 0$
(f) $y^* \geq 0$

In this example the Kuhn-Tucker conditions are necessary and sufficient to identify an optimal solution of the original problem, namely, $y^0 = 1$. As we shall see later this is always the case when the objective function is convex (concave) for minimization (maximization) problems over linear constraints.

The next example shows that the Kuhn-Tucker conditions can be satisfied by some (y^*, x^*), but this is not sufficient to ensure that y^* solves the original nonlinear minimization problem.

EXAMPLE 7.2

Consider the problem

$$\begin{aligned} \text{minimize} \quad & -(y-1)^3 \\ \text{subject to} \quad & y \leq 2 \\ & y \geq 0 \end{aligned}$$

then the Lagrangian function of the saddle value problem is

$$\phi(y, x) = -(y-1)^3 + x(y-2)$$

Clearly, the optimal solution of the problem is $y^0 = 2$. Consider the vector $(y^*, x^*) = (1, 0)$. Then from the Kuhn-Tucker conditions we have

(a) $(y^* - 2)x^* = 0$
(b) $(y^* - 2) \leq 0$
(c) $x^* \geq 0$
(d) $-3(y^* - 1)^2 y^* + y^* x^* = 0$
(e) $-3(y^* - 1)^2 + x^* \geq 0$
(f) $y^* \geq 0$

The vector $(y^*, x^*) = (1, 0)$ satisfies the Kuhn-Tucker conditions, but $y^* = 1$ is not an optimal solution of the original problem.

The next theorem is used to establish some sufficient conditions to ensure that $\mathbf{y}^0$ of the saddle value solution solves the original problem.

THEOREM 7.2. *The conditions of Theorem 7.1 and*

(*g*) $\phi(\mathbf{y}^0, \mathbf{x}) \leq \phi(\mathbf{y}^0, \mathbf{x}^0) + \phi'_{\mathbf{x}}(\mathbf{y}^0, \mathbf{x}^0)(\mathbf{x} - \mathbf{x}^0)$ *on* Q_n^+
(*h*) $\phi(\mathbf{y}, \mathbf{x}^0) \geq \phi(\mathbf{y}^0, \mathbf{x}^0) + \phi'_{\mathbf{y}}(\mathbf{y}^0, \mathbf{x}^0)(\mathbf{y} - \mathbf{y}^0)$ *on* Q_m^+

are sufficient to ensure that $(\mathbf{y}^0, \mathbf{x}^0)$ *is a saddle point of* $\phi(\mathbf{y}, \mathbf{x})$ *on* $Q_m^+ \times Q_n^+$.

Proof (sufficiency). With condition (*a*) in Theorem 7.1, the conditions of Lemma 7.2 are conditions (*b*) and (*g*) when $f(\mathbf{z}) = \phi(\mathbf{y}^0, \mathbf{x})$ and $L = Q_n^+$ (with $\leq$ on L). Hence, by Lemma 7.2, $\phi(\mathbf{y}^0, \mathbf{x}) \leq \phi(\mathbf{y}^0, \mathbf{x}^0)$ on Q_n^+, similarly for $\mathbf{y}$. Hence the additional conditions (*c*) and (*f*) imply $(\mathbf{y}^0, \mathbf{x}^0)$ is a saddle point of $\phi(\mathbf{y}, \mathbf{x})$.
Alternate proof of sufficiency. From (*g*)

$$\begin{aligned}\phi(\mathbf{y}^0,\mathbf{x}) &\le \phi(\mathbf{y}^0,\mathbf{x}^0) + \phi_{\mathbf{x}}'(\mathbf{y}^0,\mathbf{x}^0)(\mathbf{x}-\mathbf{x}^0) \text{ on } Q_n^+\\ &= \phi(\mathbf{y}^0,\mathbf{x}^0) + \phi_{\mathbf{x}}'(\mathbf{y}^0,\mathbf{x}^0)\,\mathbf{x} - \phi_{\mathbf{x}}'(\mathbf{y}^0,\mathbf{x}^0)\,\mathbf{x}^0 \text{ on } Q_n^+\\ &\le \phi(\mathbf{y}^0,\mathbf{x}^0) \quad \text{from } (a) \text{ and } (b)\\ &= \phi(\mathbf{y}^0,\mathbf{x}^0) - \phi_{\mathbf{y}}'(\mathbf{y}^0,\mathbf{x}^0)\,\mathbf{y}^0 \quad \text{from } (d)\\ &\le \phi(\mathbf{y}^0,\mathbf{x}^0) + \phi_{\mathbf{y}}'(\mathbf{y}^0,\mathbf{x}^0)(\mathbf{y}-\mathbf{y}^0) \text{ on } Q_m^+ \quad \text{from } (e)\\ &\le \phi(\mathbf{y},\mathbf{x}^0) \text{ on } Q_m^+ \quad \text{from } (h)\end{aligned}$$

Let us consider the Kuhn-Tucker conditions

$$\begin{array}{lll}\phi_{\mathbf{x}}'(\mathbf{y}^0,\mathbf{x}^0)\,\mathbf{x}^0 = 0 & \phi_{\mathbf{x}}(\mathbf{y}^0,\mathbf{x}^0) \le 0 & \mathbf{x}^0 \ge 0\\ \phi_{\mathbf{y}}'(\mathbf{y}^0,\mathbf{x}^0)\,\mathbf{y}^0 = 0 & \phi_{\mathbf{y}}(\mathbf{y}^0,\mathbf{x}^0) \ge 0 & \mathbf{y}^0 \ge 0\end{array}$$

Then, the classical Kuhn-Tucker conditions and conditions (g) and (h) of Theorem 7.2 are sufficient to ensure that $(\mathbf{y}^0, \mathbf{x}^0)$ is a saddle point of $\phi(\mathbf{y},\mathbf{x})$ on $Q_m^+ \times Q_n^+$.

The following theorem establishes that if $(\mathbf{y}^0, \mathbf{x}^0)$ satisfies the Kuhn-Tucker conditions and condition (h), then this is sufficient to ensure that $\mathbf{y}^0$ solves Problem 1.

THEOREM 7.3. *In order for* $\mathbf{y}^0$ *to be a solution of Problem 1 it is sufficient that* $\mathbf{y}^0$ *and some* $\mathbf{x}^0 \ge 0$ *satisfy conditions (a)–(f) of Theorem 7.1 and* $\phi(\mathbf{y},\mathbf{x}) = F(\mathbf{y}) + \mathbf{x}'\mathbf{f}(\mathbf{y})$ *satisfies (h) of Theorem 7.2.*

Proof. For all $\mathbf{y} \ge 0$ such that $\mathbf{f}(\mathbf{y}) \le 0$ we have

$$\begin{aligned}F(\mathbf{y}) &\ge F(\mathbf{y}) + \mathbf{x}^{0\prime}\mathbf{f}(\mathbf{y}) = \phi(\mathbf{y},\mathbf{x}^0) \quad \text{for } \mathbf{x}^0 \ge 0\\ &\ge \phi(\mathbf{y}^0,\mathbf{x}^0) + \phi_{\mathbf{y}}'(\mathbf{y}^0,\mathbf{x}^0)(\mathbf{y}-\mathbf{y}^0) \text{ on } Q_m^+ \quad \text{from } (h)\\ &\ge \phi(\mathbf{y}^0,\mathbf{x}^0) \quad \text{from } (d) \text{ and } (e)\\ &= F(\mathbf{y}^0) + \mathbf{x}^{0\prime}\mathbf{f}(\mathbf{y}^0)\\ &= F(\mathbf{y}^0)\end{aligned}$$

since $\mathbf{x}^{0\prime}\mathbf{f}(\mathbf{y}^0) \equiv \phi_{\mathbf{x}}'(\mathbf{y}^0,\mathbf{x}^0)\mathbf{x}^0 = 0$ in (a). Therefore $F(\mathbf{y}) \ge F(\mathbf{y}^0)$ for all $\mathbf{y} \ge 0$ such that $\mathbf{f}(\mathbf{y}) \le 0$.

LEMMA 7.3. *Let g possess continuous first derivatives on a convex set Y. Then g is convex on Y if and only if*

$$g(\mathbf{y}_2) - g(\mathbf{y}_1) \ge \nabla g(\mathbf{y}_1)(\mathbf{y}_2 - \mathbf{y}_1)$$

where ∇g *is the gradient vector of g.*

Proof. Since g is convex, then for all $\alpha \in (0, 1)$

$$g(\alpha \mathbf{y}_2 + (1 - \alpha)\mathbf{y}_1) \leq \alpha g(\mathbf{y}_2) + (1 - \alpha) g(\mathbf{y}_1)$$

or

$$[g(\mathbf{y}_1 + \alpha(\mathbf{y}_2 - \mathbf{y}_1)) - g(\mathbf{y}_1)]/\alpha \leq g(\mathbf{y}_2) - g(\mathbf{y}_1) \tag{7.4}$$

On expanding $g(\mathbf{y}_1 + \alpha(\mathbf{y}_2 - \mathbf{y}_1))$ by Taylor's Theorem we have, for $\theta \in [0, 1]$

$$g(\mathbf{y}_1 + \alpha(\mathbf{y}_2 - \mathbf{y}_1)) = g(\mathbf{y}_1) + \alpha \nabla g(\mathbf{y}_1 + \alpha\theta(\mathbf{y}_2 - \mathbf{y}_1))(\mathbf{y}_2 - \mathbf{y}_1) \tag{7.5}$$

Substituting (7.4) in (7.5)

$$\nabla g(\mathbf{y}_1 + \alpha\theta(\mathbf{y}_2 - \mathbf{y}_1))(\mathbf{y}_2 - \mathbf{y}_1) \leq g(\mathbf{y}_2) - g(\mathbf{y}_1) \tag{7.6}$$

In particular, on taking the limit as $\alpha \to 0$ in (7.6) we have

$$\nabla g(\mathbf{y}_1)(\mathbf{y}_2 - \mathbf{y}_1) \leq g(\mathbf{y}_2) - g(\mathbf{y}_1) \qquad \forall \mathbf{y}_1, \mathbf{y}_2 \in Y$$

Conversely, let $\mathbf{y}_1, \mathbf{y}_2 \in Y$. Then since Y is convex we have

$$(1 - \alpha)\mathbf{y}_1 + \alpha \mathbf{y}_2 \in Y \quad \text{for any } \alpha \in [0, 1]$$

Therefore

$$g(\mathbf{y}_1) - g((1 - \alpha)\mathbf{y}_1 + \alpha\mathbf{y}_2) \geq \alpha \nabla g((1 - \alpha)\mathbf{y}_1 + \alpha\mathbf{y}_2)(\mathbf{y}_1 - \mathbf{y}_2)$$

$$g(\mathbf{y}_2) - g((1 - \alpha)\mathbf{y}_1 + \alpha\mathbf{y}_2) \geq -(1 - \alpha) \nabla g((1 - \alpha)\mathbf{y}_1 + \alpha\mathbf{y}_2)(\mathbf{y}_1 - \mathbf{y}_2)$$

Multiplying the first inequality by $(1 - \alpha)$ and the second inequality by α, and adding

$$(1 - \alpha) g(\mathbf{y}_1) + \alpha g(\mathbf{y}_2) \geq g((1 - \alpha)\mathbf{y}_1 + \alpha\mathbf{y}_2)$$

i.e., g is convex over Y.

COROLLARY 7.3.1. If $\phi(\mathbf{y}, \mathbf{x}^0)$ *is a convex function of* $\mathbf{y}$ *on* Q_m^+, *then*

$$\phi'_{\mathbf{y}}(\mathbf{y}^0, \mathbf{x}^0)(\mathbf{y} - \mathbf{y}^0) \leq \phi(\mathbf{y}, \mathbf{x}^0) - \phi(\mathbf{y}^0, \mathbf{x}^0) \quad \textit{on } Q_m^+$$

COROLLARY 7.3.2. *If* $(\mathbf{y}^0, \mathbf{x}^0)$ *satisfies conditions (a)–(f) of Theorem 7.1 and* $F(\mathbf{y})$ *and* $f_i(\mathbf{y})$ *are convex, then* $F(\mathbf{y}^0)$ *is the global minimum of* $F(\mathbf{y})$ *over all* $\mathbf{y} \geq 0$ *such that* $\mathbf{f}(\mathbf{y}) \leq 0$.

Proof. By hypothesis $\phi(\mathbf{y}, \mathbf{x}^0)$ is a convex function of $\mathbf{y}$ for fixed $\mathbf{x}^0$. Therefore, in view of Lemma 7.3 and Theorem 7.3, $\mathbf{y}^0$ solves Problem 1.

The following example illustrates that the Kuhn-Tucker conditions alone are not sufficient; this example is similar to Example 7.2 except $\mathbf{y}$ is unrestricted.

EXAMPLE 7.3

Consider the problem

$$\begin{aligned} \text{minimize} \quad & -y^2 \\ \text{subject to} \quad & -2 + y \leq 0 \\ & -1 - y \leq 0 \\ & y \in E^1 \end{aligned}$$

The feasible solutions y are such that $y \in [-1, 2]$. The optimal solution is seen to be $y^0 = 2$. However, consider the feasible solution $\bar{y} = -1$ and the Lagrange multipliers $\bar{x}_1 = 0$ and $\bar{x}_2 = 2$; then, from conditions (a)–(f) in Theorem 7.1 with $\phi(y, x) = -y^2 + x_1(-2 + y) + x_2(-1 - y)$,

(i) $(-2 + \bar{y}, -1 - \bar{y})' \begin{bmatrix} \bar{x}_1 \\ \bar{x}_2 \end{bmatrix} = 0$

(ii) $\begin{bmatrix} -2 + \bar{y} \\ -1 - \bar{y} \end{bmatrix} = \begin{pmatrix} -3 \\ 0 \end{pmatrix} \leq 0$

(iii) $\begin{bmatrix} \bar{x}_1 \\ \bar{x}_2 \end{bmatrix} \geq 0$

(iv) $(-2\bar{y} + \bar{x}_1 - \bar{x}_2)(-1) = 0$

(v) $(-2\bar{y} + \bar{x}_1 - \bar{x}_2) = 0$

(vi) $\bar{y} \in E^1$

Conditions (i)–(vi) satisfy conditions (a)–(f). Hence the vector $(\bar{y}, \bar{x}_1, \bar{x}_2)$ satisfies the Kuhn-Tucker conditions, but $\bar{y} = -1$ is not the optimal solution. The reader can verify that the sufficient conditions in Theorem 7.2 are violated. Hence, the Kuhn-Tucker conditions are only necessary conditions for $(\mathbf{y}^0, \mathbf{x}^0)$ to be a saddle point of $\phi(\mathbf{y}, \mathbf{x})$, $(\mathbf{y}, \mathbf{x}) \in Q_m^+ \times Q_n^+$. Consider the saddle value problem

$$\begin{aligned} & \text{find } (\mathbf{x}^0, \mathbf{y}^0) \in Q_n^+ \times Q_m^+ \\ & \text{such that } \psi(\mathbf{x}, \mathbf{y}^0) \leq \psi(\mathbf{x}^0, \mathbf{y}^0) \leq \psi(\mathbf{x}^0, \mathbf{y}) \qquad \forall (\mathbf{x}, \mathbf{y}) \in Q_n^+ \times Q_m^+ \end{aligned}$$

In this case, considering the proof of Theorem 7.1, we see that the same six conditions are necessary conditions for $(\mathbf{x}^0, \mathbf{y}^0)$ to be a saddle point solution of $\psi(\mathbf{x}, \mathbf{y})$; i.e.,

(1) $\psi_\mathbf{x}'(\mathbf{x}^0, \mathbf{y}^0)\,\mathbf{x}^0 = 0$
(2) $\psi_\mathbf{x}(\mathbf{x}^0, \mathbf{y}^0) \leq 0$
(3) $\mathbf{x}^0 \geq 0$
(4) $\psi_\mathbf{y}'(\mathbf{x}^0, \mathbf{y}^0)\,\mathbf{y}^0 = 0$
(5) $\psi_\mathbf{y}(\mathbf{x}^0, \mathbf{y}^0) \geq 0$
(6) $\mathbf{y}^0 \geq 0$

The following example illustrates an important by-product of the Kuhn-Tucker theory when considering a Lagrangian function $\psi(\mathbf{x}, \mathbf{y})$ associated with an appropriate maximization problem.

EXAMPLE 7.4

Consider the Lagrangian $\phi(\mathbf{y}, \mathbf{x})$ associated with the primal linear programming problem

$$\begin{aligned} &\text{minimize} \quad \mathbf{b}'\mathbf{y} \\ &\text{subject to} \quad \mathbf{c} - A'\mathbf{y} \leq 0 \\ &\qquad\qquad\qquad\quad \mathbf{y} \geq 0 \end{aligned}$$

Appealing to Corollary 7.3.2, the Kuhn-Tucker conditions are necessary and sufficient to characterize an optimal solution of the primal linear programming problem. In particular, since

$$\phi(\mathbf{y}, \mathbf{x}) = \mathbf{b}'\mathbf{y} + \mathbf{x}'(\mathbf{c} - A'\mathbf{y})$$

$(\mathbf{y}^0, \mathbf{x}^0)$ must satisfy

$$\begin{aligned} &\mathbf{x}^{0\prime}\phi_\mathbf{x}(\mathbf{y}^0, \mathbf{x}^0) = 0 \qquad \phi_\mathbf{x}(\mathbf{y}^0, \mathbf{x}^0) \leq 0 \qquad \mathbf{x}^0 \geq 0 \\ &\mathbf{y}^{0\prime}\phi_\mathbf{y}(\mathbf{y}^0, \mathbf{x}^0) = 0 \qquad \phi_\mathbf{y}(\mathbf{y}^0, \mathbf{x}^0) \geq 0 \qquad \mathbf{y}^0 \geq 0 \end{aligned}$$

or

$$\begin{aligned} &\mathbf{x}^{0\prime}(\mathbf{c} - A'\mathbf{y}^0) = 0 \qquad \mathbf{c} - A'\mathbf{y}^0 \leq 0 \qquad \mathbf{x}^0 \geq 0 \\ &\mathbf{y}^{0\prime}(\mathbf{b} - A\mathbf{x}^0) = 0 \qquad \mathbf{b} - A\mathbf{x}^0 \geq 0 \qquad \mathbf{y}^0 \geq 0 \end{aligned}$$

i.e., the identical conditions needed in Chapter 4 to establish the Duality Theorem of linear programming. These conditions are also necessary and sufficient for $(\mathbf{x}^0, \mathbf{y}^0)$ to be a saddle point solution of $\psi(\mathbf{x}, \mathbf{y}) = \mathbf{c}'\mathbf{x} + \mathbf{y}'(\mathbf{b} - A\mathbf{x})$, namely, the Lagrangian associated with the following linear maximization problem

$$\begin{aligned} \text{maximize} \quad & \mathbf{c}'\mathbf{x} \\ \text{subject to} \quad & \mathbf{b} - A\mathbf{x} \geq 0 \\ & \mathbf{x} \geq 0 \end{aligned}$$

LEMMA 7.4. *Consider the problem*

$$\begin{aligned} &\textit{find } \mathbf{x}^0 \textit{ such that } G(\mathbf{x}^0) \geq G(\mathbf{x}) \qquad \forall \mathbf{x} \geq 0 \\ &\textit{subject to } \mathbf{g}(\mathbf{x}) \geq 0 \end{aligned}$$

and its associated Lagrangian function $\psi(\mathbf{x}, \mathbf{y}) = G(\mathbf{x}) + \mathbf{y}'\mathbf{g}(\mathbf{x})$. *Then if* $(\mathbf{x}^0, \mathbf{y}^0)$ *satisfies conditions (a)–(f) in Theorem 7.1 and condition (g) in Theorem 7.2, then* $G(\mathbf{x}^0)$ *is the global maximum over all* $\mathbf{x} \geq 0$ *such that* $\mathbf{g}(\mathbf{x}) \geq 0$.

Proof. The proof is similar to the proof of Theorem 7.3.

COROLLARY 7.4.1. *If* $G(\mathbf{x})$ *and* $g_i(\mathbf{x})$ *are concave and* $(\mathbf{x}^0, \mathbf{y}^0)$ *satisfies the Kuhn-Tucker conditions, then* $G(\mathbf{x}^0)$ *is the global maximum of* $G(\mathbf{x})$ *over all* $\mathbf{x} \geq 0$ *such that* $\mathbf{g}(\mathbf{x}) \geq 0$.

Figure 7.1 summarizes the relationship between the original problem, the saddle point problem and the Kuhn-Tucker conditions.

DEFINITION 7.1. *Let D be an* $n \times n$ *symmetric matrix. Then D is said to be positive (negative) semidefinite if and only if* $\mathbf{x}'D\mathbf{x} \geq 0$ (*resp.*, $\mathbf{x}'D\mathbf{x} \leq 0$) *for all* $\mathbf{x}$.

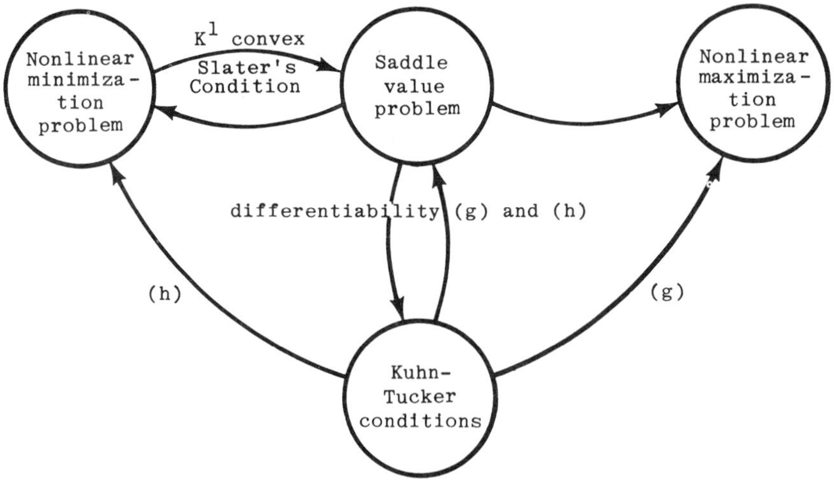

Figure 7.1. Relationship between the nonlinear programming problems, the saddle value problem, and the Kuhn-Tucker conditions.

DEFINITION 7.2. *D is said to be positive (negative) definite if and only if* $\mathbf{x}'D\mathbf{x} > 0$ *(resp.,* $\mathbf{x}'D\mathbf{x} < 0$*) for all* $\mathbf{x} \neq 0$.

LEMMA 7.5. *The function* $f(\mathbf{x}) = \mathbf{c}'\mathbf{x} + \mathbf{x}'D\mathbf{x}$ *is concave if D is negative semidefinite or definite.*

Proof. We wish to show, given any two vectors $\bar{\mathbf{x}}_1$ and $\bar{\mathbf{x}}_2$ and any $\alpha \in [0, 1]$, that

$$f(\alpha\bar{\mathbf{x}}_1 + (1 - \alpha)\bar{\mathbf{x}}_2) \geq \alpha f(\bar{\mathbf{x}}_1) + (1 - \alpha)f(\bar{\mathbf{x}}_2)$$

But

$$\begin{aligned} f(\alpha\bar{\mathbf{x}}_1 + (1 - \alpha)\bar{\mathbf{x}}_2) &= \mathbf{c}'(\alpha\bar{\mathbf{x}}_1 + (1 - \alpha)\bar{\mathbf{x}}_2) \\ &\quad + (\alpha\bar{\mathbf{x}}_1 + (1 - \alpha)\bar{\mathbf{x}}_2)'D(\alpha\bar{\mathbf{x}}_1 + (1 - \alpha)\bar{\mathbf{x}}_2) \\ &= \alpha\mathbf{c}'\bar{\mathbf{x}}_1 + (1 - \alpha)\mathbf{c}'\bar{\mathbf{x}}_2 + \bar{\mathbf{x}}_2'D\bar{\mathbf{x}}_2 \\ &\quad + \alpha\bar{\mathbf{x}}_2'D(\bar{\mathbf{x}}_1 - \bar{\mathbf{x}}_2) \\ &\quad + \alpha(\bar{\mathbf{x}}_1 - \bar{\mathbf{x}}_2)'D\bar{\mathbf{x}}_2 + \alpha^2(\bar{\mathbf{x}}_1 - \bar{\mathbf{x}}_2)'D(\bar{\mathbf{x}}_1 - \bar{\mathbf{x}}_2) \\ &= \alpha\mathbf{c}'\bar{\mathbf{x}}_1 + (1 - \alpha)\mathbf{c}'\bar{\mathbf{x}}_2 + \bar{\mathbf{x}}_2'D\bar{\mathbf{x}}_2 \\ &\quad + 2\alpha(\bar{\mathbf{x}}_1 - \bar{\mathbf{x}}_2)'D\bar{\mathbf{x}}_2 \\ &\quad + \alpha^2(\bar{\mathbf{x}}_1 - \bar{\mathbf{x}}_2)'D(\bar{\mathbf{x}}_1 - \bar{\mathbf{x}}_2) \\ &\geq \alpha\mathbf{c}'\bar{\mathbf{x}}_1 + (1 - \alpha)\mathbf{c}'\bar{\mathbf{x}}_2 + \alpha\bar{\mathbf{x}}_1'D\bar{\mathbf{x}}_1 \\ &\quad + (1 - \alpha)\bar{\mathbf{x}}_2'D\bar{\mathbf{x}}_2 \quad \text{since } \alpha^2 \leq \alpha \\ &= \alpha f(\bar{\mathbf{x}}_1) + (1 - \alpha)f(\bar{\mathbf{x}}_2) \end{aligned}$$

COROLLARY 7.5.1. *If D is positive semidefinite or definite, then* $\mathbf{x}'D\mathbf{x}$ *and* $\mathbf{c}'\mathbf{x} + \mathbf{x}'D\mathbf{x}$ *are convex functions of* $\mathbf{x}$.

EXAMPLE 7.5

Consider the problem

$$\begin{aligned} \text{maximize} \quad & x_1 + x_2 + x_1^2 + 2x_1x_2 + x_2^2 \\ \text{subject to} \quad & x_1 + x_2 \leq 1 \\ & x_1, \quad x_2 \geq 0 \end{aligned}$$

Then the associated Lagrangian function is

$$\psi(\mathbf{x}, \mathbf{y}) = \mathbf{c}'\mathbf{x} + \mathbf{x}'D\mathbf{x} + y(b - A\mathbf{x})$$

where $\mathbf{c}' = (1, 1)$, $D = \begin{bmatrix} 1 & 1 \\ 1 & 1 \end{bmatrix}$, $b = 1$, and $A = [1, 1]$. Then a set of necessary and sufficient conditions that $\mathbf{x}^0$ is a solution of the above problem is that there exists $y^0 \geq 0$ such that

$$\psi'_{\mathbf{x}}(\mathbf{x}^0, y^0)\,\mathbf{x}^0 = 0$$
$$\psi_{\mathbf{x}}(\mathbf{x}^0, y^0) \leq 0$$
$$\mathbf{x}^0 \geq 0$$
$$\psi'_y(\mathbf{x}^0, y^0)\,y^0 = 0$$
$$\psi_y(\mathbf{x}^0, y^0) \geq 0$$
$$y^0 \geq 0$$

In particular, $(\mathbf{x}^0, y^0)$ must satisfy the following system

$$\mathbf{c}'\mathbf{x}^0 + 2\mathbf{x}^{0\prime}D\mathbf{x}^0 - y^0A\mathbf{x}^0 = 0$$
$$\mathbf{c} + 2D\mathbf{x}^0 - y^0A \leq 0$$
$$\mathbf{x}^0 \geq 0$$
$$y^0(b - A\mathbf{x}^0) = 0$$
$$b - A\mathbf{x}^0 \geq 0$$
$$y^0 \geq 0$$

Since $y^0A\mathbf{x}^0 = y^0b$ it follows that $\mathbf{c}'\mathbf{x}^0 + 2\mathbf{x}^{0\prime}D\mathbf{x}^0 = y^0b$, and for any optimal solution $\mathbf{x}^0$ of the original problem there exists $y^0 \geq 0$ such that

$$y^0b = \mathbf{c}'\mathbf{x}^0 + 2\mathbf{x}^{0\prime}D\mathbf{x}^0$$
$$A'y^0 \geq \mathbf{c}' + 2D\mathbf{x}^0$$

Hence for $\mathbf{x}^{0\prime} = (1/2, 1/2)$, a sufficient value of y^0 to guarantee that $\mathbf{x}^0$ is indeed optimal is 3.

7.2 Examples of Problems Applicable to Quadratic Programming

Quadratic programming is concerned with maximizing or minimizing quadratic functions subject to linear inequality constraints. Problems of this type arise in various contexts; in this section we discuss two problems which can be classified as quadratic programming problems.

Imagine a security portfolio formed by making selections in proper proportions from n different stocks, mutual funds, or bonds. Most investors would wonder what portfolio to buy; the following outlines one way.

Assume the rate of return on investment in the ith security is distributed with expected rate of return μ_i and variance σ_{ii}^2.

The variance is a measure of the risk that the realized rate of return deviates from the expected rate of return; i.e., a security with a high variance σ_{ii}^2 in returns is risky. To arrive at a reasonable estimate of μ_i and σ_{ii}^2 consider the following:

Assume that historical data are available giving a record of the price and dividend behavior of the securities for m years. Define the return $r_j(k)$ from security j in year k to be

$$r_j(k) = \frac{p_j(k) - p_j(k-1) + d_j(k)}{p_j(k-1)} \tag{7.7}$$

where $p_j(k)$ is the average price in year k and $d_j(k)$ denotes dividends received in year k. For each j we can compute a measure of central tendency in two reasonable ways, i.e., by the arithmetic mean

$$\bar{\mu}_j = (1/m)\sum_k r_j(k)$$

or by the geometric mean

$$\mu_j = [(1 + r_j(1))(1 + r_j(2)) \cdots (1 + r_j(m))]^{1/m} - 1 \tag{7.8}$$

Without loss of generality let the one selected be denoted by μ_j. Then a measure of variation of the return is

$$\sigma_j^2 = (1/m)\sum_k [r_j(k) - \mu_j]^2 \tag{7.9}$$

and for each pair ij, the covariance is

$$\sigma_{ij} = (1/m)\sum_k [r_i(k) - \mu_i][r_j(k) - \mu_j] \tag{7.10}$$

Hence if one invests in two securities i and j, the covariance σ_{ij} of their returns provides a measure of the correlation between the rates of return on securities i and j. A positive covariance implies that the returns of the securities tend to go up and down together; therefore, a good rule of thumb for rational investors is to not invest heavily in a set of securities that seem to move together. Hence it would be advisable to diversify the portfolio somewhat, sacrificing return for risk reduction; and measuring covariance between securities allows us to achieve this required diversification.

Among the many aims of investors it is assumed that (i) among securities with equal expected rates of returns, investors prefer the one with the smallest variance, and (ii) among securities with equal variances of returns, investors prefer the one with the greatest rate of return. These two statements are essentially equivalent, and imply that investors consider only efficient portfolios. Hence a portfolio is not efficient if another portfolio can be selected with higher expected rate of return and the same variance, or with lower variance and the same expected rate of return.

To tie all the above concepts together, consider the following six securities and the values, given in Table 7.1, of each security over the last eleven intervals. Year 1 represents the initial amount that would have to be invested to purchase 100 shares of each security.

Suppose we would like to minimize our risk while determining a portfolio with an expected return of at least 20 percent on our total capital.

For convenience let $d_j(k) = 0$ for all j, k; then

$$r_1(1) = (16{,}257 - 11{,}019)/11{,}019 = .475$$

$$r_1(2) = (18{,}628 - 16{,}257)/16{,}257 = .145$$

$$\vdots$$

represents the return for each security for each of the ten years.

TABLE 7.1. Table of Values

Time Interval	Security					
	One	Two	Three	Four	Five	Six
1	$11,019	$ 16,308	$ 8,000	$10,000	$ 7,683	$10,328
2	16,257	22,501	10,000	13,313	10,396	17,475
3	18,628	25,235	15,201	15,439	12,744	25,200
4	20,315	33,345	19,524	14,908	14,206	30,106
5	28,588	49,701	26,703	19,744	18,080	39,107
6	22,827	35,784	21,355	16,937	14,249	29,840
7	27,093	41,141	33,846	23,266	16,141	34,485
8	29,294	48,769	40,634	26,729	17,054	37,984
9	38,722	68,806	57,221	39,965	23,986	55,141
10	37,245	74,035	55,779	41,124	24,546	49,203
11	74,757	126,955	82,077	59,867	42,963	83,915

To establish what yearly yield would produce an increase from $11,-019 to $74,757 for security One in the ten-year interval, using (7.8) we have

$$\hat{\mu}_1 = .211$$

This implies that \$11,019 invested at 21.1 percent compounded would yield \$74,757 in ten years. Hence, security One qualifies as a candidate for our specified portfolio. However, another security, or a combination of securities, may give the desired expected return at less risk than security One alone. Computing $\hat{\mu}_j$ for all j we have

	One	Two	Three	Four	Five	Six
Yield	.211	.228	.262	.196	.188	.233

Now to calculate a measure of risk of each security and to achieve diversification in our portfolio, we determine the following covariance matrix S, using (7.10).

TABLE 7.2. Covariance Matrix S

	One	Two	Three	Four	Five	Six
One	.102	.075	.044	.052	.076	.082
Two	.075	.067	.038	.040	.059	.063
Three	.044	.038	.055	.036	.039	.048
Four	.052	.040	.036	.042	.042	.047
Five	.076	.059	.039	.042	.061	.067
Six	.082	.063	.048	.047	.067	.089

To minimize risk arising from variability within a fund and risk from an undiversified portfolio, we form the following objective function

$$\text{minimize} \quad \mathbf{x}'S\mathbf{x}$$

Terms of the type $x_i^2 s_{ii}$ account for variability within a fund and terms of the type $x_i x_j s_{ij}$ account for the covariance between funds.

Adding the constraints: to express our determination to get at least a 20 percent return, to cause 100 percent of our capital to be invested, and positivity restrictions on our activities, we have the following quadratic programming problem

$$\begin{aligned}
&\text{minimize} \quad \mathbf{x}'S\mathbf{x} \\
&\text{subject to} \\
&\quad .211x_1 + .228x_2 + .262x_3 + .196x_4 + .188x_5 + .233x_6 \geq .2 \\
&\quad x_1 + x_2 + x_3 + x_4 + x_5 + x_6 = 1 \\
&\quad x_i \geq 0 \quad i = 1, 2, \ldots, 6
\end{aligned}$$

Upon solving this system we have

$$\mathbf{x}^0 = \begin{bmatrix} 0 \\ .0362 \\ .2339 \\ .7299 \\ 0 \\ 0 \end{bmatrix}$$

The expected yield is 21.25 percent. When the expected yield is higher, say 25 percent, then the expected total variance (risk) would be higher at an expected return of 25 percent than at 21.25 percent.

Another example of quadratic programming arises from regression. In regression, least squares theory leads to estimates $\hat{b}_0, \hat{b}_1, \ldots, \hat{b}_p$ which minimize

$$\sum_i (y_i - b_0 - b_1 x_{1i} - \cdots - b_p x_{p_i})^2$$

i.e., estimate the regression parameters for the model $y = b_0 + b_1 x_1 + \cdots + b_p x_p$. This problem is nothing more than a quadratic programming problem with the regression parameters unrestricted. In this case it is easy to obtain estimates of β by deriving and solving the well-known normal equations. However, this procedure may produce parameter estimators that seem improbable on a priori convictions. Provided these convictions are sufficiently strong (parameter estimates outside certain limits will not be believed anyway) it may be advisable to impose certain limits on these parameters, i.e., $-1.0 \leq b_0 \leq 10$. Then the quadratic programming problem cannot be readily solved by the general least squares theory, and we usually resort to the classical Kuhn-Tucker theory. Here the problem takes the form

$$\begin{aligned} &\text{minimize} \quad (\mathbf{Y} - X\beta)'(\mathbf{Y} - X\beta) \\ &\text{subject to} \quad A'\beta \geq \mathbf{b} \end{aligned} \tag{7.11}$$

where β is the vector of unknowns.

Expression (7.11) can be rewritten in a more familiar form since $\mathbf{Y}'\mathbf{Y}$ is a known scalar; it is equivalently expressed

$$\begin{aligned} &\text{minimize} \quad \beta' D\beta + \mathbf{c}'\beta \\ &\text{subject to} \quad A'\beta \geq \mathbf{b} \end{aligned} \tag{7.12}$$

where $D = (1/2)X'X$ and $\mathbf{c}' = -\mathbf{Y}'X$. If D is positive semidefinite, we can solve this problem iteratively as a quadratic programming problem.

In certain cases we can solve for β in closed form, in particular, when the matrix under consideration is of full row rank and $X'X$ is positive definite and therefore nonsingular. First consider the following lemma.

LEMMA 7.6. *Let $\bar{D}$ be an $n \times n$ positive definite matrix and A' any matrix of size $m \times n$. Then $A'\bar{D}A$ is positive semidefinite.*

Proof. Since $\bar{D}$ is positive definite, there exists a nonsingular matrix Q such that $\bar{D} = Q'Q$. Therefore, for any vector $\mathbf{x}$ we have

$$\begin{aligned} \mathbf{x}'A'\bar{D}A\mathbf{x} &= \mathbf{x}'A'Q'QA\mathbf{x} \\ &= \mathbf{y}'\mathbf{y} \quad \text{letting } \mathbf{y} = QA\mathbf{x} \\ &\geq 0 \end{aligned}$$

Hence $A'\bar{D}A$ is positive semidefinite.

LEMMA 7.7. *Let $\bar{D}$ be an $n \times n$ positive definite matrix and A' a matrix of size $m \times n$ of rank m. Then $A'\bar{D}A$ is positive definite.*

Proof. From Lemma 7.6, $A'\bar{D}A$ is positive semidefinite for arbitrary A'. Hence it suffices to show that $\mathbf{x}'A'\bar{D}A\mathbf{x} > 0$ for all $\mathbf{x} \neq 0$, or equivalently, $\mathbf{y}'\mathbf{y} > 0$ for all $\mathbf{y} \neq 0$, when A' is of full row rank.

However, if there exists $\mathbf{y}$ such that $\mathbf{y}'\mathbf{y} = 0$, then necessarily $\mathbf{y} = QA\mathbf{x} = 0$ or $A\mathbf{x} = 0$. The columns of A are independent implying that all components of $\mathbf{x}$ are zero, and likewise for $\mathbf{y}$. Therefore

$$\mathbf{x}'A'\bar{D}A\mathbf{x} > 0 \quad \text{for all } \mathbf{x} \neq 0$$

Now considering the Lagrangian associated with (7.12), we have

$$\phi(\beta, \lambda) = (1/2)\beta'X'X\beta - \beta'X'\mathbf{Y} + \lambda'(\mathbf{b} - A'\beta)$$

and the solution of the saddle value problem associated with $\phi(\beta, \lambda)$ is restricted to the following four Kuhn-Tucker conditions, i.e., determining a vector (β^0, λ^0) such that

$$\phi_\beta = 0 \tag{7.13}$$

$$\phi_\lambda \leq 0 \qquad \lambda'\phi_\lambda = 0 \qquad \lambda \geq 0 \tag{7.14}$$

Hence (β^0, λ^0) is a saddle point solution if it satisfies (7.13) and (7.14); moreover, β^0 will be an optimal solution of (7.12). This implies (β^0, λ^0) must satisfy the following system of equations

$$X'X\beta - X'\mathbf{Y} - A\lambda = 0 \tag{7.15}$$

$$A'\beta \geq \mathbf{b} \tag{7.16}$$

$$\lambda'(\mathbf{b} - A'\beta) = 0 \tag{7.17}$$

$$\lambda \geq 0 \tag{7.18}$$

Let us for a moment consider the least squares estimator $\hat{\beta}$, obtained by solving the normal equations $X'X\beta = X'\mathbf{Y}$. Here, $\hat{\beta} = (X'X)^{-1}X'\mathbf{Y}$. Therefore the restricted least squares estimator $\tilde{\beta}$, in view of (7.15), can be expressed as

$$\tilde{\beta} = \hat{\beta} + (X'X)^{-1}A\lambda \tag{7.19}$$

If $\hat{\beta}$ satisfies (7.16), then clearly $(\beta^0, \lambda^0) = (\hat{\beta}, 0)$ is a saddle point solution. However, normally $A'\hat{\beta} \not\geq \mathbf{b}$. Let us assume without loss of generality that the first m^* constraints are satisfied and the last $m - m^*$ are violated. Then we have that

$$A_1'\hat{\beta} \geq \mathbf{b}_1 \qquad A_2'\hat{\beta} < \mathbf{b}_2$$

where $A = [A_1, A_2]$ and A_1 is of size $p \times m^*$.

Now considering (7.17) we have $\lambda_i = 0$ (for $i = 1, 2, \ldots, m^*$). Also, the remaining components of λ, say $\bar{\lambda}$, are obtained by solving the following system

$$\mathbf{b}_2 - A_2'\tilde{\beta} = 0$$

or $\mathbf{b}_2 - A_2'\hat{\beta} - A_2'(X'X)^{-1}A_2\bar{\lambda} = 0$. Hence if A_2' is of full row rank, in view of Lemma 7.7

$$\bar{\lambda} = [A_2'(X'X)^{-1}A_2]^{-1}(\mathbf{b}_2 - A_2'\hat{\beta})$$

Therefore $\tilde{\beta} = \hat{\beta} + (X'X)^{-1}A_2[A_2'(X'X)^{-1}A_2]^{-1}(\mathbf{b}_2 - A_2'\hat{\beta})$. From the discussion, if the corresponding matrix A_2' is of full row rank we have the following

$$\tilde{\beta} = \hat{\beta} \quad \text{if } A'\hat{\beta} \geq \mathbf{b} \tag{7.20a}$$

$$\tilde{\beta} = \hat{\beta} + (X'X)^{-1}A_2[A_2'(X'X)^{-1}A_2]^{-1}(\mathbf{b}_2 - A_2'\hat{\beta}) \quad \text{if } A_2'\hat{\beta} < \mathbf{b}_2 \tag{7.20b}$$

$$\tilde{\beta} = \hat{\beta} + (X'X)^{-1}A_1[A_1'(X'X)^{-1}A_1]^{-1}(\mathbf{b}_1 - A_1'\hat{\beta}) \quad \text{if } A_1'\hat{\beta} < \mathbf{b}_1 \tag{7.20c}$$

$$\tilde{\beta} = \hat{\beta} + (X'X)^{-1}A[A'(X'X)^{-1}A]^{-1}(\mathbf{b} - A'\hat{\beta}) \quad \text{if } A'\hat{\beta} < \mathbf{b} \tag{7.20d}$$

At this point it may appear that the above closed form expressions will determine the restricted least squares estimator by simply modifying the unrestricted least squares estimator $\hat{\beta}$. However, we must establish some conditions on the restrictions so we indeed have a feasible solution to the original problem. For instance, it could happen that $A'\tilde{\beta} \ngeq \mathbf{b}$. Let us consider the original constraints evaluated at $\tilde{\beta}$ for each closed form given in (7.20).

If (7.20a) holds, then clearly the unrestricted least squares estimator $\hat{\beta}$ solves (7.12).

If (7.20d) holds, we have

$$A'\tilde{\beta} = A'\hat{\beta} + [A'(X'X)^{-1}A][A'(X'X)^{-1}A]^{-1}(\mathbf{b} - A'\hat{\beta}) = \mathbf{b}$$

Hence the closed form given by (7.20d) yields an optimal solution of (7.12) provided $\lambda = [A'(X'X)^{-1}A]^{-1}(\mathbf{b} - A'\hat{\beta}) \geq 0$.

Now consider the other two closed forms; without loss of generality consider (7.20b). Here we have

$$A_2'\tilde{\beta} = A_2'\hat{\beta} + [A_2'(X'X)^{-1}A_2][A_2'(X'X)^{-1}A_2]^{-1}(\mathbf{b}_2 - A_2'\hat{\beta}) = \mathbf{b}_2$$
$$A_1'\tilde{\beta} = A_1'\hat{\beta} + [A_1'(X'X)^{-1}A_2][A_2'(X'X)^{-1}A_2]^{-1}(\mathbf{b}_2 - A_2'\hat{\beta}) \geq \mathbf{b}_1$$

if the covariance $(A_1'\hat{\beta}, A_2'\hat{\beta}) \geq 0$, which is equivalent to a nonnegative submatrix $A_1'(X'X)^{-1}A_2$ in the matrix $A'(X'X)^{-1}A$.

Therefore a sufficient condition to ensure that the closed form expressions (7.20b) and (7.20c) are feasible, and hence optimal in view of the Kuhn-Tucker conditions, is that $A_1'(X'X)^{-1}A_2 = \text{covariance } (A_1'\hat{\beta}, A_2'\hat{\beta})$ contains all nonnegative components and $\bar{\lambda} = [A_2'(X'X)^{-1}A_2]^{-1}(\mathbf{b}_2 - A_2'\hat{\beta})$ is nonnegative in (7.20b); likewise in (7.20c).

EXAMPLE 7.6

(i) Consider the linear model $y = \beta_0 + \beta_1 x$ where the slope $\beta_1 \geq 1$ and (x, y) are (1, 1), (2, 3), and (3, 2). Then

$$X = \begin{bmatrix} 1 & 1 \\ 1 & 2 \\ 1 & 3 \end{bmatrix} \qquad X'X = \begin{bmatrix} 3 & 6 \\ 6 & 14 \end{bmatrix} \qquad (X'X)^{-1} = \begin{bmatrix} 7/3 & -1 \\ -1 & 1/2 \end{bmatrix}$$

$$\hat{\beta} = (X'X)^{-1}X'\mathbf{Y} = \begin{bmatrix} 1 \\ 1/2 \end{bmatrix}$$

Now $A'\beta \geq b$ is defined as follows: $A' = [0, 1]$ and $b = 1$; therefore $A'\hat{\beta} = \hat{\beta}_1 \ngeq 1$ and the restricted least squares estimator is

$$\begin{aligned} \tilde{\beta} &= \hat{\beta} + (X'X)^{-1}A[A'(X'X)^{-1}A]^{-1}(1 - \hat{\beta}_1) \\ &= \begin{bmatrix} 1 \\ 1/2 \end{bmatrix} + \begin{bmatrix} 7/3 & -1 \\ -1 & 1/2 \end{bmatrix} \begin{bmatrix} 0 \\ 1 \end{bmatrix} (1/2)^{-1}(1/2) \\ &= \begin{bmatrix} 1 \\ 1/2 \end{bmatrix} + \begin{bmatrix} -1 \\ 1/2 \end{bmatrix} = \begin{bmatrix} 0 \\ 1 \end{bmatrix} \end{aligned}$$

(ii) Consider the same model and observations, but restrict the parameters so that $\beta_0 \geq 1/2$ and $0 \leq \beta_1 \leq 1/4$. This system can be expressed equivalently as

$$\beta_0 \geq 1/2 \qquad \beta_1 \geq 0 \qquad -\beta_1 \geq -1/4$$

Here

$$A' = \begin{bmatrix} 1 & 0 \\ 0 & 1 \\ 0 & -1 \end{bmatrix}$$

and clearly A' is of rank 2. However, $A'\hat{\beta}$ satisfies the first two constraints of the feasible region, but not the third. Therefore

$$\begin{aligned} \tilde{\beta} &= \hat{\beta} + (X'X)^{-1} \begin{bmatrix} 0 \\ -1 \end{bmatrix} [A_3'(X'X)^{-1}A_3]^{-1}(-1/4 + \hat{\beta}_1) \\ &= \begin{bmatrix} 1 \\ 1/2 \end{bmatrix} + \begin{bmatrix} 1 \\ -1/2 \end{bmatrix} (1/2)^{-1}(-1/4 + 1/2) \\ &= \begin{bmatrix} 1 \\ 1/2 \end{bmatrix} + \begin{bmatrix} 1/2 \\ -1/4 \end{bmatrix} = \begin{bmatrix} 3/2 \\ 1/4 \end{bmatrix} \end{aligned}$$

Note in this example that not all of the components of $A_1'(X'X)^{-1}A_2$ are nonnegative where $A_1' = \begin{bmatrix} 1 & 0 \\ 0 & 1 \end{bmatrix}$ and $A_2 = \begin{bmatrix} 0 \\ -1 \end{bmatrix}$. In particular, $A_1'(X'X)^{-1}A_2 = \begin{bmatrix} 1 \\ -1/2 \end{bmatrix}$. However, since the second constraint is redundant after it is determined that $\hat{\beta}_1 \geq 0$, it suffices to consider only

$$\bar{A}_1' = [1, 0] \qquad A_2 = \begin{bmatrix} 0 \\ -1 \end{bmatrix}$$

Then, clearly we have Cov $(\bar{A}_1'\hat{\beta}, A_2'\hat{\beta}) = 1$. Therefore, as the third constraint evaluated at $\tilde{\beta}$ is made feasible, then necessarily the other constraints evaluated at $\tilde{\beta}$ will remain feasible.

(iii) The following example illustrates that the closed forms (7.20b) and (7.20c) can yield an incorrect solution if some condition is not imposed on $A_1'(X'X)^{-1}A_2$. Let the inequality restrictions for the above problem be $\beta_0 \leq 5/4$ and $\beta_1 \leq 1/4$ or, equivalently, $-\beta_0 \geq -5/4$ and $-\beta_1 \geq -1/4$. Then clearly $-\hat{\beta}_1 \ngeq -1/4$ whereas $-\hat{\beta}_0 \geq -5/4$. Now appealing to (7.20b) we have

$$\tilde{\beta} = \hat{\beta} + (X'X)^{-1}A_2[A_2'(X'X)^{-1}A_2]^{-1}(\mathbf{b}_2 - A_2'\hat{\beta})$$

$$= \begin{bmatrix} 1 \\ 1/2 \end{bmatrix} + \begin{bmatrix} 1/2 \\ -1/4 \end{bmatrix} = \begin{bmatrix} 3/2 \\ 1/4 \end{bmatrix}$$

However, $\tilde{\beta}$ is now an infeasible solution. Note that $A_1'(X'X)^{-1}A_2 = -1$ and $A_1'\tilde{\beta}$ is made infeasible even though it was previously a feasible constraint.

7.3 Dual Quadratic Programs

The derivation of dual quadratic programs is presented appealing to the results in Section 7.1. Consider

Problem I

$$\text{maximize } \mathbf{x}'D\mathbf{x} + \mathbf{c}'\mathbf{x}$$
$$\text{subject to } \quad \mathbf{x} \in \Omega_I = \{\mathbf{x} \mid \mathbf{x} \geq 0, \mathbf{b} - A\mathbf{x} \geq 0\}$$

Problem II

$$\text{minimize } \quad -\mathbf{x}'D\mathbf{x} + \boldsymbol{\lambda}'\mathbf{b}$$
$$\text{subject to } \quad (\mathbf{x}, \boldsymbol{\lambda}) \in \Omega_{II} = \{(\mathbf{x}, \boldsymbol{\lambda}) \mid \boldsymbol{\lambda} \geq 0, 2D\mathbf{x} - A'\boldsymbol{\lambda} + \mathbf{c} \leq 0\}$$

LEMMA 7.8 (*Weak Duality*). *If there exists* $\bar{\mathbf{x}} \in \Omega_I$ *and* $(\hat{\mathbf{x}}, \hat{\boldsymbol{\lambda}}) \in \Omega_{II}$, *then*

$$\bar{\mathbf{x}}'D\bar{\mathbf{x}} + \mathbf{c}'\bar{\mathbf{x}} \leq -\hat{\mathbf{x}}'D\hat{\mathbf{x}} + \hat{\boldsymbol{\lambda}}'\mathbf{b}$$

Proof. For $(\hat{\mathbf{x}}, \hat{\boldsymbol{\lambda}}) \in \Omega_{II}$, we have $2D\hat{\mathbf{x}} - A'\hat{\boldsymbol{\lambda}} + \mathbf{c} \leq 0$. Since $\bar{\mathbf{x}} \geq 0$

$$2\bar{\mathbf{x}}'D\hat{\mathbf{x}} - \bar{\mathbf{x}}'A'\hat{\boldsymbol{\lambda}} + \bar{\mathbf{x}}'\mathbf{c} \leq 0 \tag{7.21}$$

Adding $\hat{\boldsymbol{\lambda}}'\mathbf{b}$ to both sides of (7.21) we have

$$2\bar{\mathbf{x}}'D\hat{\mathbf{x}} + \mathbf{c}'\bar{\mathbf{x}} + \hat{\boldsymbol{\lambda}}'(\mathbf{b} - A\bar{\mathbf{x}}) \leq \hat{\boldsymbol{\lambda}}'\mathbf{b} \tag{7.22}$$

However, D is negative semidefinite; therefore

$$(\bar{\mathbf{x}} - \hat{\mathbf{x}})'D(\bar{\mathbf{x}} - \hat{\mathbf{x}}) \leq 0 \qquad \text{or} \qquad \bar{\mathbf{x}}'D\bar{\mathbf{x}} + \hat{\mathbf{x}}'D\hat{\mathbf{x}} \leq 2\bar{\mathbf{x}}'D\hat{\mathbf{x}}$$

and from (7.22)

$$\bar{\mathbf{x}}'D\bar{\mathbf{x}} + \mathbf{c}'\bar{\mathbf{x}} + \hat{\boldsymbol{\lambda}}'(\mathbf{b} - A\bar{\mathbf{x}}) \leq -\hat{\mathbf{x}}'D\hat{\mathbf{x}} + \hat{\boldsymbol{\lambda}}'\mathbf{b}$$

and since $\hat{\boldsymbol{\lambda}}'(\mathbf{b} - A\bar{\mathbf{x}}) \geq 0$, the result follows.

COROLLARY 7.8.1. *If* $\bar{\mathbf{x}} \in \Omega_I$, $(\hat{\mathbf{x}}, \hat{\boldsymbol{\lambda}}) \in \Omega_{II}$, *and the optima are equal, then* $\bar{\mathbf{x}}$ *and* $(\hat{\mathbf{x}}, \hat{\boldsymbol{\lambda}})$ *solve Problems I and II, respectively.*

Let $\psi(\mathbf{x}, \boldsymbol{\lambda}) = \mathbf{x}'D\mathbf{x} + \mathbf{c}'\mathbf{x} + \boldsymbol{\lambda}'(\mathbf{b} - A\mathbf{x})$; i.e., the Lagrangian function associated with Problem I. Then, appealing to Theorem 6.2, we have the following lemma associated with Problem I.

LEMMA 7.9. *If* $\mathbf{x}^0$ *solves Problem I, then there exists* $\boldsymbol{\lambda}^0 \geq 0$ *such that* $(\mathbf{x}^0, \boldsymbol{\lambda}^0)$ *is a saddle point of* $\psi(\mathbf{x}, \boldsymbol{\lambda})$, $(\mathbf{x}, \boldsymbol{\lambda}) \in Q_n^+ \times Q_m^+$.

THEOREM 7.4. *If* $\mathbf{x}^0$ *solves Problem I, then there exists* $\boldsymbol{\lambda}^0$ *such that*

(*a*) $(\mathbf{x}^0, \boldsymbol{\lambda}^0)$ *solves Problem II*
(*b*) $\mathbf{x}^{0\prime}D\mathbf{x}^0 + \mathbf{c}'\mathbf{x}^0 = -\mathbf{x}^{0\prime}D\mathbf{x}^0 + \boldsymbol{\lambda}^{0\prime}\mathbf{b}$

Proof. Appealing to Lemma 7.9, there exists $\boldsymbol{\lambda}^0 \geq 0$ such that $(\mathbf{x}^0, \boldsymbol{\lambda}^0)$ is a saddle point of $\psi(\mathbf{x}, \boldsymbol{\lambda})$, i.e., by Theorem 7.1, such that $\psi_{\mathbf{x}}(\mathbf{x}^0, \boldsymbol{\lambda}^0) \leq 0$, or $2D\mathbf{x}^0 - A'\boldsymbol{\lambda}^0 + \mathbf{c} \leq 0$. Also, since $\boldsymbol{\lambda}^0 \geq 0$ it follows that $(\mathbf{x}^0, \boldsymbol{\lambda}^0) \in \Omega_{II}$. Now $(\mathbf{x}^0, \boldsymbol{\lambda}^0)$, a saddle point of $\psi(\mathbf{x}, \boldsymbol{\lambda})$, implies, appealing to (*a*) and (*d*) in Theorem 7.1, that

$$2\mathbf{x}^{0\prime}D\mathbf{x}^0 - \mathbf{x}^{0\prime}A'\boldsymbol{\lambda}^0 + \mathbf{x}^{0\prime}\mathbf{c} = 0 \tag{7.23}$$

$$\boldsymbol{\lambda}^{0\prime}\mathbf{b} - \boldsymbol{\lambda}^{0\prime}A\mathbf{x}^0 = 0 \tag{7.24}$$

Therefore

$$\psi(\mathbf{x}^0, \boldsymbol{\lambda}^0) = \mathbf{c}'\mathbf{x}^0 + \mathbf{x}^{0\prime}D\mathbf{x}^0 \tag{7.25}$$

In particular, (7.23), (7.24), and (7.25) imply that

$$\psi(\mathbf{x}^0, \boldsymbol{\lambda}^0) = -\mathbf{x}^{0\prime}D\mathbf{x}^0 + \boldsymbol{\lambda}^{0\prime}\mathbf{b}$$

thus (*b*) holds and, moreover, $(\mathbf{x}^0, \boldsymbol{\lambda}^0)$ solves Problem II in view of Corollary 7.8.1.

THEOREM 7.5. *If* $(\mathbf{x}^0, \boldsymbol{\lambda}^0)$ *solves Problem II, then there exists* $\mathbf{v}^0 \geq 0$ *such that*

(*i*) $D\mathbf{v}^0 = D\mathbf{x}^0$
(*ii*) $\mathbf{v}^{0\prime}D\mathbf{v}^0 + \mathbf{c}'\mathbf{v}^0 = -\mathbf{x}^{0\prime}D\mathbf{x}^0 + \boldsymbol{\lambda}^{0\prime}\mathbf{b}$
(*iii*) $\mathbf{v}^0$ *solves Problem I*

Proof. Consider the associated Lagrangian function for Problem II, i.e.

$$\phi(\mathbf{x}, \boldsymbol{\lambda}, \mathbf{v}) = -\mathbf{x}'D\mathbf{x} + \boldsymbol{\lambda}'\mathbf{b} + \mathbf{v}'(2D\mathbf{x} - A'\boldsymbol{\lambda} + \mathbf{c})$$
$$(\mathbf{x}, \boldsymbol{\lambda}, \mathbf{v}) \in E^n \times Q_m^+ \times Q_n^+$$

From (*d*), (*e*), (*a*), and (*f*) in Theorem 7.1 with $\mathbf{x} \in E^n$ and $(\mathbf{x}, \boldsymbol{\lambda}) = \mathbf{y}$

$$(\mathbf{b} - A\mathbf{v}^0)'\boldsymbol{\lambda}^0 = 0 \tag{7.26a}$$

$$-D\mathbf{x}^0 + D\mathbf{v}^0 = 0 \tag{7.26b}$$

$$\mathbf{b} - A\mathbf{v}^0 \geq 0 \tag{7.27}$$

$$(2D\mathbf{x}^0 - A'\boldsymbol{\lambda}^0 + \mathbf{c})'\mathbf{v}^0 = 0 \tag{7.28}$$

$$\mathbf{v}^0 \geq 0 \tag{7.29}$$

Equations (7.27) and (7.29) imply that $\mathbf{v}^0 \in \Omega_I$. To prove (*iii*) it suffices to show (*ii*); i.e., the optima of Problems I and II are equal. Hence

$$\begin{aligned}(\mathbf{b}'\boldsymbol{\lambda}^0 - \mathbf{x}^{0\prime}D\mathbf{x}^0) - (\mathbf{v}^{0\prime}D\mathbf{v}^0 + \mathbf{c}'\mathbf{v}^0)& \\ &= \mathbf{b}'\boldsymbol{\lambda}^0 - 2\mathbf{x}^{0\prime}D\mathbf{v}^0 - \mathbf{c}'\mathbf{v}^0 \quad \text{from (7.26b) and with } D = D' \\ &= \boldsymbol{\lambda}^{0\prime}A\mathbf{v}^0 - 2\mathbf{x}^{0\prime}D\mathbf{v}^0 - \mathbf{c}'\mathbf{v}^0 \quad \text{from (7.26a)} \\ &= 0 \quad \text{from (7.28)}\end{aligned}$$

Therefore $\mathbf{v}^0$ solves Problem I, and from (7.26b), $\mathbf{v}^0$ is such that $D\mathbf{v}^0 = D\mathbf{x}^0$.

Note that quadratic duality holds without qualification since the constraints are linear, as discussed in Chapter 4. In particular, if the objective function is a concave function defined for all $\mathbf{x} \geq 0$ with a finite maximum over Q_n^+, then Condition A in Chapter 4 is satisfied.

If D is null in Problems I and II, then it is seen that linear duality is a special case of the above results.

The following section presents algorithms for solving quadratic programming problems.

EXAMPLE 7.7

Consider the problem

$$\begin{aligned}\text{maximize} \quad & x_1 + x_1^2 + 2x_1x_2 + x_2^2 \\ \text{subject to} \quad & x_1 + x_2 \leq 1 \\ & x_1, \quad x_2 \geq 0\end{aligned}$$

Then, letting $\mathbf{c}' = (1, 0)$, $D = \begin{bmatrix} 1 & 1 \\ 1 & 1 \end{bmatrix}$, $A = [1, 1]$, and $b = 1$, this problem can be expressed equivalently as

$$\begin{aligned}\text{maximize} \quad & \mathbf{c}'\mathbf{x} + \mathbf{x}'D\mathbf{x} \\ \text{subject to} \quad & A\mathbf{x} \leq b \\ & \mathbf{x} \geq 0\end{aligned}$$

Its corresponding dual problem is

$$\begin{aligned}\text{minimize} \quad & -\mathbf{x}'D\mathbf{x} + \lambda b \\ \text{subject to} \quad & 2D\mathbf{x} - A'\lambda + \mathbf{c} \leq 0 \\ & \lambda \geq 0\end{aligned}$$

or

$$\begin{aligned}\text{minimize} \quad & -x_1^2 - 2x_1x_2 - x_2^2 + \lambda \\ \text{subject to} \quad & 2x_1 + 2x_2 - \lambda + 1 \leq 0 \\ & 2x_1 + 2x_2 - \lambda \leq 0 \\ & \lambda \geq 0\end{aligned}$$

Applying the Kuhn-Tucker conditions we have $\mathbf{x}^0 = \begin{bmatrix} 1 \\ 0 \end{bmatrix}$ solving the original problem and $(\mathbf{x}^0, \boldsymbol{\lambda}^0)' = (1, 0, 3)$, solving the dual problem, as can be verified by the reader.

7.4 Quadratic Programming Algorithms

Let us again consider Problem I. We have the following quadratic programming problem

$$\begin{aligned} \text{maximize} \quad & \mathbf{x}'D\mathbf{x} + \mathbf{c}'\mathbf{x} \\ \text{subject to} \quad & A\mathbf{x} \leq \mathbf{b} \\ & \mathbf{x} \geq 0 \end{aligned} \tag{7.30}$$

As before, A is an $m \times n$ matrix, $\mathbf{b}$ is an m-component vector, $\mathbf{c}$ is an n-component vector, and D is a symmetric negative semidefinite $n \times n$ matrix, all with known elements; and $\mathbf{x}$ is the vector of unknowns. Any quadratic form can be expressed as a symmetric matrix. This is shown in the following lemma.

LEMMA 7.10. *Any quadratic form can be expressed as a symmetric matrix.*

Proof. Consider the form $\mathbf{x}'B\mathbf{x}$ where B is any $n \times n$ matrix.
Let $D = (B + B')/2$, then

$$\begin{aligned} \mathbf{x}'D\mathbf{x} &= \mathbf{x}'[(B + B')/2]\mathbf{x} \\ &= (1/2)\,\mathbf{x}'B\mathbf{x} + (1/2)\,\mathbf{x}'B'\mathbf{x} \\ &= \mathbf{x}'B\mathbf{x} \end{aligned}$$

since $\mathbf{x}'B'\mathbf{x}$ is a scalar and $\mathbf{x}'B'\mathbf{x} = (\mathbf{x}'B'\mathbf{x})' = \mathbf{x}'B\mathbf{x}$.

7.4.1 Wolfe's Algorithm (1959)

To characterize the optimal solution of (7.30), let us consider the Lagrangian function of (7.30) and the Kuhn-Tucker conditions. If we let

$$\frac{\partial \psi}{\partial x} = -\mathbf{v} \qquad \frac{\partial \psi}{\partial y} = \mathbf{x}_s$$

then

$$\frac{\partial \psi}{\partial x} = -\mathbf{v} = \mathbf{c} + 2D\mathbf{x} - A'\mathbf{y} \tag{7.31}$$

$$\frac{\partial \psi}{\partial y} = \mathbf{x}_s = \mathbf{b} - A\mathbf{x} \tag{7.32}$$

With these, the Kuhn-Tucker conditions are expressed as:

(a) $2D\mathbf{x} - A'\mathbf{y} + \mathbf{v} = -\mathbf{c}$
(b) $A\mathbf{x} + \mathbf{x}_s = \mathbf{b}$
(c) $\mathbf{x} \geq 0, \mathbf{x}_s \geq 0, \mathbf{y} \geq 0, \mathbf{v} \geq 0$
(d) $\mathbf{x}'\mathbf{v} = 0 = \mathbf{y}'\mathbf{x}_s$

Conditions (a)–(c) form a linear system. This suggests that we could use the simplex method to find an appropriate solution of (7.30) by making certain that at each iteration a vector is allowed to enter the basis only if condition (d) is satisfied. This is indeed the case even though condition (d) is nonlinear.

A basic solution of the linear system will have no more than $m + n$ positive variables and the remaining variables will be zero. Here we are defining a feasible solution of (a)–(c) to be a basic feasible solution if (d) is also satisfied. Furthermore, if a vector $(\mathbf{x}^0, \mathbf{x}_s^0, \mathbf{y}^0, \mathbf{v}^0)$ is a feasible solution of the following linear constraints

$$\begin{bmatrix} 2D & 0_{n\times m} & -A' & I_n \\ A & I_m & 0_{m\times m} & 0_{m\times n} \end{bmatrix} \begin{bmatrix} \mathbf{x} \\ \mathbf{x}_s \\ \mathbf{y} \\ \mathbf{v} \end{bmatrix} = \begin{bmatrix} -\mathbf{c} \\ \mathbf{b} \end{bmatrix} \tag{7.33}$$

$(\mathbf{x}, \mathbf{x}_s, \mathbf{y}, \mathbf{v} \geq 0)$ and if it satisfies condition (d), then from the Kuhn-Tucker theory, $\mathbf{x}^0$ solves the quadratic programming problem (7.30).

The system of linear equations expressed in (7.33) is of the familiar form

$$\text{maximize} \quad 0'\mathbf{z} \tag{7.34}$$
$$\text{subject to} \quad \bar{A}\mathbf{z} = \bar{\mathbf{b}} \tag{7.35}$$
$$\mathbf{z} \geq 0 \tag{7.36}$$

Using a modification of the simplex algorithm as described in Chapter 2, we can solve (7.30) by incorporating the following two additional rules to ensure that $\mathbf{x}'\mathbf{v} = 0 = \mathbf{y}'\mathbf{x}_s$:

(1) If a variable x_j (or v_j) is currently in the basis at a positive level, do not consider v_j (or x_j) as a candidate for entry into the basis. If x_j (or v_j) is currently in the basis at a zero level, then v_j (or x_j) may enter the basis only if x_j (or v_j) remains at a zero level.
(2) If a variable y_i (or x_{s_i}) is currently in the basis at a positive level,

do not consider x_{s_i} (or y_i) as a candidate for entry into the basis. If y_i (or x_{s_i}) is in the basis at a zero level, x_{s_i} (or y_i) may enter the basis only if y_i (or x_{s_i}) remains at a zero level.

The foregoing algorithm, developed by Wolfe, always yields an optimal solution if (7.30) has a feasible solution and D is negative definite. If D is negative semidefinite, the solution could be unbounded. This case will not cause difficulties in practice if one considers the matrix $D - \epsilon I$ for any $\epsilon > 0$.

LEMMA 7.11. *Let D be any $n \times n$ negative semidefinite matrix. Then for any $\epsilon > 0$ the matrix $D - \epsilon I$ is negative definite.*

Proof. The expression $\mathbf{x}'(D - \epsilon I)\mathbf{x} = \mathbf{x}'D\mathbf{x} - \epsilon\mathbf{x}'I\mathbf{x} < 0$ since $-\epsilon\mathbf{x}'I\mathbf{x}$ is negative definite. Thus $D - \epsilon I$ is negative definite.

Although a modified version of the simplex algorithm can be used to solve quadratic programming problems, note that this does not imply the optimal solution will lie on an extreme point as in linear programming; it may lie on an edge or in the interior of the feasible region, see Exercise 6 at the end of this chapter.

EXAMPLE 7.8

Consider the problem

$$\begin{aligned} \text{maximize} \quad & x_1 + x_2 - (1/2)x_1^2 + x_1x_2 - x_2^2 \\ \text{subject to} \quad & x_1 + x_2 \le 3 \\ & 2x_1 + 3x_2 \ge 6 \\ & x_1, \quad x_2 \ge 0 \end{aligned}$$

Here

$$D = \begin{bmatrix} -1/2 & 1/2 \\ 1/2 & -1 \end{bmatrix} \quad \mathbf{c} = \begin{bmatrix} 1 \\ 1 \end{bmatrix} \quad \mathbf{b} = \begin{bmatrix} 3 \\ -6 \end{bmatrix} \quad A = \begin{bmatrix} 1 & 1 \\ -2 & -3 \end{bmatrix}$$

D is negative definite since all principal minor determinants of $-D$ are positive; i.e., $|1/2| = 1/2$ and

$$\begin{vmatrix} 1/2 & -1/2 \\ -1/2 & 1 \end{vmatrix} = 1/2 - 1/4 = 1/4$$

Equation (7.33) implies that the initial tableau is

TABLEAU 1

$\mathbf{c}_B$	Basis	$\mathbf{P}_0$	$\mathbf{P}_1$	$\mathbf{P}_2$	$\mathbf{P}_3$	$\mathbf{P}_4$	$\mathbf{P}_5$	$\mathbf{P}_6$	$\mathbf{P}_7$	$\mathbf{P}_8$	$\mathbf{A}_1$	$\mathbf{A}_2$	$\mathbf{A}_3$	θ
−1	$\mathbf{A}_1$	1	1	−1	0	0	1	−2	−1	0	1	0	0	−1
−1	$\mathbf{A}_2$	1	−1	[2]	0	0	1	−3	0	−1	0	1	0	1/2
0	$\mathbf{P}_3$	3	1	1	1	0	0	0	0	0	0	0	0	3
−1	$\mathbf{A}_3$	6	2	3	0	−1	0	0	0	0	0	0	1	2
	$\mathbf{z} - \mathbf{c}$	−8	−2	−4	0	1	−2	5	1	1	0	0	0	

Before introducing appropriate artificial variables to establish an initial solution, we must multiply each row by −1 if its right-hand component is negative since the simplex algorithm requires that the constraint vector be greater than or equal to zero. Also, in Tableau 1 only three artificial variables are introduced since the vector $\mathbf{P}_3$ is in the initial basic solution; the three artificial variables can be assigned any arbitrary negative cost, for convenience say −1.

We now introduce any vector with a corresponding negative $z_j - c_j$ to enter the solution. Choosing $\mathbf{P}_2$ to enter, we see that $\mathbf{A}_2$ leaves the basis since $\theta_2 = 1/2 = \min_i \{\theta_i \mid \theta_i \geq 0\}$. This results in Tableau 2. Now one can delete $\mathbf{A}_2$ from the tableau since it will never be considered as a candidate for reentry into the basis.

TABLEAU 2

$\mathbf{c}_B$	Basis	$\mathbf{P}_0$	$\mathbf{P}_1$	$\mathbf{P}_2$	$\mathbf{P}_3$	$\mathbf{P}_4$	$\mathbf{P}_5$	$\mathbf{P}_6$	$\mathbf{P}_7$	$\mathbf{P}_8$	$\mathbf{A}_1$	$\mathbf{A}_2$	$\mathbf{A}_3$	θ
−1	$\mathbf{A}_1$	3/2	1/2	0	0	0	3/2	−7/2	−1	−1/2	1	1/2	0	3
0	$\mathbf{P}_2$	1/2	−1/2	1	0	0	1/2	−3/2	0	−1/2	0	1/2	0	−1
0	$\mathbf{P}_3$	5/2	3/2	0	1	0	−1/2	3/2	0	1/2	0	−1/2	0	5/3
−1	$\mathbf{A}_3$	9/2	[7/2]	0	0	−1	−3/2	9/2	0	3/2	0	−3/2	1	9/7
	$\mathbf{z} - \mathbf{c}$	−6	−4	0	0	1	0	1	1	−1	0	2	0	

The vector $\mathbf{P}_1$ can enter the basis without violating rules (1) or (2); hence when $\mathbf{P}_1$ enters the basis then $\mathbf{A}_3$ will leave the basis. Thus we have the following tableau.

TABLEAU 3

$\mathbf{c}_B$	Basis	$\mathbf{P}_0$	$\mathbf{P}_1$	$\mathbf{P}_2$	$\mathbf{P}_3$	$\mathbf{P}_4$	$\mathbf{P}_5$	$\mathbf{P}_6$	$\mathbf{P}_7$	$\mathbf{P}_8$	$\mathbf{A}_1$	$\mathbf{A}_2$	$\mathbf{A}_3$	θ
−1	$\mathbf{A}_1$	6/7	0	0	0	1/7	12/7	−29/7	−1	−5/7	1	5/7	−1/7	6
0	$\mathbf{P}_2$	8/7	0	1	0	−1/7	2/7	−6/7	0	−2/7	0	2/7	1/7	
0	$\mathbf{P}_3$	4/7	0	0	1	[3/7]	1/7	−3/7	0	−1/7	0	1/7	−3/7	4/3
0	$\mathbf{P}_1$	9/7	1	0	0	−2/7	- 3/7	9/7	0	3/7	0	−3/7	2/7	
	$\mathbf{z} - \mathbf{c}$	−6/7	0	0	0	−1/7	−12/7	29/7	1	5/7	0	2/7	8/7	

The vectors $\mathbf{P}_4$ and $\mathbf{P}_5$ are candidates for entry into the basis; however if $\mathbf{P}_5$ enters, rule (2) will be violated. Therefore, let $\mathbf{P}_4$ enter the basis and $\mathbf{P}_3$ leaves the basis. The corresponding tableau follows.

TABLEAU 4

$\mathbf{c}_B$	Basis	$\mathbf{P}_0$	$\mathbf{P}_1$	$\mathbf{P}_2$	$\mathbf{P}_3$	$\mathbf{P}_4$	$\mathbf{P}_5$	$\mathbf{P}_6$	$\mathbf{P}_7$	$\mathbf{P}_8$	$\mathbf{A}_1$	$\mathbf{A}_2$	$\mathbf{A}_3$	θ
−1	$\mathbf{A}_1$	2/3	0	0	−1/3	0	[5/3]	−4	−1	−2/3	1	2/3	0	2/5
0	$\mathbf{P}_2$	4/3	0	1	1/3	0	1/3	−1	0	−1/3	0	1/3	0	4
0	$\mathbf{P}_4$	4/3	0	0	7/3	1	1/3	−1	0	−1/3	0	1/3	−1	4
0	$\mathbf{P}_1$	5/3	1	0	2/3	0	−1/3	1	0	1/3	0	−1/3	0	
	$\mathbf{z} - \mathbf{c}$	−2/3	0	0	1/3	0	−5/3	4	1	2/3	0	1/3	1	

Now $\mathbf{P}_5$ is eligible to enter the basis without violating rule (2), and $\mathbf{A}_1$ will leave the basis. The next tableau yields an optimal solution

$$\begin{bmatrix} x_1^0 \\ x_2^0 \end{bmatrix} = \begin{bmatrix} 9/5 \\ 6/5 \end{bmatrix}$$

since all the components $z_j - c_j$ are nonnegative.

TABLEAU 5

$\mathbf{c}_B$	Basis	$\mathbf{P}_0$	$\mathbf{P}_1$	$\mathbf{P}_2$	$\mathbf{P}_3$	$\mathbf{P}_4$	$\mathbf{P}_5$	$\mathbf{P}_6$	$\mathbf{P}_7$	$\mathbf{P}_8$
0	$\mathbf{P}_5$	2/5	0	0	−1/5	0	1	−12/5	−3/5	−2/5
0	$\mathbf{P}_2$	6/5	0	1	2/5	0	0	−1/5	1/5	−1/5
0	$\mathbf{P}_4$	6/5	0	0	12/5	1	0	−1/5	1/5	−1/5
0	$\mathbf{P}_1$	9/5	1	0	3/5	0	0	1/5	−1/5	1/5
	$\mathbf{z} - \mathbf{c}$	0	0	0	0	0	0	0	0	0

7.4.2 *Hildreth's Algorithm (1957)*

The algorithm presented in this section for solving quadratic programming problems is an asymptotic method making direct use of the duality concepts introduced previously. The steps at each iteration are extremely simple and make this algorithm highly suitable for computer computation.

For simplicity assume the quadratic model is structured as a minimization problem and, if nonnegativity is assumed on any activities, these restrictions are incorporated into the linear system $A\mathbf{x} \leq \mathbf{b}$. Thus the model is essentially the same as (7.33) where we change the criterion of the objective function from a maximization problem to a minimization problem. (As noted previously, any problem can be changed from maximization to minimization without affecting the optimal solution.) Thus we have the following model

$$\begin{aligned} &\text{minimize} \quad -(\mathbf{c}'\mathbf{x} + \mathbf{x}'D\mathbf{x}) \\ &\text{subject to} \quad A\mathbf{x} \le \mathbf{b} \end{aligned} \tag{7.37}$$

Now assume $\mathbf{x}^0$ is the optimal solution of (7.37); then from the Kuhn-Tucker theory we know that there exists a vector $\mathbf{y}^0$ such that $(\mathbf{x}^0, \mathbf{y}^0)$ is a saddle point solution of $\phi(\mathbf{x}, \mathbf{y}) = -\mathbf{c}'\mathbf{x} - \mathbf{x}'D\mathbf{x} + \mathbf{y}'(A\mathbf{x} - \mathbf{b})$. Note that we are relating the quadratic model to the saddle value problem as outlined in Chapter 6. Since $\phi(\mathbf{x}, \mathbf{y})$ is a convex function in $\mathbf{x}$ for a fixed $\mathbf{y}$, then $(\mathbf{x}^0, \mathbf{y}^0)$ must also satisfy the Kuhn-Tucker conditions. Moreover, since $\mathbf{x}$ is unrestricted, the six Kuhn-Tucker conditions can be reduced to four, i.e.

$$\phi_{\mathbf{x}} = 0 \tag{7.38}$$

$$\phi_{\mathbf{y}} \le 0 \qquad \phi_{\mathbf{y}}'\mathbf{y}^0 = 0 \qquad \mathbf{y}^0 \ge 0 \tag{7.39}$$

Therefore $(\mathbf{x}^0, \mathbf{y}^0)$ must satisfy the following conditions

$$\mathbf{c} + 2D\mathbf{x}^0 - A'\mathbf{y}^0 = 0 \tag{7.40}$$

$$A\mathbf{x}^0 - \mathbf{b} \le 0 \tag{7.41}$$

$$\mathbf{y}^{0\prime}(A\mathbf{x}^0 - \mathbf{b}) = 0 \tag{7.42}$$

$$\mathbf{y}^0 \ge 0 \tag{7.43}$$

Since D is negative definite, then the determinant of D is nonzero; i.e., D is nonsingular and D^{-1} exists. Then from (7.40)

$$\mathbf{x}^0 = -(1/2)D^{-1}(\mathbf{c} - A'\mathbf{y}^0) \tag{7.44}$$

Substituting (7.44) into equations (7.41) and (7.42), we have the following equivalent necessary and sufficient conditions

$$-\mathbf{b} - (1/2)AD^{-1}(\mathbf{c} - A'\mathbf{y}^0) \le 0 \tag{7.45}$$

$$\mathbf{y}^{0\prime}[-\mathbf{b} - (1/2)AD^{-1}(\mathbf{c} - A'\mathbf{y}^0)] = 0 \tag{7.46}$$

$$\mathbf{y}^0 \ge 0 \tag{7.47}$$

Let $\varphi_{\mathbf{y}} = \mathbf{b} + (1/2)AD^{-1}(\mathbf{c} - A'\mathbf{y}^0)$, then there exists some dual problem such that

$$-\varphi_{\mathbf{y}} \leq 0 \qquad \varphi_{\mathbf{y}}'\mathbf{y}^0 = 0 \qquad \mathbf{y}^0 \geq 0 \tag{7.48}$$

are necessary and sufficient to ensure $\mathbf{y}^0$ is its corresponding optimal solution. An appropriate Lagrangian function of this dual problem is

$$\varphi(\mathbf{y}) = \mathbf{y}'\mathbf{b} + (1/2)\mathbf{y}'[AD^{-1}(\mathbf{c} - (1/2)A'\mathbf{y})]$$

and its associated dual problem would be

$$\begin{aligned} &\text{minimize} \quad \mathbf{y}'\mathbf{b} + (1/2)\mathbf{y}'AD^{-1}\mathbf{c} - (1/4)\mathbf{y}'AD^{-1}A'\mathbf{y} \\ &\text{subject to} \quad \mathbf{y} \geq 0 \end{aligned} \tag{7.49}$$

The above formulation gives the following lemma.

LEMMA 7.12. *Given the problem*

$$\begin{aligned} &\textit{minimize} \quad -(\mathbf{c}'\mathbf{x} + \mathbf{x}'D\mathbf{x}) \\ &\textit{subject to} \quad A\mathbf{x} \leq \mathbf{b} \end{aligned}$$

a dual problem is

$$\begin{aligned} &\textit{minimize} \quad \mathbf{y}'\mathbf{b} + (1/2)\mathbf{y}'AD^{-1}\mathbf{c} - (1/4)\mathbf{y}'AD^{-1}A'\mathbf{y} \\ &\textit{subject to} \quad \mathbf{y} \geq 0 \end{aligned}$$

COROLLARY 7.12.1. *An optimal solution of (7.37) is* $\mathbf{x}^0$ *if and only if* $\mathbf{x}^0 = -(1/2)D^{-1}(\mathbf{c} - A'\mathbf{y}^0)$ *where* $\mathbf{y}^0$ *is an optimal solution of (7.49).*

Note that the objective function in the dual problem is convex; in particular, the quadratic term is convex since $AD^{-1}A'$ is negative semidefinite (see Lemma 7.7).

7.4.3. *Solution of the Dual Problem*

Let $\mathbf{h} = (1/2)AD^{-1}\mathbf{c} + \mathbf{b}$ and $G = -(1/4)AD^{-1}A'$; then (7.49) can be expressed equivalently as

$$\begin{aligned} &\text{minimize} \quad \mathbf{h}'\mathbf{y} + \mathbf{y}'G\mathbf{y} \\ &\text{subject to} \quad \mathbf{y} \geq 0 \end{aligned} \tag{7.50}$$

Now

$$\frac{\partial(\mathbf{h}'\mathbf{y} + \mathbf{y}'G\mathbf{y})}{\partial y_i} = h_i + 2\sum_{j=1}^{m} g_{ij}y_j \geq 0 \qquad i = 1, 2, \ldots, m$$

Thus (7.50) can be solved iteratively by considering m linear programming problems; i.e.

$$\begin{array}{llll} \text{minimize} & \sum_{j=1}^{m} g_{1j} y_j & \ldots\text{minimize} & \sum_{j=1}^{m} g_{mj} y_j \\ \text{subject to} & \sum_{j=1}^{m} g_{1j} y_j \geq -h_1/2 & \text{subject to} & \sum_{j=1}^{m} g_{mj} y_j \geq -h_m/2 \\ & y_j \geq 0 & & y_j \geq 0 \\ & j = 1, 2, \ldots, m & & j = 1, 2, \ldots, m \end{array}$$

where the restrictions in each programming model require $y_i = 0$ or $\varphi_{y_i} = 0$.

Hildreth proved that the algorithm converges using Gauss-Seidel's iterative procedure. For example, choose an arbitrary feasible solution, possibly $\bar{\mathbf{y}} = 0$. Then, in each linear programming problem, hold all but one variable fixed and determine a new value for this variable. This involves nothing more than determining the new value of $y_i^{\ell+1}$ where

$$y_i^{\ell+1} = \max\left\{0, (-1/g_{ii})\left(h_i/2 + \sum_{j<i} g_{ij} y_j^{\ell+1} + \sum_{j\geq i} g_{ij} y_j^{\ell}\right)\right\}$$
$$i = 1, 2, \ldots, m \qquad \ell = 0, 1, 2, \ldots$$

where $\ell + 1$ denotes the $(\ell + 1)$th iteration. The procedure is continued until

$$|y_i^{\ell+1} - y_i^{\ell}| \leq \text{some predetermined value} \qquad \forall i = 1, 2, \ldots, m$$

Example 7.9

Consider the problem

$$\begin{array}{ll} \text{minimize} & -x_1 - 2x_2 + (1/2)x_1^2 + (1/2)x_2^2 \\ \text{subject to} & 2x_1 + 3x_2 \leq 6 \\ & x_1 + 4x_2 \leq 5 \\ & x_1, \quad x_2 \geq 0 \end{array}$$

Therefore

$$A = \begin{bmatrix} 2 & 3 \\ 1 & 4 \\ -1 & 0 \\ 0 & -1 \end{bmatrix} \qquad \mathbf{b} = \begin{bmatrix} 6 \\ 5 \\ 0 \\ 0 \end{bmatrix} \qquad \mathbf{c} = \begin{bmatrix} 1 \\ 2 \end{bmatrix}$$

$$D = \begin{bmatrix} -1/2 & 0 \\ 0 & -1/2 \end{bmatrix} \qquad D^{-1} = \begin{bmatrix} -2 & 0 \\ 0 & -2 \end{bmatrix}$$

To solve the dual problem we must determine the vector

$$\mathbf{h} = (1/2)AD^{-1}\mathbf{c} + \mathbf{b} = \begin{bmatrix} -2 \\ -4 \\ 1 \\ 2 \end{bmatrix}$$

and the matrix

$$G = -(1/4)AD^{-1}A' = \begin{bmatrix} 13/2 & 7 & -1 & -3/2 \\ 7 & 17/2 & -1/2 & -2 \\ -1 & -1/2 & 1/2 & 0 \\ -3/2 & -2 & 0 & 1/2 \end{bmatrix}$$

The four linear programming problems yield the following four iterative equations

$$\begin{aligned} y_1^\ell &= \max\{0, (-2/13)[-2/2 + 7y_2^{\ell-1} - y_3^{\ell-1} - (3/2)y_4^{\ell-1}]\} \\ y_2^\ell &= \max\{0, -(2/17)[7y_1^\ell - 4/2 - (1/2)y_3^{\ell-1} - 2y_4^{\ell-1}]\} \\ y_3^\ell &= \max\{0, -2[-y_1^\ell - (1/2)y_2^\ell + 1/2 + 0]\} \\ y_4^\ell &= \max\{0, -2[-(3/2)y_1^\ell - 2y_2^\ell + 0 + 2/2]\} \end{aligned}$$

Since $\bar{\mathbf{y}} = 0$ is a feasible solution, let us start the iterative procedure with this vector. At the end of the first iteration

$$y_1^1 = 2/13 \qquad y_2^1 = 24/221 \qquad y_3^1 = 0 \qquad y_4^1 = 0$$

Proceeding in this manner an iteration log is developed in Table 7.3 which, at the end of the fourth iteration, yields the optimal solution. Thus the solution to our original problem is $x_1^0 = 13/17$, $x_2^0 = 18/17$, by using equation (7.44).

TABLE 7.3. Iteration Log

Iteration	y_1	y_2	y_3	y_4
1	2/13	24/221	0	0
2	106/2,873	10,008/48,841	0	0
3	0	4/17	0	0
4	0	4/17	0	0

Exercises

1. Find a dual problem of the problem

$$\begin{aligned} \text{maximize} \quad & x_1 + x_2 - (1/2)x_1^2 + x_1x_2 - x_2^2 \\ \text{subject to} \quad & x_1 + x_2 \leq 3 \\ & 2x_1 + 3x_2 \geq 6 \\ & x_1, \quad x_2 \geq 0 \end{aligned}$$

2. The primal problem expressed in Exercise 1 is solved in Section 7.4. Resolve this problem and convince yourself that selecting any vector not violating rules (1) and (2) to enter the basis will yield the same optimal solution.
3. Consider the problem

$$\begin{aligned} \text{minimize} \quad & (1/3)(x_1 + 1)^3 + x_2 \\ \text{subject to} \quad & x_1 \geq 1 \\ & x_2 \geq 0 \end{aligned}$$

   a) Derive the optimal solution from the Kuhn-Tucker conditions.
   b) Why is one allowed to assume that the Kuhn-Tucker conditions in this example are sufficient to ensure that $\mathbf{x}^{0'} = (1, 0)$ is indeed the optimal solution?
4. Consider the problem

$$\begin{aligned} \text{maximize} \quad & -(x_1 + 1)^2 - x_2^2 \\ \text{subject to} \quad & x_1^2 + x_2^2 \leq 1 \\ & x_1, \quad x_2 \geq 0 \end{aligned}$$

   a) Derive the optimal solution from the Kuhn-Tucker conditions.
   b) What are the Kuhn-Tucker conditions if the positivity restrictions are removed? How would this affect Exercise 3?
5. Consider the following quadratic programming problem

$$\begin{aligned} \text{maximize} \quad & \mathbf{c}'\mathbf{x} + \mathbf{x}'D\mathbf{x} \\ \text{subject to} \quad & A\mathbf{x} \leq \mathbf{b} \\ & \mathbf{x} \geq 0 \end{aligned}$$

   where

$$D = \begin{bmatrix} -2 & 1 \\ 1 & -4 \end{bmatrix} \quad A = \begin{bmatrix} 1 & 1 \\ -2 & 1 \end{bmatrix} \quad \mathbf{b} = \begin{bmatrix} 300 \\ 1000 \end{bmatrix} \quad \mathbf{c} = \begin{bmatrix} -1 \\ -3 \end{bmatrix}$$

   a) Determine a feasible solution. Is it optimal?
   b) Determine the optimum of its associated saddle value problem, and determine the optimal solution of its dual problem.
6. Give an example showing that if one minimizes a convex function subject to linear restrictions, it is not necessarily true that the optimal solution lies on an extreme point.
7. Consider the following problem

$$\begin{aligned}&\text{minimize} \quad (\mathbf{Y} - X\beta)'(\mathbf{Y} - X\beta)\\ &\text{subject to} \quad A'\beta \geq \mathbf{b}\end{aligned}$$

where A' is a $2 \times p$ matrix of rank 2. Let $\hat{\beta}$ be the unrestricted least squares estimator; i.e., $\hat{\beta} = (X'X)^{-1}X\mathbf{Y}$.

a) Show that if both constraints are violated; i.e., $A'\hat{\beta} \ngeq \mathbf{b}$ and $A'(X'X)^{-1}A = D$, and

(1) if $d_{12} < 0$, then $\tilde{\beta} = \hat{\beta} + (X'X)^{-1}A[A'(X'X)^{-1}A]^{-1}(\mathbf{b} - A'\hat{\beta})$

(2) if $d_{12} > 0$ with

$$d_{12}/d_{22} \leq (b_1 - A_1'\hat{\beta})/(b_2 - A_2'\hat{\beta}) \leq d_{11}/d_{22}$$

then $\tilde{\beta} = \hat{\beta} + (X'X)^{-1}A[A'(X'X)^{-1}A]^{-1}(\mathbf{b} - A'\hat{\beta})$.

(3) if $d_{12} > 0$ with

$$d_{12}/d_{22} \geq (b_1 - A_1'\hat{\beta})/(b_2 - A_2'\hat{\beta})$$

then $\tilde{\beta} = \hat{\beta} + (X'X)^{-1}A_2(1/d_{22})(b_2 - A_2'\hat{\beta})$

(4) if $d_{12} > 0$ with

$$(b_1 - A_1'\hat{\beta})/(b_2 - A_2'\hat{\beta}) \geq d_{11}/d_{12}$$

then $\tilde{\beta} = \hat{\beta} + (X'X)^{-1}A_1(1/d_{11})(b_1 - A_1'\hat{\beta})$

b) Show that if $A_1'\hat{\beta} \geq b_1$ and $A_2'\hat{\beta} \ngeq b_2$

(1) and $d_{12} > 0$, then $\tilde{\beta} = \hat{\beta} + (X'X)^{-1}A_2(1/d_{22})(b_2 - A_2'\hat{\beta})$.

(2) and $d_{12} < 0$ with

$$d_{12}/d_{22} \geq (b_1 - A_1'\hat{\beta})/(b_2 - A_2'\hat{\beta})$$

then $\tilde{\beta} = \hat{\beta} + (X'X)^{-1}A_2(1/d_{22})(b_2 - A_2'\hat{\beta})$

(3) and $d_{12} < 0$ with

$$d_{12}/d_{22} \leq (b_1 - A_1'\hat{\beta})/(b_2 - A_2'\hat{\beta}) \qquad d_{11}/d_{12} \leq (b_1 - A_1'\hat{\beta})/(b_2 - A_2'\tilde{\beta})$$

then $\tilde{\beta} = \hat{\beta} + (X'X)^{-1}A[A'(X'X)^{-1}A]^{-1}(\mathbf{b} - A'\hat{\beta})$.

c) Consider (iii) of Example 7.6. Show that (2) and (3) are not satisfied for this particular problem.

8. Consider the following problem

$$\begin{aligned}&\text{minimize} \quad (\mathbf{Y} - X\beta)'(\mathbf{Y} - X\beta)\\ &\text{subject to} \quad b_0 \leq A'\beta \leq b_1\end{aligned}$$

where A' is $1 \times p$.

a) Show that if the unrestricted least squares estimator $\hat{\beta}$ is such that $A'\hat{\beta} \notin [b_0, b_1]$, then necessarily only one inequality is violated.

b) Assume $A'\hat{\beta} \geq b_0$ and $-A'\hat{\beta} \ngeq -b_1$. Show that here no conditions are needed to estimate the restricted least squares estimator and it can be expressed as

$$\tilde{\beta} = \hat{\beta} - (X'X)^{-1}A[A'(X'X)^{-1}A]^{-1}(-b_1 + A'\hat{\beta})$$

References

Abrams, R. A., and Ben-Israel, Adi. 1969. A duality theorem for complex quadratic programming. *J. Optim. Theory Appl.* 4:244–52.

Dorn, W. S. 1960. Duality in quadratic programming. *Q. Appl. Math.* 18:155–62.

Hadley, G. 1964. *Nonlinear and Dynamic Programming.* Addison-Wesley, Reading, Mass.

Hildreth, C. 1957. A quadratic programming procedure. *Nav. Res. Logist. Q.* 4: 79–85.

Klemm, B.; Sposito, V. A.; and Pyne, D. 1974. Useful least squares solutions over inequality restrictions. *J. Stat. Res.* vol. 8, no. 2. In press.

Kuhn, H. W., and Tucker, A. W. 1950. Nonlinear programming. 2nd Berkeley Symp. Proc. Math. Stat. Probab., pp. 481–92. Univ. Calif. Press, Berkeley.

Kunzi, Hans, and Krelle, Wilhelm. 1966. *Nonlinear Programming.* Blaisdell, Waltham, Mass.

Lowell, M. C., and Prescott, E. 1970. Multiple regression with inequality constraints: Pretesting bias, hypothesis testing and efficiency. *J. Am. Stat. Assoc.* 65:913–25.

Luenberger, D. 1973. *Introduction to Linear and Nonlinear Programming.* Addison-Wesley, Reading, Mass.

McMillan, Claude, Jr. 1970. *Mathematical Programming.* Wiley, New York.

Mangasarian, O. L. 1969. *Nonlinear Programming.* McGraw-Hill, New York.

Markowitz, Harry M. 1959. Portfolio selection. *Cowles Foundation Monograph 16.* Wiley, New York.

O'Hagen, A. 1973. Bayes estimation of a convex quadratic. *Biometrika* 60:565–71.

Pyne, D. 1972. Duality in abstract mathematical programming with applications to statistical problems. Diss. Iowa State Univ., Ames.

Shah, B. V.; Buehler, R. J.; and Kempthorne, O. 1964*a*. Methods of parallel tangents. *Chem. Eng. Prog. Symp. Ser.* 60:1–7.

———. 1964*b*. Some algorithms for minimizing a function of several variables. *SIAM J. Appl. Math.* 12:74–92.

Slater, M. 1951. Lagrange multipliers revisited: a contribution to nonlinear programming. Rand Corp. Rept. RM-676. Santa Monica, Calif.

Sposito, V. A., and Soults, D. J. 1969. An algorithm for the optimization of a quadratic form subject to linear restrictions: Zorilla. *Numer. Anal. Program. Ser. 9.* Stat. Lab., Iowa State Univ., Ames.

Wolfe, P. 1959. The simplex algorithm for quadratic programming. *Econometrica* 27:382–98.

Zellner, A. 1961. Linear regression with inequality constraints on the coefficients: An application of quadratic programming and linear decision rules. *Int. Cent. Manage. Sci. Rept. 6109 (MSN09).* Rotterdam.

C H A P T E R   E I G H T

GEOMETRIC PROGRAMMING

If one uses the classical optimization techniques where the function to be minimized (maximized) is convex (concave), a global solution is often obtained by solving a system of linear equations. In this section, we shall reduce an optimization problem involving an arbitrary function to a problem of solving a system of linear equations. This technique will always produce a global minimum when

$$\text{number of terms} - \text{number of unknowns} - 1 = 0$$

provided a certain linear system of equations is consistent.

The original investigation of this optimization procedure was by C. Zener. Zener observed in 1961 that the minimum of certain objective functions could be determined almost by inspection. In 1962, R. J. Duffin learned of Zener's work and jointly they extended the early ideas of Zener to more general problems. Duffin, E. L. Peterson (a student of Zener), and Zener published the first text in geometric programming summarizing their results in 1967. This procedure is still in its infancy and this chapter presents only a brief description of geometric programming.

This method provides a systematic way of solving a class of nonlinear optimization problems by first finding the optimal value of the objective function and then finding the unknown variables which yield this minimum.

Consider the following (primal) problem

$$\begin{aligned} &\text{minimize} \quad g_0(\mathbf{x}) \\ &\text{subject to} \quad x_j > 0 \qquad j = 1, 2, \ldots, n \end{aligned} \tag{8.1}$$

The function $g_0(\mathbf{x})$ is assumed to be of the form $g_0 = u_1 + u_2 + \cdots + u_{m_0}$ where

$$u_i = c_i x_1^{a_{i1}} \cdots x_n^{a_{in}} = c_i \prod_{j=1}^{n} x_j^{a_{ij}}$$

the coefficients c_i are positive, the exponents a_{ij} are real constants, and the variables x_j are positive. Such functions are called positive polynomials or *posynomials*.

8.1 The Arithmetic and Geometric Means

Given m_0 terms $U_1, U_2, \ldots, U_{m_0}$, their arithmetic mean is

$$(1/m_0)(U_1 + U_2 + \cdots + U_{m_0}) \tag{8.2}$$

and their geometric mean is

$$(U_1 U_2 \cdots U_{m_0})^{1/m_0} \tag{8.3}$$

The arithmetic mean of a group of nonnegative numbers is always greater than or equal to their geometric mean, i.e., (8.2) $\geq$ (8.3). For instance, if $m_0 = 2$, then $(U_1 - U_2)^2 \geq 0$, or

$$U_1^2 - 2U_1U_2 + U_2^2 \geq 0$$

Adding $4U_1U_2$ to both sides we have

$$U_1^2 + 2U_1U_2 + U_2^2 \geq 4U_1U_2$$

Taking the square root of both sides and dividing by 2

$$(1/2)U_1 + (1/2)U_2 \geq (U_1U_2)^{1/2} \tag{8.4}$$

A generalization of (8.4) is obtained by considering four variables $U_1, \ldots, U_4$; then

$$(1/4)U_1 + \cdots + (1/4)U_4 \geq [(U_1 + U_2)/2]^{1/2}[(U_3 + U_4)/2]^{1/2} \tag{8.5a}$$

Using (8.4) twice on the right side of (8.5a) we have

$$(1/4)U_1 + \cdots + (1/4)U_4 \geq (U_1U_2U_3U_4)^{1/4} \tag{8.5b}$$

And if $U_1 = \cdots = U_4$, the inequality (8.5b) becomes an equality.

Furthermore, a weighted mean is obtained from (8.5a) by letting $U_2 = U_3 = U_4$, consequently

$$(1/4)U_1 + (3/4)U_2 \geq U_1^{1/4}U_2^{3/4} \tag{8.6}$$

The following set of examples shows how the greatest lower bound on $g_0(\mathbf{x})$ is achieved when an appropriate weighted mean is used.

EXAMPLE 8.1

Let $g(x) = 4x + 1/x = (1/2)(8x + 2/x)$. Using (8.4)

$$g(x) \geq (8x)^{1/2}(2/x)^{1/2} = 4$$

Note by letting the partial derivative be zero, we have $4 - 1/x^2 = 0$ or $x = \pm 1/2$, and $g(1/2) = 4$. Thus 4 is the greatest lower bound.

EXAMPLE 8.2

Let $g(x_1, x_2) = 4x_1 + x_1/x_2^2 + 4x_2/x_1$, and using the geometric inequality

$$(1/4)U_1 + (1/4)U_2 + (2/4)U_3 \geq U_1^{1/4}U_2^{1/4}U_3^{1/2}$$

we have

$$g(x_1, x_2) \geq (16x_1)^{1/4}(4x_1/x_2^2)^{1/4}(8x_2/x_1)^{1/2} = 8$$

Since $g(1/2, 1/2) = 8$, the lower bound is achieved and 8 is the greatest lower bound.

EXAMPLE 8.3

Let $g(x) = 2x^3 + 6/x = (1/4)(8x^3) + (3/4)(8/x)$; then using (8.6)

$$g(x) \geq (8x^3)^{1/4}(8/x)^{3/4} = 8$$

and for $x^0 = 1, g(x^0) = 8$; i.e., 8 is the greatest lower bound.

Note in these three examples that the exponents of x indicate which geometric mean to use. For example, if we had used (8.4) in the last example, then

$$g(x) \geq (4x^3)^{1/2}(12/x)^{1/2} = 4(3)^{1/2}x$$

and we do not gain anything by using this inequality since it is a function of x.

We now show how the desired geometric inequality is obtained and the greatest lower bound is achieved.

8.2 The General Relationship

From the last section it is apparent that we can form a variety of geometric inequalities. By considering different weights δ_i we obtain different inequalities. The general form, however, is

$$\delta_1 U_1 + \cdots + \delta_{m_0} U_{m_0} \geq U_1^{\delta_1} \cdots U_{m_0}^{\delta_{m_0}} \tag{8.7}$$

where $U_i, \delta_i > 0$ for all i and $\sum_{i=1}^{m_0} \delta_i = 1$.

Letting $u_i = \delta_i U_i$ $(i = 1, 2, \ldots, m_0)$, then from (8.7)

$$u_1 + \cdots + u_{m_0} \geq (u_1/\delta_1)^{\delta_1} \cdots (u_{m_0}/\delta_{m_0})^{\delta_{m_0}} \tag{8.8}$$

This relationship is a special case of Jensen's inequality (see Exercise 2); i.e.

$$E[g(X)] \leq g(E[X])$$

where $g(X)$ is a concave function. Hence, taking g to be the natural log function, then for N points

$$E[g(X)] = \sum_{i=1}^{N} \delta_i \ln X_i = \sum_{i=1}^{N} \ln X_i^{\delta_i} = \ln \prod_{i=1}^{N} X_i^{\delta_i}$$

and $g(E[X]) = \ln \sum_{i=1}^{N} \delta_i X_i$. Therefore,

$$\ln \prod_{i=1}^{N} X_i^{\delta_i} \leq \ln \sum_{i=1}^{N} \delta_i X_i$$

or $\prod_{i=1}^{N} X_i^{\delta_i} \leq \sum_{i=1}^{N} \delta_i X_i$, which can be equivalently expressed as (8.7) by letting $U_i = X_i$.

8.3 Conditions for an Extreme

Denote the left and right sides of (8.8) by g and V respectively, then $g \geq V$.

DEFINITION 8.1. *Let g be denoted as the primal function and V be denoted as the predual function. The predual function V is a product function of the terms u_i divided by appropriate weights δ_i and then raised to the power δ_i.*

Now, if the primal function g is a posynomial to be minimized, then each u_i term is of the form $c_i x_1^{a_{i1}} \cdots x_n^{a_{in}}$ such that $c_i > 0$ and a_{ij} are real constants. Moreover

$$V(\delta, \mathbf{x}) = (c_1/\delta_1)^{\delta_1} \cdots (c_{m_0}/\delta_{m_0})^{\delta_{m_0}} x_1^{D_1} \cdots x_n^{D_n}$$

where $D_j = \sum_{i=1}^{m_0} \delta_i a_{ij}$ $(j = 1, 2, \ldots, n)$.

Note that in the previous section we desired a geometric inequality such that the lower bound is not dependent on the unknown variables. Therefore if we select the weights δ_i such that $\sum_{i=1}^{m_0} \delta_i = 1$ and $\delta_i > 0$, $D_j = 0$ for all i and j, then

$$V(\delta) = \prod_{i=1}^{m_0} (c_i/\delta_i)^{\delta_i} \tag{8.9}$$

We define (8.9) to be the *dual function.*

The weight vector $\boldsymbol{\delta}$ is said to satisfy the *normality* and *orthogonality* conditions, respectively, if $\sum_{i=1}^{m_0} \delta_i = 1$ and $D_j = 0$ $(j = 1, 2, \ldots, n)$.

LEMMA 8.1. *The maximum of the dual function is equal to the minimum of the primal function.*

Proof. Assume $\mathbf{x}^{0'} = (x_1^0, x_2^0, \ldots, x_n^0)$ is the minimizing point of $g(\mathbf{x})$ where $x_i^0 > 0$ for all i. Then at the point $\mathbf{x}^0$, the derivative of $g(\mathbf{x})$ with respect to each variable vanishes; i.e., we obtain n equations of the form

$$0 = x_j^0 \frac{\partial g(\mathbf{x}^0)}{\partial x_j} = \sum_{i=1}^{m_0} x_j^0 \frac{\partial u_i(\mathbf{x}^0)}{\partial x_j} = \sum_{i=1}^{m_0} u_i(\mathbf{x}^0) a_{ij}$$

Dividing these equations by $g(\mathbf{x}^0)$ and letting

$$\bar{\delta}_i = u_i(\mathbf{x}^0)/g(\mathbf{x}^0) > 0 \qquad i = 1, 2, \ldots, m_0 \tag{8.10}$$

then we have

$$0 = \sum_{i=1}^{m_0} \bar{\delta}_i a_{ij} \qquad j = 1, 2, \ldots, n \tag{8.11}$$

i.e., $\bar{\delta}$ satisfies the condition $D_j = 0$ $(j = 1, 2, \ldots, n)$. Now summing over the terms in (8.10) and noting $\Sigma\, u_i = g$, we have that the $\bar{\delta}_i$ satisfy the condition $\sum_{i=1}^{m_0} \bar{\delta}_i = 1$.

We shall now show that $V(\bar{\delta}) = g(\mathbf{x}^0)$. From (8.10)

$$u_i(\mathbf{x}^0) = \bar{\delta}_i g(\mathbf{x}^0) \quad \text{for all } i$$

or

$$c_i x_1^{0^{a_{i1}}} \cdots x_n^{0^{a_{in}}} = \bar{\delta}_i g(\mathbf{x}^0) \quad \text{for all } i$$

Hence, for any $\bar{\delta}_{i'}$ $(i' \in \{i \mid i = 1, 2, \ldots, m_0\})$

$$(c_{i'})^{\bar{\delta}_{i'}} \prod_{j=1}^{n} (x_j^0)^{a_{i'j}\bar{\delta}_{i'}} = (\bar{\delta}_{i'} g(\mathbf{x}^0))^{\bar{\delta}_{i'}}$$

or

$$(c_{i'}/\bar{\delta}_{i'})^{\bar{\delta}_{i'}} \prod_{j=1}^{n} (x_j^0)^{\bar{\delta}_{i'}a_{i'j}} = g(\mathbf{x}^0)^{\bar{\delta}_{i'}}$$

Therefore, multiplying all m_0 such equations together, we have

$$\prod_{i=1}^{m_0} (c_i/\bar{\delta}_i)^{\bar{\delta}_i} \prod_{j=1}^{n} (x_j^0)^{\sum_{i=1}^{m_0} \bar{\delta}_i a_{ij}} = g(\mathbf{x}^0)^{\Sigma \bar{\delta}_i} = g(\mathbf{x}^0)$$

But from (8.11), $\sum_{i=1}^{m_0} \bar{\delta}_i a_{ij} = 0$; therefore $V(\bar{\delta}) = g(\mathbf{x}^0)$.

8.4 Determination of the Minimizing Point

From the fact that $g(\mathbf{x}^0)\bar{\delta}_i = u_i(\mathbf{x}^0)$ $(i = 1, 2, \ldots, m_0)$ and, from the last lemma, $g(\mathbf{x}^0) = V(\bar{\delta})$, we have

$$V(\bar{\delta})\bar{\delta}_i = u_i(\mathbf{x}^0) \qquad i = 1, 2, \ldots, m_0$$

This implies a necessary condition for the geometric inequality to be an equality is that $u_i(\mathbf{x}^0)/\bar{\delta}_i$ be a constant, independent of i. Say

$$u_i(\mathbf{x}^0)/\bar{\delta}_i = k \tag{8.12}$$

Then

$$\begin{aligned} k = \sum_{i=1}^{m_0} k\bar{\delta}_i &= \sum_{i=1}^{m_0} u_i(\mathbf{x}^0) \quad \text{from (8.12) and with } \Sigma\bar{\delta}_i = 1 \\ &= g(\mathbf{x}^0) \quad \text{from the definition of } g(\mathbf{x}^0) \\ &= V(\bar{\delta}) \quad \text{by Lemma 8.1} \end{aligned}$$

Hence, $\mathbf{x}^0$ is such that

$$u_i(\mathbf{x}^0)/\bar{\delta}_i = V(\bar{\delta}) \qquad i = 1, 2, \ldots, m_0 \tag{8.13}$$

i.e., the optimal solution of the primal problem is determined by solving a system of equations. The following example illustrates that this system of equations can be reduced to a system of linear equations.

EXAMPLE 8.4

Let $g(x_1, x_2, x_3) = 40/x_1x_2x_3 + 40x_2x_3 + 20x_1x_3 + 10x_1x_2$. Then the dual function is

$$V = (40/\delta_1)^{\delta_1}(40/\delta_2)^{\delta_2}(20/\delta_3)^{\delta_3}(10/\delta_4)^{\delta_4}$$

Now to determine the appropriate weights. The variables δ_i must satisfy the following system of equations

$$\begin{aligned}
D_1 &= -\delta_1 \phantom{{}+\delta_2} + \delta_3 + \delta_4 = 0\\
D_2 &= -\delta_1 + \delta_2 \phantom{{}+\delta_3} + \delta_4 = 0\\
D_3 &= -\delta_1 + \delta_2 + \delta_3 \phantom{{}+\delta_4} = 0\\
&\phantom{=-{}}\delta_1 + \delta_2 + \delta_3 + \delta_4 = 1
\end{aligned}$$

Solving this system, $\delta_1 = 2/5$, $\delta_2 = \delta_3 = \delta_4 = 1/5$, and $V(\delta) = 100$. Now we will determine the $\mathbf{x}^0$ that achieves this greatest lower bound. From (8.13), this vector must satisfy the system $V(\delta)\,\delta_i = u_i(\mathbf{x}^0)$ $(i = 1, 2, 3, 4)$; i.e.

$$\begin{aligned}
(100)\,2/5 &= 40x_1^{-1}x_2^{-1}x_3^{-1}\\
(100)\,1/5 &= 40x_2x_3\\
(100)\,1/5 &= 20x_1x_3\\
(100)\,1/5 &= 10x_1x_2
\end{aligned}$$

Taking logarithms of each side of each equation, we have, letting $z_i = \ln(x_i)$

$$\begin{aligned}
0 &= -z_1 - z_2 - z_3\\
-\ln 2 &= z_2 + z_3\\
0 &= z_1 + z_3\\
\ln 2 &= z_1 + z_2
\end{aligned}$$

and $\mathbf{z}^{0\prime} = (\ln 2, 0, -\ln 2)$ or $\mathbf{x}^{0\prime} = (2, 1, 1/2)$.

LEMMA 8.2. *Let $g(\mathbf{x})$ be a posynomial with T terms; then any stationary point $\mathbf{x}^0$ is a minimizing point for $g(\mathbf{x})$.*

Proof. Let $e^{u_j} = x_j$ $(j = 1, 2, \ldots, n)$. Each of the terms e^{u_j} is convex with respect to u_j and the sum $\sum_{i=1}^{T} c_i \exp\left(\sum_{j=1}^{n} a_{ij}u_j\right)$ is convex. Hence for any stationary point

$$\left.\frac{\partial \sum_i c_i \exp\left(\sum_j a_{ij}u_j\right)}{\partial u_j}\right|(\mathbf{u} = \mathbf{u}^0) = 0 \qquad \forall j$$

Therefore $\mathbf{x}^0$ must be a global minimum of $g(\mathbf{x})$.

8.5 Inequality Constraints

We now consider optimization problems involving inequality constraints, namely

$$\begin{aligned} &\text{minimize} \quad g_0(\mathbf{x}) \\ &\text{subject to} \quad \sigma_k(1 - g_k(\mathbf{x})) \geq 0 \qquad k = 1, 2, \ldots, m \end{aligned} \tag{8.14}$$

For each constraint $g_k(\mathbf{x}) \geq$ (resp., $\leq$) 1 we introduce a signum function σ_k that can be either 1 or -1 depending on whether $g_k(\mathbf{x})$ is ≤ 1 or ≥ 1. As in the unconstrained case, the weights δ_i are introduced for the terms in the objective function

$$\delta_i = u_i(\mathbf{x}^0)/g_0(\mathbf{x}^0) \qquad i = 1, 2, \ldots, T \tag{8.15}$$

Let the weights for the objective function be denoted by δ_{0i} $(i = 1, 2, \ldots, T_0)$. The weights for the constraint terms are simply the values of the terms themselves, since if the constraint (8.14) is satisfied exactly then the terms must sum to unity; i.e.

$$\delta_{ki} = c_{ki} \prod_{\ell=1}^{n} x_\ell^{a_{ki\ell}} \qquad k = 1, 2, \ldots, m \qquad i = 1, 2, \ldots, T_k \tag{8.16}$$

where T_k is the number of terms in the kth constraint.

Let $e^{u_\ell} = x_\ell$ $(\ell = 0, 1, \ldots, n)$ where $x_0 = g_0(\mathbf{x})$. Taking natural logarithms, we can transform the original problem, with σ_k in (8.16), to

$$
\begin{array}{lll}
\text{minimize} & u_0 & \\
\text{subject to} & -u_0 + \sum_{\ell=1}^{n} a_{0i\ell} u_\ell = \ln(\delta_{0i}/c_{0i}) & i = 1, 2, \ldots, T_0 \\
& \sigma_k \sum_{\ell=1}^{n} a_{ki\ell} u_\ell = \sigma_k \ln(\delta_{ki}/c_{ki}) & k = 1, 2, \ldots, m \\
& & i = 1, 2, \ldots, T_k \\
& \sum_{i=1}^{T_0} \delta_{0i} = 1 & \\
& f_k \equiv \sigma_k(1 - \sum_{i=1}^{T_k} \delta_{ki}) \geq 0 & \quad (8.17)
\end{array}
$$

To construct a Lagrangian function, (8.17) needs special consideration. Define Lagrange multipliers λ_k such that if $f_k > 0$ then $\lambda_k = 0$, and if $f_k = 0$ then λ_k is nonnegative. This condition is called the Kuhn-Tucker complementary slackness condition: $\lambda_k f_k = 0$. Hence this condition can be added to the Lagrangian function without changing its value. An appropriate function for the above problem is

$$
\begin{aligned}
\mathcal{L}(\mathbf{u}, \boldsymbol{\delta}, \boldsymbol{\lambda}, \mathbf{w}) = {} & u_0 - \lambda_0 \left(1 - \sum_{i=1}^{T_0} \delta_{0i}\right) \\
& - \sum_{i=1}^{T_0} w_{0i} \left[\ln(\delta_{0i}/c_{0i}) + u_0 - \sum_{\ell=1}^{n} a_{0i\ell} u_\ell\right] \\
& - \sum_{k=1}^{m} \sum_{i=1}^{T_k} w_{ki} \sigma_k \left[\ln(\delta_{ki}/c_{ki}) - \sum_{\ell=1}^{n} a_{ki\ell} u_\ell\right] \\
& - \sum_{k=1}^{m} \lambda_k f_k
\end{aligned}
\tag{8.18}
$$

LEMMA 8.3. *Given the optimal Lagrange multipliers* $\mathbf{w}$ *in (8.18), then*

(a) $\lambda_0 = 1$

(b) $\delta_{0i} = w_{0i} \qquad i = 1, 2, \ldots, T_0$

(c) $\delta_{ki} = w_{ki}/\sum_{i=1}^{T_k} w_{ki} \quad \text{if } \lambda_k \neq 0 \qquad k = 1, 2, \ldots, m$

Proof. To show (*a*) and (*b*)

$$
\frac{\partial \mathcal{L}}{\partial u_0} = 1 - \sum_{i=1}^{T_0} w_{0i} = 0
$$

$$
\frac{\partial \mathcal{L}}{\partial \delta_{0i}} = \lambda_0 - \frac{w_{0i}}{\delta_{0i}} = 0 \qquad i = 1, 2, \ldots, T_0 \tag{8.19}
$$

$$
w_{0i} = \lambda_0 \delta_{0i} \qquad i = 1, 2, \ldots, T_0 \quad \text{from (8.19)} \tag{8.20}
$$

and summing over i, $1 = \sum_{i=1}^{T_0} w_{0i} = \lambda_0 \sum_{i=1}^{T_0} \delta_{0i} = \lambda_0$. Hence $\lambda_0 = 1$, and (b) holds by substituting $\lambda_0 = 1$ in (8.20).

Next consider the derivative with respect to δ_{ki} ($k \neq 0$)

$$\frac{\partial \mathcal{L}}{\partial \delta_{ki}} = -\frac{w_{ki}\sigma_k}{\delta_{ki}} + \lambda_k \sigma_k = 0$$

or

$$w_{ki} = \lambda_k \delta_{ki} \tag{8.21}$$

and summing over i we have $\sum_{i=1}^{T_k} w_{ki} = \lambda_k$. Hence, from (8.21)

$$\delta_{ki} = w_{ki}/\lambda_k \quad \text{if } \lambda_k \neq 0$$

The weights w_{ki} are determined by differentiating with respect to u_ℓ ($\ell \neq 0$)

$$\frac{\partial \mathcal{L}}{\partial u_\ell} = \sum_{k=0}^{m} \sum_{i=1}^{T_k} w_{ki}\sigma_k a_{ki\ell} = 0 \quad \text{with } \sigma_0 = 1.$$

Note that the variables are the Lagrange multipliers w_{ki}, and the multipliers $w_{0i} = \delta_{0i}$ are subject to $\Sigma \delta_{0i} = 1$.

LEMMA 8.4. *The objective function of the associated dual problem of (8.14) is*

$$V(\delta) = \prod_{k=0}^{m} \prod_{i=1}^{T_k} \left(c_{ki} w_{ki}^{-1} \sum_{\ell=1}^{T_k} w_{k\ell} \right)^{\sigma_k w_{ki}}$$

Proof. Consider the Lagrangian function (8.18) with Lagrange multipliers $(u_0 - 1), u_1, \ldots, u_m$ and λ_k ($k = 1, 2, \ldots, m$). Then

$$\begin{aligned} \mathcal{L}(\mathbf{u}, \delta, \lambda, \mathbf{w}) &= L(\mathbf{w}, \mathbf{u}, \mathbf{f}) \\ &= \sum_{k=0}^{m} \sum_{i=1}^{T_k} \sigma_k w_{ki} \ln \left(c_{ki} w_{ki}^{-1} \sum_{\ell=1}^{T_k} w_{k\ell} \right) \\ &\quad + (u_0 - 1)\left(1 - \sum_{i=1}^{T_0} w_{0i}\right) \\ &\quad + \sum_{\ell=1}^{n} u_\ell \sum_{k=0}^{m} \sum_{i=1}^{T_k} \sigma_k a_{ki\ell} w_{ki} - \sum_{j=1}^{m} f_j \lambda_j \end{aligned}$$

This Lagrangian function can be considered as another constrained maximization problem with a new objective function and subject to the following normality and orthogonality restrictions

$$\begin{aligned}
\text{maximize} \quad & z(\mathbf{w}) = \textstyle\sum_{k=0}^{m}\sum_{i=1}^{T_k} \sigma_k w_{ki} \ln (c_{ki} w_{ki}^{-1} \sum_{\ell=1}^{T_k} w_{k\ell}) \\
\text{subject to} \quad & \textstyle\sum_{i=1}^{T_0} w_{0i} = 1 \\
& \textstyle\sum_{k=0}^{m}\sum_{i=1}^{T_k} \sigma_k a_{ki\ell} w_{ki} = 0 \qquad l = 1, 2, \ldots, n \\
& \lambda_k = \textstyle\sum_{i=1}^{T_k} w_{ki} \geq 0 \qquad k = 1, 2, \ldots, m
\end{aligned}$$

Hence, in view of the exponential transformation, the optimum of this problem is equal to the optimum of (8.14) so that

$$V(\delta) = \exp[z(\mathbf{w})] = \prod_{k=0}^{m} \prod_{i=1}^{T_k} \left(c_{ki} w_{ki}^{-1} \sum_{\ell=1}^{T_k} w_{k\ell} \right)^{\sigma_k w_{ki}}$$

EXAMPLE 8.5

Consider the problem

$$\begin{aligned}
\text{minimize} \quad & 40/x_1 x_2 x_3 + 40 x_2 x_3 \\
\text{subject to} \quad & x_1 x_3/2 + x_1 x_2/4 \leq 1 \\
& x_j > 0 \qquad j = 1, 2, 3
\end{aligned}$$

Note that $\sigma_1 = 1$ and the Lagrange multipliers must satisfy the following linear system

$$\begin{aligned}
-w_{01} \qquad\qquad + w_{11} + w_{12} &= 0 \\
-w_{01} + w_{02} \qquad\qquad + w_{12} &= 0 \\
-w_{01} + w_{02} + w_{11} \qquad\qquad &= 0 \\
w_{01} + w_{02} \qquad\qquad\qquad &= 1
\end{aligned}$$

Hence $\mathbf{w}' = (2/3, 1/3, 1/3, 1/3)$ and therefore, from Lemmas 8.3 and 8.4

$$\begin{aligned}
\delta' &= (2/3, 1/3, 1/2, 1/2) \\
g_0(\mathbf{x}) &= \left(\frac{40}{2/3}\right)^{2/3} \left(\frac{40}{1/3}\right)^{1/3} [(1/2)(3/1)(2/3)]^{1/3} [(1/4)(3/1)(2/3)]^{1/3} \\
&= (60)^{2/3} (120)^{1/3} (1/2)^{1/3} = 60
\end{aligned}$$

To determine the optimal solution, applying (8.15) and (8.16)

$$\begin{aligned}
&40 x_1^{-1} x_2^{-1} x_3^{-1} = (2/3)60 = 40 \\
&40 x_2 x_3 = (1/3)60 = 20 \\
&x_1 x_3/2 = 1/2 \qquad x_1 x_2/4 = 1/2
\end{aligned}$$

or

$$x_1x_2x_3 = 1 \qquad x_2x_3 = 1/2 \qquad x_1x_3 = 1 \qquad x_1x_2 = 2$$

Hence $\mathbf{x}^{0\prime} = (2, 1, 1/2)$.

Note that another way of writing $g_0(\mathbf{x})$, in view of Lemmas 8.3 and 8.4, is

$$\begin{aligned} g_0(\mathbf{x}) &= \left(\frac{40}{w_{01}}\right)^{w_{01}}\left(\frac{40}{w_{02}}\right)^{w_{02}}\left(\frac{1/2}{w_{11}}\right)^{w_{11}}\left(\frac{1/4}{w_{12}}\right)^{w_{12}}(w_{11}+w_{12})^{w_{11}}(w_{11}+w_{12})^{w_{12}} \\ &= \left(\frac{40}{w_{01}}\right)^{w_{01}}\left(\frac{40}{w_{02}}\right)^{w_{02}}\left(\frac{1}{2w_{11}}\right)^{w_{11}}\left(\frac{1}{4w_{12}}\right)^{w_{12}}(w_{11}+w_{12})^{(w_{11}+w_{12})} \\ &= (60)^{2/3}(120)^{1/3}(3/2)^{1/3}(3/4)^{1/3}(2/3)^{2/3} \\ &= 60 \end{aligned}$$

This form is sometimes used in the literature.

Note that in our presentation we assume the constraints are tight when evaluated at $\mathbf{x}^0$, i.e., $g_k(\mathbf{x}^0) = 1$ $(k = 1, 2, \ldots, m)$. In view of this assumption

$$\begin{aligned} \delta_{ki} &= c_{ki}\prod_{\ell=1}^{n} x_\ell^{a_{ki\ell}} \qquad k = 1, 2, \ldots, m \qquad i = 1, 2, \ldots, T_k \\ &= w_{ki}\left(\sum_{\ell=1}^{T_k} w_{k\ell}\right)^{-1} \end{aligned}$$

with $\sum_{\ell=1}^{T_k} w_{k\ell} \neq 0$, and

$$\begin{aligned} V(\delta) &= \prod_{k=0}^{m}\prod_{i=1}^{T_k}\left(c_{ki}w_{ki}^{-1}\sum_{\ell=1}^{T_k} w_{k\ell}\right)^{w_{ki}} \\ &= \prod_{k=0}^{m}\prod_{i=1}^{T_k}(c_{ki}/\delta_{ki})^{w_{ki}} \\ &= \prod_{i=1}^{T_0}(c_{0i}/\delta_{0i})^{\delta_{0i}}\prod_{k=1}^{m}\prod_{i=1}^{T_k}(c_{ki}/u_{ki})^{w_{ki}} \end{aligned}$$

Moreover, under the above assumption, δ_{ki}^0 are positive for $i = 1, 2, \ldots, T_k$ if and only if the kth constraint is binding at $\mathbf{x}^0$. If the kth constraint is binding at $\mathbf{x}^0$, then $g_k(\mathbf{x}^0) = 1$ or $\sum_{i=1}^{T_k} u_{ki} = 1$. Hence, it follows that

$$\sum_{i=1}^{T_k} \delta_{ki}^0 = 1 \qquad \delta_{ki}^0 = c_{ki} \prod_{j=1}^{n} (x_j^0)^{a_{kij}} > 0 \qquad \text{for all } i$$

Conversely, suppose δ_{ki}^0 are all positive. Then necessarily

$$\sum_{i=1}^{T_k} \delta_{ki}^0 = \sum_{i=1}^{T_k} w_{ki}^0 \left(\sum_{\ell=1}^{T_k} w_{k\ell}^0 \right)^{-1} = 1$$

and the kth constraint evaluated at $\mathbf{x}^0$ is binding.

Therefore it follows that if any constraint is not binding at $\mathbf{x}^0$, then some of its associated w_{ki} must be zero. However, since $u_{ki} > 0$ for $i = 1, 2, \ldots, T_k$ at $\mathbf{x}^0$, then $(u_{ki})^{w_{ki}} = 1$. This implies that if any constraint is loose, then $V(\delta)$ can be computed by observing that

$$1 = (c_{ki}/u_{ki})^{w_{ki}} = \left(c_{ki} w_{ki}^{-1} \sum_{\ell=1}^{T_k} w_{k\ell} \right)^{w_{ki}} \qquad \text{for } w_{ki} = 0$$

EXAMPLE 8.6

Consider the problem

$$\begin{aligned} \text{minimize} \quad & 2/x_1 x_2 + 50 x_1 x_2 \\ \text{subject to} \quad & x_1 \geq 2 \\ & x_2 > 0 \end{aligned}$$

The weights w_{ki} must satisfy the following system of equations

$$\begin{aligned} -w_{01} + w_{02} - w_{11} &= 0 \\ -w_{01} + w_{02} &= 0 \\ w_{01} + w_{02} &= 1 \end{aligned}$$

Hence $\mathbf{w}' = (1/2, 1/2, 0)$, implying that

$$g_0(\mathbf{x}) = \left(\frac{2}{1/2}\right)^{1/2} \left(\frac{50}{1/2}\right)^{1/2} = 20$$

$$2x_1^{-1}x_2^{-1} = (1/2)20 = 10$$

$$50 x_1 x_2 = 10$$

or $x_1^0 x_2^0 = 1/5$. Hence there exist many solutions of this problem, in particular, any value of $x_1^0 \geq 2$ and $x_2^0 = 1/5\, x_1^0$.

In this example there exist alternate solutions; any one is a global minimum. Wilde and Beightler (1967) point out that in such situations when solutions exist that satisfy

$$u_{0i} = \delta_{0i} V(\delta) \qquad i = 1, 2, \ldots, n$$

then the set of solutions of this system need not be optimal solutions; in particular, one must examine every member of the set to determine the optimal solution.

A procedure is given later that yields an equivalent primal problem such that all constraints are necessarily binding at the optimal solution; hence $\delta_{ki}^0 > 0$ for all k and i.

8.6 Main Theorem of Geometric Programming

When the objective function and constraint functions of a problem are posynomials (functions with arbitrary real exponents, positive coefficients, and positive independent variables) restricted by positive upper bound coefficients, geometric programming provides a powerful procedure for determining its solution. Here, the general (prototype) geometric programming problem GP is defined as

$$\begin{aligned} &\text{minimize} \quad g_0(\mathbf{x}) \\ &\text{subject to} \quad g_k(\mathbf{x}) \le 1 \qquad k = 1, 2, \ldots, m \\ &\qquad\qquad\quad x_j > 0 \qquad j = 1, 2, \ldots, n \end{aligned}$$

where $g_k(\mathbf{x}) \triangleq \sum_{i=1}^{T_k} c_{ki} \prod_{j=1}^{n} x_j^{a_{kij}}$ $(k = 0, 1, \ldots, m)$, and each $c_{ki} > 0$ $(i = 1, 2, \ldots, T_k,\ k = 0, 1, \ldots, m)$ and T_k denotes the number of terms in $g_k(\mathbf{x})$.

Associated with the primal prototype geometric problem is a dual problem. This transformed problem is easier to solve than the primal since the constraints are linear. The dual of GP is GPD

$$\begin{aligned} &\text{maximize} \quad V(\delta) = \prod_{k=0}^{m} \prod_{i=1}^{T_k} (c_{ki} w_{k0}/w_{ki})^{w_{ki}} \\ &\text{subject to} \quad \sum_{i=1}^{T_0} w_{0i} = 1 \quad \text{normality} \\ &\qquad \sum_{k=0}^{m} \sum_{i=1}^{T_k} a_{kij} w_{ki} = 0 \qquad j = 1, 2, \ldots, n \quad \text{orthogonality} \\ &\qquad w_{ki} \ge 0 \qquad k = 0, 1, \ldots, m \qquad i = 1, 2, \ldots, T_k \quad \text{positivity} \end{aligned}$$

where $w_{k0} = \sum_{i=1}^{T_k} w_{ki}$.

Clearly, if the total number of terms (w_{ki} associated with each) − total number of variables (one equation for each) − 1 (a normality equation) = 0, the solution is unique. This expression, called the *degree of difficulty*, gives an index of the difficulty in determining an optimal solution.

The main theorem of geometric programming relating the optimal solutions of GP and GPD with an arbitrary degree of difficulty is now given.

LEMMA 8.5. *Let* $\mathbf{x}^0$ *satisfy the constraints of the primal problem and let* $\mathbf{w}^0$ *satisfy the constraints of the dual problem; then* $\mathbf{x}^0$ *and* $\mathbf{w}^0$ *are optimal solutions, i.e.,* $g_0(\mathbf{x}^0) = V(\mathbf{w}^0)$ *if*

$$w_{0i}^0 = u_{0i}(\mathbf{x}^0)/g_0(\mathbf{x}^0) \qquad i = 1, 2, \ldots, T_0 \tag{8.22}$$

and for $w_{k0}^0 > 0$

$$w_{ki}^0/w_{k0}^0 = u_{ki}(\mathbf{x}^0) \qquad k = 1, 2, \ldots, m \qquad i = 1, 2, \ldots, T_k \tag{8.23}$$

This lemma is used later in establishing optimality conditions for condensed (posy)monomial geometric problems. Also, observe that conditions (8.22) and (8.23) were given previously in (8.10), (8.15), and (8.16).

Proof. Suppose $\mathbf{x}^0$ and $\mathbf{w}^0$ satisfy (8.22) and (8.23) in addition to the constraints of the primal and dual problems, GP and GPD, respectively. Then (8.22) implies

$$\begin{aligned} g_0(\mathbf{x}^0) &= u_{0i}(\mathbf{x}^0)/w_{0i}^0 \qquad i = 1, 2, \ldots, T_0 \\ &= \prod_{i=1}^{T_0} \left(c_{0i} \sum_{\ell=1}^{T_0} w_{0\ell}^0/w_{0i}^0 \right)^{w_{0i}^0} \prod_{j=1}^{n} (x_j^0)^{a_{0ij}\, w_{0i}^0} \end{aligned}$$

since $\sum_{i=1}^{T_0} w_{0i}^0 = 1$. Also, (8.23) implies

$$u_{ki}(\mathbf{x}^0) = w_{ki}^0/w_{k0}^0 = \delta_{ki}^0 > 0$$

Therefore, since

$$1 = \frac{u_{ki}(x^0)}{u_{ki}(x^0)}$$

and $u_{ki}(\mathbf{x}^0) = c_{ki}\prod_{j=1}^{n}(x_j^0)^{a_{kij}}$, it follows that for any w_{ki}^0 in (8.23)

$$1 = \left(u_{ki}(\mathbf{x}^0) \sum_{\ell=1}^{T_k} w_{k\ell}^0 / w_{ki}^0\right)^{w_{ki}^0} = \left(c_{ki} \sum_{\ell=1}^{T_k} w_{k\ell}^0 / w_{ki}^0\right)^{w_{ki}^0} \prod_{j=1}^{n} (x_j^0)^{a_{kij} w_{ki}^0}$$

Therefore, for $k = 1, 2, \ldots, m$ and $i = 1, 2, \ldots, T_k$

$$1 = \prod_{k=1}^{m} \prod_{i=1}^{T_k} \left(c_{ki} \sum_{\ell=1}^{T_k} w_{k\ell}^0 / w_{ki}^0\right)^{w_{ki}^0} \prod_{j=1}^{n} (x_j^0)^{\Sigma\Sigma a_{kij} w_{ki}^0}$$

Multiplying the above by $g_0(\mathbf{x}^0)$ and since $\mathbf{w}^0$ satisfies the dual constraints

$$g_0(\mathbf{x}^0) = \prod_{k=0}^{m} \prod_{i=1}^{T_k} \left(c_{ki} \sum_{\ell=1}^{T_k} w_{k\ell}^0 / w_{ki}^0\right)^{w_{ki}^0} = V(\mathbf{w}^0)$$

THEOREM 8.1. *Suppose the primal problem satisfies Slater's Condition and* $\mathbf{x}^0$ *solves the primal problem; then there exists a vector* $\mathbf{y}^0$ *such that*

$$\phi(\mathbf{x}^0, \mathbf{y}) \le \phi(\mathbf{x}^0, \mathbf{y}^0) \le \phi(\mathbf{x}, \mathbf{y}^0) \qquad \forall \mathbf{x} > 0 \qquad \forall \mathbf{y} \ge 0$$

where $\phi(\mathbf{x}, \mathbf{y}) = g_0(\mathbf{x}) + \mathbf{y}'(\mathbf{g}(\mathbf{x}) - 1)$. *Moreover, there is a maximizing vector* $\mathbf{w}$ *for the dual problem whose components are*

$$w_{ki} = \frac{c_{0i} \prod_{j=1}^{n} (x_j^0)^{a_{0ij}}}{g_0(\mathbf{x}^0)} \qquad k = 0 \qquad i = 1, 2, \ldots, T_0$$

$$= \frac{y_k^0 c_{ki} \prod_{j=1}^{n} (x_j^0)^{a_{kij}}}{g_0(\mathbf{x}^0)} \qquad k = 1, 2, \ldots, m \qquad i = 1, 2, \ldots, T_k$$

Furthermore, $w_{k0} = y_k^0 / g_0(\mathbf{x}^0)$.

Proof. Let $x_j = e^{z_j}$; then $g_k(\mathbf{z}) = \Sigma c_{ki} \exp(\Sigma a_{kij} z_j)$ is a convex function provided each c_{ki} is positive. Hence under this transformation the primal problem is a convex programming problem. Therefore, from the equivalence theorem of nonlinear programming, Theorem 6.2, under Slater's Condition there exists a vector $\mathbf{y}^0$ which solves the associated saddle value problem.

Since the saddle value solution holds for $\mathbf{z}$ unrestricted and the Lagrangian is differentiable, we have

$$\frac{\partial \phi(\mathbf{z}^0, \mathbf{y}^0)}{\partial z_q} = 0 \qquad q = 1, 2, \ldots, n$$

or

$$\Sigma a_{0iq} c_{0i} \exp(\Sigma a_{0ij} z_j^0) + \sum_{k=1}^{m} y_k^0 [\Sigma a_{kiq} c_{ki} \exp(\Sigma a_{kij} z_j^0)] = 0$$

After dividing this expression by $g_0(\mathbf{z}^0)$ we have the components w_{ki}^0 which will satisfy the orthogonality condition of the dual problem

$$\begin{aligned} w_{ki}^0 &= \frac{c_{0i} \exp(\Sigma a_{0ij} z_j^0)}{g_0(\mathbf{z}^0)} & k = 0 \quad & i = 1, 2, \ldots, T_0 \\ &= \frac{y_k^0 c_{ki} \exp(\Sigma a_{kij} z_j^0)}{g_0(\mathbf{z}^0)} & k = 1, 2, \ldots, m \quad & i = 1, 2, \ldots, T_k \end{aligned}$$

Consider w_{ki}^0 when $k = 0$ and summing from $i = 1, 2, \ldots, T_0$, then $\mathbf{w}^0$ satisfies the normality condition. Hence the dual problem is feasible.

Now let $w_{k0}^0 = y_k^0 g_k(\mathbf{z}^0)/g_0(\mathbf{z}^0)$ for $k = 1, 2, \ldots, m$ and note that at $(\mathbf{z}^0, \mathbf{y}^0)$ we have $y_k^0(g_k(\mathbf{z}^0) - 1) = 0$; then

$$w_{k0}^0 = y_k^0 / g_0(\mathbf{z}^0) \qquad k = 1, 2, \ldots, m$$

Hence

$$\begin{aligned} w_{ki}^0 &= \frac{c_{0i} \exp(\Sigma a_{0ij} z_j^0)}{g_0(\mathbf{z}^0)} & k = 0 \quad & i = 1, 2, \ldots, T_0 \\ &= w_{k0}^0 c_{ki} \exp(\Sigma a_{kij} z_j^0) & k = 1, 2, \ldots, m \quad & i = 1, 2, \ldots, T_k \end{aligned}$$

Hence from the main lemma of geometric programming (Lemma 8.5), $g_0(\mathbf{x}^0) = V(\mathbf{w}^0)$; thus $\mathbf{x}^0$ and $\mathbf{w}^0$ are optimal solutions.

8.7 Shortcomings

The presentation up to this point has been based on some strong assumptions, namely,

(i) the number of terms − the number of unknowns − 1 = 0
(ii) associated with each x_j is a positive exponent as well as a negative exponent
(iii) $c_i > 0$ for all i

Wilde and Passy show how to overcome shortcomings (ii) and (iii), as outlined in Section 8.8. The problem of more terms than variables could yield solutions which are local optima, and in this situation the necessary conditions presented so far are not sufficient. Techniques to handle the cases when there are more terms than variables are given by Duffin, Peterson, and Zener (1967). The following three examples are presented to demonstrate the problems that arise when assumptions (i), (ii), and (iii) do not hold.

EXAMPLE 8.7. (more terms than variables)

Consider the problem

$$\begin{aligned} &\text{minimize} \quad x_1^2 + x_2^2 - x_1x_2 + 2x_1^{-1} + 2x_2 \\ &\text{subject to} \quad x_1, x_2 > 0 \end{aligned}$$

The normality and orthogonality conditions that determine the appropriate weights are

$$\begin{aligned} 2\delta_1 \qquad\quad + \delta_3 - \delta_4 \qquad\quad &= 0 \\ 2\delta_2 + \delta_3 \qquad\quad + \delta_5 &= 0 \\ \delta_1 + \delta_2 + \delta_3 + \delta_4 + \delta_5 &= 1 \end{aligned}$$

i.e., a system of three equations with five unknowns. Hence we have a system with many solutions and the selection of the ideal weights is not immediately apparent.

EXAMPLE 8.8. (only positive exponents)

Consider the problem

$$\begin{aligned} &\text{minimize} \quad (1/3)x^3 + (1/2)x^2 - 2x \\ &\text{subject to} \quad x > 0 \end{aligned}$$

The appropriate weights must satisfy the following system of equations

$$\begin{aligned} 3\delta_1 + 2\delta_2 + \delta_3 &= 0 \\ \delta_1 + \delta_2 + \delta_3 &= 1 \end{aligned}$$

but this system of equations is inconsistent. Hence there does not exist an appropriate vector δ, even though this problem has an optimal solution $x^0 = 1$.

EXAMPLE 8.9. (negative coefficients)

Assume we have the problem

$$\begin{aligned}&\text{minimize} \quad 4x - 1/x\\ &\text{subject to} \quad x > 0\end{aligned}$$

The normality and orthogonality conditions are satisfied by $\delta_1 = 1/2$ and $\delta_2 = 1/2$ as can be easily verified. However, to determine the optimum we must take the square root of a negative quantity, specifically

$$V(\delta) = (8)^{1/2}(-2)^{1/2}$$

We now outline Wilde and Passy's procedure which overcomes the problems involved with negative coefficients and all positive (or negative) exponents. Also, in Section 8.8 we introduce signum multipliers which can take on only two values, ± 1.

The theoretical derivation of the procedure is omitted; refer to Passy and Wilde (1967).

8.8 Wilde-Passy Method

The general constrained nonlinear programming problem is expressed as

$$\begin{aligned}&\text{minimize} \quad g_0(\mathbf{x}) = \textstyle\sum_{i=1}^{T_0} c_{0i} \prod_{\ell=1}^{n} x_\ell^{a_{0i\ell}}\\ &\text{subject to} \quad g_k = \textstyle\sum_{i=1}^{T_k} c_{ki} \prod_{\ell=1}^{n} x_\ell^{a_{ki\ell}}\\ &\qquad\qquad 0 < \sigma_k g_k^{\sigma_k} \le 1 \qquad \sigma_k = \pm 1 \qquad k = 1, 2, \ldots, m\end{aligned}$$

(A) The w_{ki} must satisfy the orthogonality conditions

$$\sum_{k=0}^{m} \sum_{i=1}^{T_k} \sigma_{ki} a_{ki\ell} w_{ki} = 0 \qquad \ell = 1, 2, \ldots, n$$

where each $\sigma_{ki} = 1$ and has the same sign as the coefficient c_{ki}. In a sense the new weights are now $\sigma \mathbf{w}$.

(B) The second condition is the normality condition, now modified to avoid the problem associated with negative coefficients. In particular, determine $\sigma \mathbf{w}$ so that there exists $\mathbf{w} \ge 0$ for either $\sigma = 1$ or $\sigma = -1$ in $\sum_{i=1}^{T_0} \sigma_{0i} w_{0i} = \sigma$.

(C) After determining the vector $\mathbf{w}$, determine the multiplier λ_{k0} from

$$\lambda_{k0} = \sigma_k \sum_{i=1}^{T_k} \sigma_{ki} w_{ki} \qquad k = 1, 2, \ldots, m$$

(D) Determine optimum

$$g_{\min} = \sigma \left[\prod_{k=0}^{m} \prod_{i=1}^{T_k} (c_{ki}\lambda_{k0}/\sigma_{ki}w_{ki})^{\sigma_{ki}w_{ki}} \right]^{\sigma}$$

where $\lambda_{00} = 1$.

(E) Determine the vector $\mathbf{x}^0$ by solving the following system of equations

$$c_{0i} \prod_{\ell=1}^{n} x_\ell^{a_{0i\ell}} = \sigma_{ki} w_{ki} \sigma g_{\min} \qquad i = 1, 2, \ldots, T_0$$

$$c_{ki} \prod_{\ell=1}^{n} x_\ell^{a_{ki\ell}} = (w_{ki}/\lambda_{k0})\sigma_{ki} \qquad k = 1, 2, \ldots, m \qquad i = 1, 2, \ldots, T_k$$

To illustrate the procedure, assume we have the following problem

$$\begin{aligned} \text{minimize} \quad & -2x_1^3 + x_2 x_3^3 \\ \text{subject to} \quad & -2x_1x_2 + 2x_2x_3 \geq 2 \\ & x_1, x_2, x_3 > 0 \end{aligned}$$

The constraint is expressed equivalently as $-1(x_1x_2 - x_2x_3)^{-1} \leq 1$. Hence $\sigma_1 = -1$. The remaining signum multipliers are

$$\sigma_{01} = -1 \qquad \sigma_{02} = 1 \qquad \sigma_{11} = 1 \qquad \sigma_{12} = -1$$

determined by considering the signs of c_{01}, c_{02}, c_{11}, and c_{12}, respectively. Note that the signum multipliers are incorporated so each term of the geometric mean is transformed to a positive quantity. For example, the first term of the geometric mean has been changed from $(-2x_1^3/\delta_{01})^{\delta_{01}}$ to $(2x_1^3/w_{01})^{-w_{01}}$.

With respect to the orthogonality and normality conditions, the $\mathbf{w}$ vector must satisfy the following system

$$\begin{aligned} -3w_{01} \qquad\qquad & + w_{11} \qquad\qquad = 0 \\ w_{02} & + w_{11} - w_{12} = 0 \\ 3w_{02} & \qquad\quad\; - w_{12} = 0 \\ -w_{01} + w_{02} & \qquad\qquad\qquad = \sigma \end{aligned}$$

The **w** vector is negative for $\sigma = -1$; and $w_{01} = 2$, $w_{02} = 3$, $w_{11} = 6$, and $w_{12} = 9$ for $\sigma = 1$.

Steps C and D of the procedure are now easily computed as

$$\begin{aligned}\lambda_{10} &= \sigma_1(\sigma_{11}w_{11} + \sigma_{12}w_{12}) = 3\\ g_{\min} &= \sigma[(2/w_{01})^{-w_{01}}(1/w_{02})^{w_{02}}(\lambda_{10}/w_{11})^{w_{11}}(\lambda_{10}/w_{12})^{-w_{12}}]^{\sigma}\\ &= 10.95\end{aligned}$$

If any component of **w** were negative, then we would again encounter the dilemma of taking the square root of a negative quantity.

It remains to determine the optimal solution $\mathbf{x}^0$. From step E, $\mathbf{x}^0$ is determined by solving the following system of equations

$$\begin{aligned}-2x_1^3 &= \sigma_{01}w_{01}\sigma\,(10.95) = -21.9\\ x_2x_3^3 &= \sigma_{02}w_{02}\sigma\,(10.95) = 32.8\\ x_1x_2 &= \sigma_{11}(w_{11}/\lambda_{10}) = 2\\ -x_2x_3 &= \sigma_{12}(w_{12}/\lambda_{10}) = -3\end{aligned}$$

and we have $x_1^0 = 2.16$, $x_2^0 = .925$, and $x_3^0 = 3.29$.

8.9 Nongeometric Problems

A nongeometric programming problem can be transformed into a geometric programming problem. This is illustrated by the following example.

EXAMPLE 8.10

$$\begin{aligned}&\text{minimize} \quad g(\mathbf{x})\\ &\text{subject to} \quad \mathbf{x} > 0\end{aligned}$$

where $g(x_1, x_2) = x_1^{-1/2}x_2^{1/8} + (4/5\,x_1^{1/2}x_2^{2/3} + 2/5\,x_1^{1/3}x_2)^{1/2}x_1^{1/4}x_2^{-1/2}$. This expression is not a posynomial, but it can be transformed to an equivalent geometric problem

$$\begin{aligned}&\text{minimize}\; x_1^{-1/2}x_2^{1/8} + x_3^{1/2}x_1^{1/4}x_2^{-1/2}\\ &\text{subject to} \quad 4/5\,x_3^{-1}x_1^{1/2}x_2^{2/3} + 2/5\,x_3^{-1}x_1^{1/3}x_2 \le 1\\ &\qquad\qquad x_j > 0 \qquad j = 1, 2, 3\end{aligned}$$

by letting $x_3 \ge 4/5\,x_1^{1/2}x_2^{2/3} + 2/5\,x_1^{1/3}x_2$.

8.10 Relationship to Linear Programming

Let us consider a special class of geometric programming problems in which the objective function and each constraint has only a single term, i.e., of the form

$$\begin{array}{lll} \text{minimize} & g_0(\mathbf{x}) & \\ \text{subject to} & g_k(\mathbf{x}) \le 1 & k = 1, 2, \ldots, m \\ & x_j > 0 & j = 1, 2, \ldots, n \end{array} \tag{8.24}$$

where $g_k(\mathbf{x}) = c_k \prod_{j=1}^n x_j^{a_{kj}}$ and $c_k > 0$. Letting $C_k = \ln c_k$ and $z_j = \ln x_j$, problem (8.24) can be expressed as

$$\begin{array}{lll} \text{minimize} & C_0 + \sum_{j=1}^n a_{0j} z_j & \\ \text{subject to} & \sum_{j=1}^n a_{kj} z_j \le -C_k & k = 1, 2, \ldots, m \end{array} \tag{8.25}$$

It is clearly a typical linear programming problem and the solution of either of these two problems yields an optimal solution of the other.

Now consider the dual of (8.24) with w_k defined to be w_{ki}

$$\begin{array}{lll} \text{maximize} & c_0 \prod_{k=1}^m (c_k)^{w_k} & \\ \text{subject to} & \sum_{k=0}^m a_{kj} w_k = 0 & j = 1, 2, \ldots, n \\ & w_0 = 1 & \\ & w_k \ge 0 & k = 1, 2, \ldots, m \end{array} \tag{8.26}$$

Note that maximizing $c_0 \prod_{k=1}^m (c_k)^{w_k} \equiv$ maximizing $C_0 + \sum_{k=1}^m w_k \ln c_k$ over the above orthogonality, normality, and positivity constraints. But this problem is clearly the dual of (8.25). Hence, if any one of the above problems has an optimal solution, then all four problems have optimal solutions.

EXAMPLE 8.11

Consider the (posy)monomial geometric problem

$$\begin{array}{ll} \text{minimize} & x_1^{-1} x_2 \\ \text{subject to} & x_1^2 \le 1 \\ & x_2 \le 1 \\ & x_1 x_2^{-1} \le 1 \\ & x_1, x_2 \ge 0 \end{array} \tag{8.27}$$

This problem clearly has an optimal solution $\mathbf{x}^0 = \begin{bmatrix} 1 \\ 1 \end{bmatrix}$; moreover, its associated linear programming problem, under the natural log transformation, has an optimal solution $\mathbf{z}^0 = \begin{bmatrix}0\\0\end{bmatrix}$. In particular, consider the problem

$$\begin{array}{lrl} \text{minimize} & -z_1 + z_2 & \\ \text{subject to} & 2z_1 & \leq 0 \\ & z_2 & \leq 0 \\ & z_1 - z_2 & \leq 0 \end{array} \tag{8.28}$$

The dual of (8.28) is

$$\begin{array}{lrl} \text{maximize} & 0'\mathbf{y} & \\ \text{subject to} & 2y_1 \quad + y_3 & = 1 \\ & y_2 - y_3 & = -1 \\ & y_1, y_2, y_3 & \geq 0 \end{array} \tag{8.29}$$

Considering the geometric dual of (8.27), we have

$$\begin{array}{lrl} \text{maximize} & \Pi_{i=0}^{3} (c_i)^{w_i} & \\ \text{subject to} & -w_0 + 2w_1 \qquad\quad + w_3 & = 0 \\ & w_0 \qquad\quad + w_2 - w_3 & = 0 \\ & w_0 & = 1 \\ & w_1, \quad w_2, \quad w_3 & \geq 0 \end{array} \tag{8.30}$$

Letting $w_i = y_i$ ($i = 1, 2, 3$), we see that the feasible regions given in (8.29) and (8.30) are equivalent.

The above relationship can be used to solve any (posy)monomial geometric program. In the remainder of this chapter we consider how to condense or decompose any geometric problem into an equivalent monomial problem.

First consider how each coefficient c_{ki} (assumed to be positive) can be normalized.

8.11 Normalization of Coefficients

Both notationally and computationally it is advantageous to normalize all coefficients to unity.

It is shown below how each c_{ki} can be replaced with a new variable $x_{n+1}^{a_{ki}} > 0$ for an appropriate selection of a_{ki} when x_{n+1} is restricted to satisfy a new constraint of the form

$$x_{n+1} = C \tag{8.31}$$

for an appropriate selection of C.

Note that if x_{n+1} is fixed, then there exists a_{ki} such that

$$c_{ki} = x_{n+1}^{a_{ki}} \quad \text{or} \quad \ln c_{ki} = a_{ki} \ln x_{n+1} \tag{8.32}$$

Also, since $x_{n+1} = C$ for some C, then in view of (8.32)

$$a_{ki} \ln C = \ln c_{ki} \quad \text{or} \quad a_{ki} = \ln c_{ki} / \ln C$$

Given any geometric problem we can normalize each c_{ki} to unity by replacing each coefficient with $x_{n+1}^{a_{ki}}$ subject to the new restriction that $x_{n+1}/C \leq 1$. When this constraint is tight, the normalized problem is equivalent to the original problem. However, in view of the consistency needed in the dual problem, we initially consider the orthogonality restrictions. For $j = n + 1$ we require that there exists $\mathbf{w}^*$ such that

$$\sum_{k=0}^{m} \sum_{i=1}^{T_k} a_{kij} w_{ki}^* = 0$$

This condition, together with the normality restriction, reveals that there must exist both a strictly positive a_{ki} and a strictly negative a_{ki} for x_{n+1}. Otherwise $w_{ki}^* = 0$ for all k, i and the dual problem is inconsistent in view of the normality condition. Hence, to guarantee both a negative and positive exponent for x_{n+1}, C should be selected accordingly:

(i) if $c_{ki} > 1$ for all i and k, let $C \in (0, 1)$
(ii) if $c_{ki} < 1$ for all i and k, let $C \in (1, \infty)$
(iii) otherwise, $C \in (0, \infty)(C \neq 1)$

8.12 Objective Function Decomposition

The decomposition of the objective function into (posy)monomial form is easily accomplished using the following well-known transformation where it is known that $g_0(\mathbf{x}^0) > 0$

$$\text{minimize} \quad g_0(\mathbf{x}) \Longleftrightarrow \text{minimize} \quad x_0$$
$$\text{subject to} \quad x_0^{-1} g_0(\mathbf{x}) \leq 1$$

This transformation simply replaces the objective function with an additional independent variable x_0 constrained to be at least as large as the objective function. We show later how the above constraint is condensed into monomial form.

Let us first consider the situation where the objective function is to be maximized. Assume in the following problem that the objective function is a signomial subject to prototype posynomial constraints (the difference of two posynomials is a *signomial*).

Problem A

$$\begin{aligned} &\text{maximize} && g_0(\mathbf{x}) && && \\ &\text{subject to} && g_k(\mathbf{x}) \le 1 && k = 1, 2, \ldots, m && (8.33) \\ & && x_j > 0 && j = 1, 2, \ldots, n && (8.34) \\ & && g_k(\mathbf{x}) \triangleq \textstyle\sum_{i=1}^{T_k} c_{ki} \prod_{j=1} x_j^{a_{kij}} && k = 1, 2, \ldots, m && (8.35) \end{aligned}$$

where the exponents a_{kij} denote arbitrary real numbers and the coefficients c_{ki} for $k \neq 0$ are assumed to be positive. Also, let

$$g_0(\mathbf{x}) \triangleq g_{01}(\mathbf{x}) - g_{02}(\mathbf{x}) \qquad (8.36)$$

denote the difference of two posynomials of the type in (8.35). Furthermore, assume that Problem A has an optimal solution with positive program value denoted by M_A.

Problem A differs from a prototype geometric problem in that the objective function is to be maximized. A procedure, different from the method developed by Duffin and Peterson (1973), is now presented whereby Problem A can be expressed as an equivalent prototype geometric problem.

Problem B

Using the following transformation we have

$$\text{maximize} \quad g_0(\mathbf{x}) \Longleftrightarrow \begin{array}{ll} \text{minimize} & x_0 \\ \text{subject to} & g_0(\mathbf{x}) \ge x_0^{-1} \end{array} \qquad (8.37)$$

Consider now a reformulation of the lower-bound inequality (8.37) and note that it can be expressed equivalently, in view of (8.36), as

$$g_{01}(\mathbf{x}) \ge x_0^{-1} + g_{02}(\mathbf{x}) \qquad (8.38)$$

This constraint is clearly binding at the optimum. Under our assumption that $0 < M_A$, we can proceed to condense (8.38) into equivalent prototype constraints.

Assumption I. There exists a posynomial function of the primal variables denoted as

$$\theta(\mathbf{x}) \triangleq c \prod_{j=1}^{n} x_j^{b_j}$$

where b_j are real numbers such that $\Theta(\mathbf{x})$ is an upper bound on (8.38) for all $(x_0, \mathbf{x})$ in the feasible region of Problem B; hence $\Theta(\mathbf{x})$ is such that

$$x_0^{-1} + g_{02}(\mathbf{x}) \leq g_{01}(\mathbf{x}) \leq \Theta(\mathbf{x}) \quad \text{for all } (x_0, \mathbf{x}) \tag{8.39}$$

subject to (8.33), (8.34), and (8.37).

Consider now the following system

$$g_{01}(\mathbf{x})/\Theta(\mathbf{x}) \leq 1 \tag{8.40}$$

$$1/x_0\Theta(\mathbf{x}) + g_{02}(\mathbf{x})/\Theta(\mathbf{x}) \leq 1 \tag{8.41}$$

LEMMA 8.6. *Assume Assumption I is satisfied. If* $(x_0, \mathbf{x})$ *is a feasible solution of Problem B, then* $(x_0, \mathbf{x})$ *satisfies (8.40) and (8.41).*

Proof. Assume that $(x_0, \mathbf{x})$ is a feasible solution of Problem B. Then there exists a positive quantity x_{n+1}^* such that

$$(x_0^*)^{-1} + g_{02}(\mathbf{x}^*) \leq g_{01}(\mathbf{x}^*) = x_{n+1}^*$$

However, under Assumption I there exists some posynomial $\Theta(\mathbf{x})$ such that $x_{n+1}^* \leq \Theta(\mathbf{x}^*)$. Therefore $(x_0, \mathbf{x})$ satisfies (8.40) and (8.41).

LEMMA 8.7. *If* $(x_0, \mathbf{x})$ *satisfies (8.33), (8.34), and (8.39), then* $(x_0, \mathbf{x})$ *satisfies (8.38).*

Proof. This follows immediately from (8.39).

LEMMA 8.8. *There exists* $\mathbf{x}$ *in the positive orthant of* E^{n+1} *and* $\Theta(\mathbf{x})$ *such that (8.40) and (8.41) hold with equality.*

Proof. If $(x_0^0, \mathbf{x}^0)$ solves Problem B, then (8.38) is binding. Hence for any $\Theta(\mathbf{x})$ such that $\Theta(\mathbf{x}^0) = g_{01}(\mathbf{x}^0)$, constraints (8.40) and (8.41) are binding; i.e., select $b_j = 0$ $(j = 2, 3, \ldots, n)$ with c and b_1 such that $c(x_1^0)^{b_1} = g_{01}(\mathbf{x}^0)$.

Consider now the following problem

Problem C

minimize x_0
subject to (8.33), (8.34), (8.40), (8.41)

LEMMA 8.9. *Assume Assumption I is satisfied. If there exists* $(\bar{x}_0, \bar{\mathbf{x}})$ *that solves Problem C and satisfies (8.39), then* $\bar{\mathbf{x}}$ *solves Problem A.*

Proof. Assume that $(\bar{x}_0, \bar{\mathbf{x}})$ is not an optimal solution of Problem B and let $(\hat{x}_0, \hat{\mathbf{x}})$ denote the optimal solution associated with Problem B. Now in view of Lemma 8.7, $(\bar{x}_0, \bar{\mathbf{x}})$ satisfies (8.38). Therefore $(\bar{x}_0, \bar{\mathbf{x}})$ is a feasible solution of Problem B and it follows that

$$\hat{x}_0 < \bar{x}_0 \tag{8.42}$$

However, if $(\hat{x}_0, \hat{\mathbf{x}})$ solves Problem B, then in view of Lemma 8.6, $(\hat{x}_0, \hat{\mathbf{x}})$ is a feasible solution of Problem C. This implies that $\bar{x}_0 \leq \hat{x}_0$, which contradicts (8.42).

Thus a reformulation of Problem A to Problem C can be made under Assumption I. The above procedure is also valid when the objective function is a posynomial. In this situation, $g_{02}(\mathbf{x})$ is deleted in the reformulation.

Also, the procedure presented here constitutes a framework within which a certain class of constrained and unconstrained, except for positivity, optimization problems can be viewed as an equivalent prototype geometric problem. For example, consider the problem

$$\begin{aligned} &\text{maximize} \quad -(1/3)x^3 - (1/2)x^2 + 2x \\ &\text{subject to} \quad x > 0 \end{aligned}$$

The optimal solution is clearly $x^0 = 1$. An ideal posynomial to select is $\Theta(x) = 2x$. Hence we have the following equivalent prototype geometric problem

$$\begin{aligned} &\text{minimize} \quad x_0 \\ &\text{subject to} \quad (1/2)x^{-1}x_0^{-1} + (1/6)x^2 + (1/4)x \leq 1 \\ &\qquad\qquad\qquad\qquad\qquad\qquad x_0, x > 0 \end{aligned}$$

which has a solution $\bar{x}_0 = 6/7$ and $\bar{x} = 1$.

The following two examples are also applicable to our presentation

$$\begin{aligned} &\text{maximize} \quad 2x_1^2 + 2x_2^2 \\ &\text{subject to} \quad x_1^2 + x_2^2 \leq 1 \\ &\qquad\qquad\quad x_1, \quad x_2 > 0 \end{aligned}$$

In this situation a sufficient $\Theta(\mathbf{x})$ is $(1/2)x_1^{-2}x_2^{-2}$. Also, for the problem

maximize x^2
subject to $x \in (0, 1]$

let $\Theta(x) = x^b$ for any $b \in [0, 2]$.

The above presentation shows that when $g_{01}(\mathbf{x})$ is a monomial, $\Theta(\mathbf{x}) = g_{01}(\mathbf{x})$ is ideal to ensure the equivalence of Problems A, B, and C.

8.13 Prototype Posynomial Decomposition

Consider the prototype posynomial constraint of the form

$$g(\mathbf{x}) = u_1 + u_2 + \cdots + u_m \leq 1 \tag{8.43}$$

where $u_i = \Pi_{j=1}^{n} x_j^{a_{ij}}$. It will be shown that a parametric condensation of (8.43) to a system of monomial constraints can be achieved with a decomposition by pairs.

In view of (8.43), there exists a new variable $x_{n+1} > 0$ such that

$$u_1 + u_2 \leq x_{n+1} + \sum_{i=3}^{m} u_i \leq 1 \tag{8.44}$$

Since each u_i is a posynomial, (8.44) can be written equivalently as

$$\frac{u_1}{x_{n+1}} + \frac{u_2}{x_{n+1}} \leq 1 \tag{8.45}$$

$$x_{n+1} + \sum_{i=3}^{m} u_i \leq 1 \tag{8.46}$$

Now consider the (posy)binomial constraint (8.45) and note that for some $\epsilon \in (0, 1)$ when this constraint is binding

$$\frac{u_1}{x_{n+1}} + \frac{u_2}{x_{n+1}} \leq 1 \Longleftrightarrow \begin{array}{l} \left(\dfrac{1}{\epsilon}\right) \dfrac{u_1}{x_{n+1}} \leq 1 \\ \left(\dfrac{1}{1-\epsilon}\right) \dfrac{u_2}{x_{n+1}} \leq 1 \end{array}$$

Hence, (8.45) is condensed into the desired (posy)monomial form for some ϵ.

Similarly, we can further decompose (8.46) intc two constraints with at least one in (posy)monomial form for some ϵ^*. Iteratively the procedure transforms a pair of positive terms into a set of monomial constraints. The procedure terminates when the remaining equation (8.46) is binomial (or monomial). The remaining constraint can be futher decomposed if necessary.

8.14 Active Constraints

In general, it is not known which constraints, if any, are tight at the optimal $\mathbf{x}^0$ prior to the solution of the problem. In the light of an algorithm to be introduced later, a procedure to ensure tightness is now given.

Consider any posynomial constraint, say

$$g(\mathbf{x}^0) \leq 1 \tag{8.47}$$

then there exists a new variable $x_{n+2} \geq 1$ such that (8.47) can be expressed equivalently as

$$x_{n+2}\, g(\mathbf{x}^0) = 1 \tag{8.48}$$

$$x_{n+2}^{-1}/\epsilon \leq 1 \qquad \text{for } \epsilon \in (0, 1] \tag{8.49}$$

Moreover, note that for some $\epsilon \in (0, 1]$, equation (8.49) is an equality. At this point (8.48) can be decomposed into an equivalent (posy)monomial form appealing to the procedure given in Section 8.13.

8.15 The Equivalent Geometric Program and Optimality

In this section, necessary and sufficient optimality conditions for the equivalent (posy)monomial geometric program are given, based in part on the main duality theorem of geometric programming. A necessary condition for convergence of any algorithm is that all constraints must be binding at the optimum. Therefore, it is shown that a monomial geometric program with all active constraints constitutes a sufficient condition for optimality.

The equivalent (condensed) geometric program EGP is stated to reflect the monomial form as follows

Program EGP

$$\begin{aligned} &\text{minimize} && x_0 \\ &\text{subject to} && g_k(\mathbf{x}, \epsilon) = w_k(\epsilon)\, \Pi_{j=0}^{N} x_j^{a_{kj}} \leq 1 \\ & && x_j > 0 \quad j = 0, 1, \ldots, N \\ & && \epsilon_k \in \Omega_{\epsilon_k} \quad k = 1, 2, \ldots, M \end{aligned}$$

Here, Ω_{ϵ_k} denotes the set of lower and upper bounds on ϵ_k, and $w_k(\cdot)$ denotes a strictly positive function where

$$\begin{aligned} w_k(\epsilon) &= 1/C \quad C > 0 \quad \text{normalization constraint} \\ &= 1 \quad \text{original tight monomial constraint} \\ &= (1/\epsilon_k) \text{ or } (1 - \epsilon_k)^{-1} \quad \text{condensed constraint} \end{aligned}$$

Investigating the dual objective function with $T_k = 1$ for all k, let $i = 0$ denote the primal objective function and $i = 1$ denote the normalization constraint; then we have

$$V(\mathbf{w}, \epsilon) = (1)^{w_0}(1/C)^{w_1} \prod_{k=2}^{M} w_k(\epsilon)^{w_k} \tag{8.50}$$

Since monotonicity of the logarithm function ensures that ln $[V(\mathbf{w})]$ has the same maximizing point as $V(\mathbf{w})$, we can express (8.50) equivalently as

$$w_1 \ln(1/C) + \sum_{k=2}^{M} \ln[w_k(\epsilon)]\, w_k \tag{8.51}$$

The dual problem EGD can therefore be stated as

Program EGD

$$\begin{aligned} &\text{maximize} && w_1 \ln(1/C) + \textstyle\sum_{k=2}^{M} \ln[w_k(\epsilon)]\, w_k \\ &\text{subject to} && w_0 = 1 \quad \text{normality} \\ & && w_k \geq 0 \quad k = 1, 2, \ldots, M \quad \text{positivity} \\ & && \textstyle\sum_{k=0}^{M} a_{kj} w_k = 0 \quad j = 0, 1, \ldots, N \quad \text{orthogonality} \end{aligned}$$

Observe that all coefficients $w_k(\epsilon)$ in EGD are isolated in the objective function. Moreover, for fixed $w_k(\epsilon)$, the dual problem is linear in the dual variables. An overview of optimality as it relates to the tightness of every constraint is now presented.

The main theorem of geometric programming (Lemma 8.5) establishes a set of sufficient conditions for optimality of EGP and EGD. For example, given that $\mathbf{x}^*$ and $\mathbf{w}^*$ satisfy the constraints of EGP and EGD, respectively, for fixed $w_k(\epsilon)$

$$\begin{aligned} g_0(\mathbf{x}^*) &= V(\mathbf{w}^*) \quad \text{if } w_0^* = g_0(\mathbf{x}^*)/g_0(\mathbf{x}^*) \\ w_k^*/w_k^* &= g_k(\mathbf{x}^*, \epsilon^*) \quad \text{for } w_k^* > 0 \qquad k = 1, 2, \ldots, M \end{aligned} \tag{8.52}$$

Hence if all constraints are binding, i.e., $g_k(\mathbf{x}^*, \epsilon^*) = 1$, $(k = 1, 2, \ldots, M)$, then (8.52) holds and optimality is guaranteed.

Since EGD can be stated as a linear programming problem for fixed weights, we can construct an algorithm based on finding the appropriate weights that satisfy the above necessary and sufficient conditions. Recall that the optimal criterion is established when, in any tableau of the simplex procedure, we have $z_j - c_j \geq 0$ for all j. Here, $z_j = \mathbf{c}_B'\mathbf{P}_j$ where $\mathbf{P}_j$ denotes the jth column in the tableau and $\mathbf{c}_B$ are the cost coefficients of the optimal solution. Furthermore, a by-product of the final tableau is the optimal solution of the associated dual linear programming problem. Here, $\mathbf{y}^{*\prime} = \mathbf{c}_B' B^{-1}$ where B is the optimal basis in the final tableau. Using these facts we can verify the following lemma

LEMMA 8.10. *Assume that in the final tableau with respect to EGD, $z_j - c_j = 0$ for all j; then all constraints of the associated dual linear programming problem evaluated at $\mathbf{y}^{*\prime} = \mathbf{c}_B' B^{-1}$ are binding.*

Proof. Let B denote the first m column vectors in A. Recall that the dual constraints have the form $A'\mathbf{y} = \mathbf{c}$. Then

$$\begin{aligned} \mathbf{y}^{*\prime} A &= \mathbf{y}^{*\prime}[B, \bar{B}] \\ &= \mathbf{c}_B' B^{-1}[B, \bar{B}] \\ &= \mathbf{c}_B'[\mathbf{P}_1, \mathbf{P}_2, \ldots, \mathbf{P}_n] \\ &= \mathbf{z}' = \mathbf{c}' \end{aligned}$$

Hence $A'\mathbf{y}^* = \mathbf{c}$; i.e., all constraints in the associated dual linear programming problem are binding when evaluated at $\mathbf{y}^*$.

In view of the results given earlier we recall that the constraints of the (posy)monomial geometric program are equivalent to the constraints of the dual of the linear programming problem EGD. Hence if one problem has all binding constraints at its optimum, then so does the other.

From these results it follows that we need a scheme to adjust the weights in some way so all the $z_j - c_j$ components are zero. The following section describes an algorithm based on the above optimality conditions.

8.15.1 Search Procedure

The optimum of the dual program EGD, for fixed $w_k(\epsilon)$, is $z = \mathbf{c}_B'\mathbf{w}_B$. We want to adjust the weights in some fashion that increases the value of the objective function $\mathbf{c}_B'\mathbf{w}_B$, restricts the weights to remain in Ω_{ϵ_k}, and adjusts the weights so that $z_j - c_j = 0$ for all j. Appealing to the method of feasible directions we denote a change in $w_k(\epsilon)$ by d_k. Then a change in z is

$$\sum_{k=1}^{N} \frac{\partial C_{Bk}}{\partial \epsilon_{Bk}} d_{Bk} w_{Bk} \tag{8.53}$$

Moreover, a change in $z_j - c_j$ is given by

$$\sum_{k=1}^{N} \frac{\partial C_{Bk}}{\partial \epsilon_{Bk}} d_{Bk} \bar{a}_{kj} - \frac{\partial C_j}{\partial \epsilon_j} \equiv d_j \tag{8.54}$$

where $\bar{a}_{kj}$ are the components of the jth nonbasic column. We want to determine a change in the weights d_{Bk} so that $z_j - c_j + d_j = 0$ for j nonbasic and $\epsilon_{Bk} + d_{Bk} \in \Omega_{\epsilon_k}$, and which will minimize an increase in $\mathbf{c}_B'\mathbf{w}_B$ in (8.53) subject to the restrictions on $\mathbf{d}$. The optimal d_{Bk} are determined from the following linear programming problem

$$\text{maximize} \quad -\textstyle\sum_{k=1}^{N} \dfrac{\partial C_{Bk}}{\partial \epsilon_{Bk}} d_{Bk} w_{Bk} \tag{8.55}$$

$$\text{subject to} \quad \textstyle\sum_{k=1}^{N} \left(\dfrac{\partial C_{Bk}}{\partial \epsilon_{Bk}} \bar{a}_{kj}\right) d_{Bk} = \dfrac{\partial C_j}{\partial \epsilon_j} - z_j + c_j$$

for j nonbasic and $\epsilon_{Bk} + d_{Bk} \in \Omega_{\epsilon_k}$. This problem can be solved via simple upper bound procedures.

Since the vector $\mathbf{d}$ only gives a feasible direction for increasing the optimum, we can compute the sum of squares of the updated $z_j - c_j$ and reduce $\mathbf{d}$ by a half until a reduction is obtained.

Summary of Procedure

(1) Select each ϵ_k so that $\epsilon_k \in \Omega_{\epsilon_k}$.
(2) Solve problem EGD for this given ϵ.
(3) Solve problem (8.55) to determine $w_k(\boldsymbol{\epsilon}^0)$. If all $\sum_j(z_j - c_j)^2$ are less than some tolerance, terminate; otherwise return to step (2).
(4) Derive the optimal $\mathbf{x}^0$ by solving the following set of equations

$$w_k(\epsilon^0) \prod_{j=1}^{N} x^{a_{kj}} = 1 \qquad k = 1, 2, \ldots, M$$

or the linear system

$$\sum_{j=1}^{N} a_{kj} \ln x_j = -\ln w_k(\epsilon^0) \qquad k = 1, 2, \ldots, M$$

EXAMPLE 8.12. (Duffin, Peterson, and Zener p. 88)

Find a vector $\mathbf{x}^0$ that minimizes

$$40x_1^{-1}x_2^{-1/2}x_3^{-1} + 20x_1x_3 + 20x_1x_2x_3 \tag{8.56}$$

$$\text{subject to} \quad (1/3)x_1^{-2}x_2^{-2} + (4/3)x_2^{1/2}x_3^{-1} \le 1$$
$$x_j > 0 \qquad j = 1, 2, 3$$

Normalization of Coefficients. Select some $x_4 > 0$ which replaces the coefficients of each term. Appealing to the criterion presented in Section 8.11, a possible selection is $x_4 = 40$. A statement of the equivalent normalized problem is given as follows

find a vector $\mathbf{x}^0$ that minimizes

$$x_1^{-1}x_2^{-1/2}x_3^{-1}x_4 + x_1x_3x_4^{a_1} + x_1x_2x_3x_4^{a_2} \tag{8.57}$$

$$\text{subject to} \quad g_1(\mathbf{x}) = (1/40)x_4 \le 1 \tag{8.58}$$
$$x_1^{-2}x_2^{-2}x_4^{a_3} + x_2^{1/2}x_3^{-1}x_4^{a_4} \le 1 \tag{8.59}$$
$$x_j > 0 \qquad j = 1, 2, 3, 4$$

Here

$$a_1 = a_2 = \ln 20/\ln 40 \sim 0.8$$
$$a_3 = \ln(1/3)/\ln 40 \sim -0.3$$
$$a_4 = \ln(4/3)/\ln 40 \sim 0.1$$

Clearly, if all constraints are binding at the optimum then equivalence has been maintained.

Objective Function Decomposition. From Section 8.12 we have the transformation

$$\text{minimize} \quad g_0(\mathbf{x}) \Longleftrightarrow \text{minimize} \quad \bar{g}_0(\mathbf{x}) = x_0 \tag{8.60}$$
$$\text{subject to} \quad g_0(\mathbf{x})x_0^{-1} \le 1 \tag{8.61}$$

Since (8.57) is a posynomial, then (8.61) is also a posynomial. According to the decomposition procedure of Section 8.13, we know there exists some x_5 $(0 < x_5 < 1)$ such that (8.61) can be expressed equivalently as

$$x_0^{-1}x_1^{-1}x_2^{-0.5}x_3^{-1}x_4x_5^{-1} + x_0^{-1}x_1x_3x_4^{0.8}x_5^{-1} \leq 1 \tag{8.62}$$

$$x_5 + x_0^{-1}x_1x_2x_3x_4^{0.8} \leq 1 \tag{8.63}$$

Condensing (8.62) into monomial form

$$\begin{aligned} g_2(\mathbf{x}, \epsilon) &= (1/\epsilon_1)x_0^{-1}x_1^{-1}x_2^{-0.5}x_3^{-1}x_4x_5^{-1} \leq 1 \\ g_3(\mathbf{x}, \epsilon) &= [1/(1 - \epsilon_1)]x_0^{-1}x_1x_3x_4^{0.8}x_5^{-1} \leq 1 \qquad \epsilon_1 \in (0, 1) \end{aligned}$$

Likewise, from (8.63) we have

$$\begin{aligned} g_4(\mathbf{x}, \epsilon) &= (1/\epsilon_2)x_5 \leq 1 \\ g_5(\mathbf{x}, \epsilon) &= [1/(1 - \epsilon_2)]x_0^{-1}x_1x_2x_3x_4^{0.8} \leq 1 \qquad \epsilon_2 \in (0, 1) \end{aligned}$$

Decomposition of Constraints. Appealing to Section 8.14 we proceed to introduce a variable $x_6 > 0$ that will force tightness in the constraint (8.59) when restricted as

$$g_6(\mathbf{x}, \epsilon) = (1/\epsilon_3)x_6^{-1} \leq 1$$

for some $\epsilon_3 \in (0, 1]$. Incorporating x_6 into (8.59) we have

$$x_1^{-2}x_2^{-2}x_4^{-0.3}x_6 + x_2^{0.5}x_3^{-1}x_4^{0.1}x_6 \leq 1 \tag{8.64}$$

Since (8.64) is already binomial, we can immediately state its equivalent

$$\begin{aligned} g_7(\mathbf{x}, \epsilon) &= (1/\epsilon_4)x_1^{-2}x_2^{-2}x_4^{-0.3}x_6 \leq 1 \\ g_8(\mathbf{x}, \epsilon) &= [1/(1 - \epsilon_4)]x_2^{0.5}x_3^{-1}x_4^{0.1}x_6 \leq 1 \qquad \epsilon_4 \in (0, 1) \end{aligned}$$

A statement of the resulting equivalent problem is given as follows

find a vector $\mathbf{x}^0$ that minimizes the function $g_0(\mathbf{x}) = x_0$
subject to

$$\begin{aligned} g_1(\mathbf{x}) &= (1/40)x_4 \leq 1 \\ g_2(\mathbf{x}, \epsilon) &= (1/\epsilon_1)x_0^{-1}x_1^{-1}x_2^{-0.5}x_3^{-1}x_4x_5^{-1} \leq 1 \\ g_3(\mathbf{x}, \epsilon) &= [1/(1 - \epsilon_1)]x_0^{-1}x_1x_3x_4^{0.8}x_5^{-1} \leq 1 \\ g_4(\mathbf{x}, \epsilon) &= (1/\epsilon_2)x_5 \leq 1 \\ g_5(\mathbf{x}, \epsilon) &= [1/(1 - \epsilon_2)]x_0^{-1}x_1x_2x_3x_4^{0.8} \leq 1 \\ g_6(\mathbf{x}, \epsilon) &= (1/\epsilon_3)x_6^{-1} \leq 1 \\ g_7(\mathbf{x}, \epsilon) &= (1/\epsilon_4)x_1^{-2}x_2^{-2}x_4^{-0.3}x_6 \leq 1 \\ g_8(\mathbf{x}, \epsilon) &= [1/(1 - \epsilon_4)]x_2^{0.5}x_3^{-1}x_4^{0.1}x_6 \leq 1 \\ x_j &\in (0, \infty) \qquad j = 0, 1, \ldots, 5 \\ \epsilon_k &\in (0, 1) \qquad k = 1, 2, 4 \\ \epsilon_3 &\in (0, 1] \end{aligned}$$

Accordingly, we have the following dual problem
find a vector $\mathbf{w}^0$ that maximizes the function

$$V(\delta, \epsilon) = \ln(1/40)w_1 + \ln(1/\epsilon_1)w_2 + \ln[1/(1 - \epsilon_1)]w_3 + \ln(1/\epsilon_2)w_4 + \ln[1/(1 - \epsilon_2)]w_5 + \ln(1/\epsilon_3)w_6 + \ln(1/\epsilon_4)w_7 + \ln[1/(1 - \epsilon_4)]w_8$$

subject to $w_i \geq 0 \quad i = 1, \ldots, 8$

From x_0 $\quad w_2 + w_3 + w_5 = 1$

from x_1 $\quad -w_2 + w_3 + w_5 - 2w_7 = 0$

from x_2 $\quad -0.5w_2 + w_5 - 2w_7 + 0.5w_8 = 0$

from x_3 $\quad -w_2 + w_3 + w_5 - w_8 = 0$

from x_4 $\quad w_1 + w_2 + 0.8w_3 + 0.8w_5 - 0.3w_7 + 0.1w_8 = 0$

from x_5 $\quad -w_2 - w_3 + w_4 = 0$

from x_6 $\quad -w_6 + w_7 + w_8 = 0$

$$\epsilon_k \in (0, 1) \quad k = 1, 2, 4 \quad \epsilon_3 \in (0, 1]$$

Utilizing the solution procedure given in Section 8.15, the following results are obtained

Iteration	$\sum_{j=1}^{8}(z_j - c_j)^2$
1	$8.17(10^8)$
2	$4.03(10^5)$
3	$9.21(10^2)$
4	$3.01(10^{-1})$
5	$7.84(10^{-4})$

The optimal solution $\mathbf{x}^{0'} = (x_0^0, x_1^0, \ldots, x_6^0)$ is

$$\mathbf{x}^{0'} = (100.01, 1, 1, 2.03, 40, 0.6, 1)$$

with program value $g_0(\mathbf{x}^0) = 100.01$. These results comply with the results obtained by Duffin, Peterson, and Zener (p. 88).

Note that the degree of difficulty associated with the original problem (8.56) is $1 = 5 - 3 - 1$ and the degree of difficulty of the condensed problem is 1. This observation is always true in general.

8.15.2 *Alternative Search Procedure*

Let the matrix A of the dual problem be of size $n^* \times k$. The nullity matrix of A is the set of vectors that spans the null space of A, i.e., the set of vectors $\bar{\delta}$ such that $A\bar{\delta} = 0$. This matrix is commonly denoted as $N(A)$.

Lemma 8.10 was used to establish an important optimality condition on the cost vector of EGD. An alternate condition is given in the following lemma.

LEMMA 8.11. *Assume that the cost vector associated with EGD is orthogonal to N(A); then*

(a) $z_j - c_j = 0$ *for all* j
(b) all the constraints of the program EGP are binding

Proof. The nullity matrix is constructed from the final tableau associated with program EGD. Without loss of generality assume that this tableau is composed of B and $\bar{B}$ where B denotes the first n^* columns of A. Then $N(A) = \left[\begin{array}{c}-\bar{B}\\ \hline I\end{array}\right]$. Moreover, $z_j = \mathbf{c}_B'\mathbf{P}_j$ where $\mathbf{P}_j$ is a column in $\bar{B}$. Letting $\bar{\mathbf{P}}_j$ be a column in $N(A)$, then by hypothesis $\mathbf{c}'N(A) = 0$ or $\mathbf{c}_B'\mathbf{P}_j - c_j = 0$ for j nonbasic, and $z_j - c_j = 0$ for all j. Result (*b*) follows from Lemma 8.10.

The above lemma gives us a necessary condition on ϵ_k. Numerical techniques such as Gauss-Seidel, or root finding procedures such as false positioning, etc., can be used to determine the ϵ_k which satisfy Lemma 8.11. These procedures are not covered here.

Consider again the dual objective function associated with (8.56). Lemma 8.11 states that the ϵ_k are such that

$$[\ln 40, \ln \epsilon_1, \ln(1 - \epsilon_1), \ln \epsilon_2, \ln(1 - \epsilon_2), \ln \epsilon_3, \ln \epsilon_4, \ln(1 - \epsilon_4)]'N(A) = 0$$

where $N(A)' = (1.65, -0.5, .25, -.25, .25, 1.5, 0.5, 1)$. As can be verified, the optimal ϵ_k are $\epsilon_1 = .333$, $\epsilon_2 = .6$, $\epsilon_3 = 1$, and $\epsilon_4 = .331$.

Exercises

1. Let $U = Ax^n + Bx^{-m}y^{-a} + Cy^b$ $(A, B, C, n, m, a, b > 0)$.
   a) What are the relative contributions of the first three terms to the minimum value of U?
   b) What is the minimum value of U?
   c) Let $A, B, C = 1$. What is the optimal solution of the problem

   minimize U
   subject to $x, y > 0$

2. Show that if f is concave, then it follows from the basic definition of a concave function that Jensen's inequality holds, i.e.

$f(E[x]) \geq E[f(x)]$

where E is the expectation operator corresponding to a distribution over a finite number of points.

Solve the following five problems (subject to $x > 0$) via geometric programming:

3. minimize $x^3 + 3/x$
4. minimize $4x_1 + x_1/x_2^2 + 4x_2/x_1$
5. maximize $-4x_1/x_2 - 10x_1^2x_2 - 4/x_1$
6. minimize $x_1x_2^{-2} + 4x_2x_1^{-1}$
   subject to $4x_1^{-1} \geq 1$
7. minimize $x_1x_2^{-2} + 4x_2x_1^{-1}$
   subject to $2x_1 \leq 1$
8. Discuss the difficulties in solving the following problem

   minimize $2x^3 + 6/x$
   subject to $x \leq 1 \quad x > 0$

9. Discuss a possible method of attacking Exercise 8 in view of Exercise 3.
10. Determine the maximum value of $x_1x_2^2x_3^3$

    subject to $x_1^3 + x_2^2 + x_3 \leq 1$
    $x_1, \quad x_2, \quad x_3 > 0$

    a) by appealing to Assumption I
    b) by using the fact that the maximum of a monomial is equal to the minimum of the reciprocal of the same monomial

References

Avriel, M., and Wilde, D. J. 1967. Optimal condenser design by geometric programming. *Ind. Eng. Chem. Process Des. Dev.* 6:256–63.

Avriel, M., and Williams, A. C. 1970. Complementary geometric programming. *SIAM J. Appl. Math.* 19:125–41.

Ben-Israel, A., and Pascual, L. D. 1971. Vector-valued criteria in geometric programming. *ORSA* 19:98–104.

Charnes, A., and Cooper, W. W. 1966. A convex approximate method for non-convex extensions of geometric programming. *Proc. Nat. Acad. Sci. USA*, vol. 56, no. 5.

Duffin, R. J. 1970. Linearizing geometric programs. *SIAM Rev.* 12:211–27.

Duffin, R. J., and Peterson, E. L. 1972a. The proximity of (algebraic) geometric programming to linear programming. *J. Math. Program.* 3:250–53.

———. 1972b. Reversed geometric programs treated by harmonic means. *Ind. Univ. Math. J.* 22:531–50.

———. 1973. Geometric programming with signomials. *J. Optim. Theory Appl.* 11:3–25.

Duffin, R. J.; Peterson, E. L.; and Zener, C. 1967. *Geometric Programming.* Wiley, New York.

McMillan, C. 1970. *Mathematical Programming.* Wiley, New York.

Oleson, G. K. 1971. Computational aspects of geometric programming. Diss. Iowa State Univ., Ames.

Pascual, L. D., and Ben-Israel, A. 1970. Constrained maximization of posynomials by geometric programming. *J. Optim. Theory Appl.* 5:73–80.

Passy, U. 1972. Condensing generalized polynomials. *J. Optim. Theory Appl.* 9:221–37.

Passy, U., and Wilde, D. J. 1967. Generalized polynomial optimization. *SIAM J. Appl. Math.* 15(5):1344–56.

Trzeciak, J. M. 1974. A parametric decomposition of geometric programs. Master's thesis. Iowa State Univ., Ames.

Wilde, D. J., and Beightler, C. S. 1967. *Foundations of Optimization.* Prentice-Hall, Englewood Cliffs, N.J.

Zener, C. 1971. *Engineering Design by Geometric Programming.* Wiley, New York.

Zoutendijk, G. 1960. *Methods of Feasible Directions.* Elsevier, Amsterdam.

APPENDIX A

LINEAR PROGRAMMING COMPUTATIONS WITH THE MPSX/MPS-III SYSTEM

In the previous chapters, we discuss how the simplex algorithm is utilized to determine the optimal solution of a linear programming model. Even though special computational techniques such as the revised simplex and bounded variable algorithms were discussed to minimize the number of calculations, it is quite apparent that as the number of constraints and variables in the model increase so will the amount of calculations. Hence to minimize the errors performed in hand calculations and to obtain the solution of the linear programming model more feasibly, we usually resort to so-called computer "package" programs.

Many packages are on the market. Described in this appendix is the IBM Mathematical Programming System, MPSX, or a system quite similar to MPSX, MPS-III, capable of giving the user all the options and features outlined in the first three chapters. For example, the options of sensitivity analysis, range analysis, and parametric programming are available in MPSX. A complete description of the MPSX System is found in Mathematical Programming System 360 (Sh20-0968-1) Linear and Separable Programming—User's Manual. Other procedures are available but are not discussed here.

Hence the purpose of this appendix is to indicate and describe some of the options and procedures of MPSX and how to set up the input data and the control program to convey the user's solution strategy to MPSX. The control program is composed of a set of procedures or commands for solving a linear programming problem in an orderly fashion.

A.1 Data Format

The input format of MPSX consists of two types of cards, indicator cards and data cards. The indicator cards specify the type of data that is to follow; as shown in Table A.1, some of these are NAME, ROWS, COLUMNS, RHS, RANGES, BOUNDS, and ENDATA. Each card should always be punched so its first character is in column 1. Each indicator card specifies a certain command and identifies a section or block of data cards to the MPSX system. Each data card consists of 80 columns and is

Table A.1. Input Format

1	Field 1 2	3		Field 2 5 — 12	Field 3 15 — 22	Field 4 25 — 36	Field 5 40 — 47	Field 6 50 — 61
N	A	M	E		data set name			
*	←				C O	M M E N T	C A R D S	→
R	O N G L E	W	S	row name				
C	O	L	U	M N S column name	row name 1	value 1	row name 2	value 2
R	H	S		rhs name	row name 1	value 1	row name 2	value 2
R	A	N	G	E S range name	row name 1	value 1	row name 2	value 2
B	O L U F M P F	U O P X I L R	N	D S bound row name	column name	value	N O T	U S E D
E	N	D	A	T A				

divided into 6 fields as indicated in Table A.1. All name fields should be left-justified and no name can begin with $ or contain blanks.

Comment cards can be placed anywhere in the data stream by putting an asterisk (*) in column 1.

A.2 Organization of Input

A NAME card must always be the first card specified in the input stream and an ENDATA must always be the last card to end processing of the data. The NAME card should also have a user-specified name in field 3.

The ROWS, COLUMNS, and RHS sections are required, whereas the RANGES and BOUNDS sections are optional and need not be declared or specified.

A.2.1 Rows

The row section defines the type of row constraints incorporated in the model. There are four indicators used in field 1 to identify the type of row in the model followed by some user-specified name in field 2:

Nb or bN	Objective function row (or no constraint until declared later)
Gb or bG	Greater than or equal to constraint
Lb or bL	Less than or equal to constraint
Eb or bE	Equality constraint

Field 1 consists of two columns, one of which must be blank (b).

A.2.2 Columns

The column section specifies the restriction coefficients and the names of each structural variable. The column name is given in field 2 and the row names are specified in field 3 (and field 5 if the user wants to place two a_{ij} on an input card). Fields 4 and 6, respectively, declare the value of a_{ij}. Since the MPSX system originally treats all coefficients in the model as zero, the user need only declare nonzero entries in the data stream.

Note that the matrix elements in the column section are specified by column. Hence, once a column name is specified all other nonzero entries in that column *must* be declared before another column is defined or mentioned.

A.2.3 RHS

The resource vector (or vectors) is specified beginning in field 2. The input format is basically the same as defined in the column section; however a unique RHS name must be declared for each resource vector defined. The additional vectors are used to solve a similar problem after an original problem is solved. We discuss this later.

A.2.4 Ranges

The range section is used to condense the input data. This is done when a constraint is both a less-than or equal-to inequality and a greater-than or equal-to inequality. The original row need be specified only once in the row and column sections with only one of its lower or upper limits $b(i)$ specified in the RHS section.

The range is used as specified in the following table, where $r(i)$ is the range on row i given in the range section

Type of row	Sign of $b(i)$	Resultant upper limit on the constraint	Resultant lower limit on the constraint
G	+	$b(i) + r(i)$	$b(i)$
L	+	$b(i)$	$b(i) - r(i)$
E	+	$b(i) + r(i)$	$b(i)$
E	−	$b(i)$	$b(i) - r(i)$

If $r(i)$ is negative for a G row or an L row, the absolute value is used.

A.2.5 Bounds

This section is used to place bounds on certain structural variables when bounds are desired. If no bounds are specified, then the entire section can be eliminated. When bounds are not declared they are automatically set at 0 and $+\infty$.

Field 1 specifies the type of bound to be imposed on an activity. The following indicators are used:

LO	Lower bound
UP	Upper bound
FX	Fixed value
MI	Lower bound of $-\infty$
PL	Upper bound of $+\infty$
FR	Free variable ($-\infty$ to ∞)

Fields 3 and 4 specify the name of the variable to be bounded and its associated finite bound, respectively. Leave field 4 blank if the bound is infinity. Field 2 identifies a name associated with a set of bounds. There can be more than one set of bounds declared in the bound section which is similar to the options for defining multiple resource vectors in the RHS section.

A.3 Sample Problem (blending problem)

Suppose we are required to blend the following ingredients in such a way that the given restrictions are satisfied.

	Analysis			
Ingredient	Protein	Fat	Fibre	Cost per pound
Barley	.115	.02	.06	$.025
Corn	.086	.038	.025	.028
Screenings	.13	.03	.07	.022
Salt				.011

Restrictions

Total weight of blend = 2,000 lb.
Total protein ≥ 200 lb.
Total fat ≥ 54 lb.
Total fibre ≤ 90 lb.
Amount of corn between 400 and 1,000 lb.
Weight of salt used = 5 lb.
Total cost of ingredients to be a minimum.

These restrictions can be expressed as

(0) MIN[.025(BARLY) + .028(CORN) + .022(SCRNX) + 0.11(SALT)]
(1) 1.0(BARLY) + 1.0(CORN) + 1.0(SCRNX) + 1.0(SALT) = 2,000
(2) .115(BARLY) + .086(CORN) + .13(SCRNX) ≥ 200
(3) .02(BARLY) + .038(CORN) + .03(SCRNX) ≥ 54
(4) .06(BARLY) + .025(CORN) + .07(SCRNX) ≤ 90
(5) 1.0(CORN) ≥ 400
(6) 1.0(CORN) ≤ 1,000
(7) 1.0(SALT) = 5

or for ease of representation

Row name	Row type	BARLY	CORN	SCRNX	SALT	RHS ZZ1
COST		.025	.028	.022	.011	
R1	=	1.0	1.0	1.0	1.0	2,000
R2	≥	.115	.086	.13		200
R3	≥	.02	.038	.03		54
R4	≤	.06	.025	.07		90
R5	≥		1.0			400
R6	≤		1.0			1,000
R7	=				1.0	5

We shall set up the input in two ways, one without the range and bound sections and the other with these sections. In the second formulation of the problem the total weight of the blend is restricted to be between 1,999 and 2,001 pounds.

The two examples for listing the input follow. Note the sections must be specified in the order shown in Table A.1.

EXAMPLE A.1

```
Column 1   5         15        30
       NAME          INPUT1
       ROWS
        N  COST
        E  R1
        G  R2
        G  R3
        L  R4
        G  R5
        L  R6
        E  R7
       COLUMNS
           BARLY     COST        .025
           BARLY     R1         1.0
           BARLY     R2          .115
           BARLY     R3          .02
           BARLY     R4          .06
           CORN      COST        .028
           CORN      R1         1.0
           CORN      R2          .086
           CORN      R3          .038
           CORN      R4          .025
           CORN      R5         1.0
           CORN      R6         1.0
           SCRNX     COST        .022
```

EXAMPLE A.1 (continued)

```
Column  1       5          15          30
        COLUMNS
                SCRNX      R1           1.0
                SCRNX      R2            .13
                SCRNX      R3            .03
                SCRNX      R4            .07
                SALT       COST          .011
                SALT       R1           1.0
                SALT       R7           1.0
        RHS
                ZZ1        R1        2000.0
                ZZ1        R2         200.0
                ZZ1        R3          54.0
                ZZ1        R4          90.0
                ZZ1        R5         400.0
                ZZ1        R6        1000.0
                ZZ1        R7           5.0
        ENDATA
```

EXAMPLE A.2

```
Column  1       5          15          30
        NAME               INPUT2
        ROWS
         N      COST
         E      R1
         G      R2
         G      R3
         L      R4
        COLUMNS
                BARLY      COST          .025
                BARLY      R1           1.0
                BARLY      R2            .115
                BARLY      R3            .02
                BARLY      R4            .06
                CORN       COST          .028
                CORN       R1           1.0
                CORN       R2            .086
                CORN       R3            .038
                CORN       R4            .025
                SCRNX      COST          .022
                SCRNX      R1           1.0
                SCRNX      R2            .13
```

EXAMPLE A.2 (continued)

```
Column 1           5          15            30
       COLUMNS cont.
                   SCRNX      R3              .03
                   SCRNX      R4              .07
                   SALT       COST            .011
                   SALT       R1             1.0
       RHS
                   ZZ1        R1          1999.0
                   ZZ1        R2           200.0
                   ZZ1        R3            54.0
                   ZZ1        R4           90.0
       RANGES
                   TOL        R1             2.0
       BOUNDS
        LO         B1         CORN         400.0
        UP         B1         CORN        1000.0
        FX         B1         SALT           5.0
       ENDATA
```

A.4 Control Program

In the previous section we outline the proper input format needed to specify completely any linear programming model. Two examples are coded. In this section the control program is discussed. The control program is a set of procedures specified by the user to define the strategy to be used in solving the model.

To exhibit a typical set of procedures, consider Example A.1 given in the last section. Table A.2 is a control program for this example.

TABLE A.2. Control Program (Ex. A.1)

```
PROGRAM
INITIALZ
MOVE(XDATA, 'INPUT1')
MOVE(XPBNAME,'SAMPLE')
CONVERT
SETUP('MIN')
MOVE(XRHS, 'ZZ1')
MOVE(XOBJ, 'COST')
PRIMAL
SOLUTION
EXIT
PEND
```

Each card defines one procedure or command and can begin in any column, provided that column 71 is the last one used.

The first command in the control program, PROGRAM, indicates to the MPSX system that the program is to follow and a listing of all coding errors, if any, is desired. If any errors in the program are found, the system will terminate before the input data is read. If the user wants to continue processing the data regardless of what undecodable statements are found, the first command should be PROGRAM('ND'). This command is a necessity when the user is defining a procedure unknown to the MPSX system. An example of this situation occurs when using MPSX's READCOMM (not discussed in this appendix).

The second command, INITIALZ, establishes initial settings of all tolerances at their standard values. The next two statements move the name of the data INPUT1 into the MPSX cell XDATA, and move the problem file name SAMPLE into the cell XPBNAME. These two cells must be defined before such procedures as CONVERT and SETUP are indicated in the control program.

The CONVERT statement instructs the MPSX system to check the input data for proper specifications and convert the data into internal representation onto the PROBFILE device with a problem name of SAMPLE. If a report of the number of elements of each row and column is desired, the command should be CONVERT('SUMMARY').

The next instruction SETUP('MIN') allocates memory space within the computer and adds appropriate slack variables. The added parameter 'MIN' specifies that the problem is to be minimized (if the objective function is to be maximized, the parameter used is 'MAX'). Since there may exist many objective functions or resource vectors in the input stream, the following two instructions specify which vectors are relevant. In this example, the resource vector ZZ1 is moved into MPSX cell XRHS and the objective function name COST is moved into the cell XOBJ.

PRIMAL instructs MPSX to apply a variant of the simplex algorithm to solve the problem. Two other commands, DUAL or OPTIMIZE, can be used in MPSX. OPTIMIZE provides an excellent strategy to solve linear

TABLE A.3. Job Control Cards

```
//JOB CARD
//STEP1 EXEC MPSX
//MPSCOMP.SYSIN DD *
(CONTROL PROGRAM)
/*
//MPSEXEC.SYSIN DD *
(DATA)
/*
```

programming models, especially if the number of structural activities is larger than 400.

The command SOLUTION is used to output the solution. After the solution is given, the commands EXIT and PEND terminate the program and turn over control to the IBM system. Hence, Table A.2 completely defines a set of commands or procedures to execute Example A.1.

Table A.3 outlines a typical set of job control cards that instruct the computer where to find the MPSX program. The control program follows the //MPSCOMP.SYSIN DD * card and the input data follows the //MPSEXEC.SYSIN DD * card.

Consider Example A.2 in which we imposed range and bound sections in our data stream. To communicate with MPSX so the input data will be properly executed we only need to add some additional parameters to the SETUP command in Table A.2, in particular

```
SETUP('MIN', 'RANGE', 'TOL', 'BOUNDS', 'B1')
```

which implies that in the range and bound sections the vectors TOL and B1 will be used in solving the problem. If these parameters are omitted the solution is obtained without these restrictions.

The above procedures are usually sufficient or at least basic in processing a linear programming model. Other commands can be added depending on the particular situation.

A.5 Additional Commands

To ensure that the proposed model is defined or coded correctly the user can use the BCDOUT and/or PICTURE procedures. These commands can appear anywhere after the CONVERT statement. The BCDOUT instruction reconverts the user-specified binary problem back to the standard input data format and lists it accordingly, two coefficients per line. PICTURE creates a picture of the magnitude of the nonzero coefficients indicated by alphabetic code. Table A.4 gives a sample printout of Example A.1 after PICTURE is called. These two commands are useful in finding coding errors before a problem is executed. Unfortunately, if a problem is originally specified improperly, there may not exist an optimal solution; i.e., the problem may be infeasible. The command TRACE is used to find binding constraints in the model. If this command is desired it will give an analysis only when an infeasible solution is reached. Upon reaching an infeasible solution the computer output will contain an asterisk (*) to the left of any constraint that cannot be satisfied.

TABLE A.4. PICTURE (Ex. A.1)

						Z
		X	X	X	X	Z
		1	2	3	4	1
COST	N	U	U	U	U	
R1	E	1	1	1	1	D
R2	G	T	U	T		C
R3	G	U	U	U		B
R4	L	U	U	U		B
R5	G		1			C
R6	L		1			C
R7	E				1	A

SUMMARY OF MATRIX

SYMBOL	RANGE			COUNT (INCL.RHS)
Z	LESS THAN		.000001	
Y	.000001	THRU	.000009	
X	.000010		.000099	
W	.000100		.000999	
V	.001000		.009999	
U	.010000		.099999	11
T	.100000		.999999	2
1	1.000000		1.000000	7
A	1.000001		10.000000	1
B	10.000001		100.000000	2
C	100.000001		1,000.000000	3
D	1,000.000001		10,000.000000	1
E	10,000.000001		100,000.000000	
F	100,000.000001		1,000,000.000000	
G		GREATER THAN	1,000,000.000000	

MINIMUM = .110000E−01 MAXIMUM = .200000E+04

As noted in Chapter 3, often one wants to solve one problem and then solve a similar problem without starting from step one of the simplex algorithm. To accomplish this use the instructions SAVE and RESTORE. The SAVE procedure saves the optimal basis and is usually placed in the control program after the PRIMAL command. The RESTORE instruction brings back in the solution the vectors saved using the SAVE command. In effect we really utilize B^{-1}. The RESTORE statement is placed after the second problem is specified and before PRIMAL is called.

TABLE A.5a. Multiple RHS

```
            PROGRAM
            INITIALZ
            MOVE(XDATA, 'INPUT2')
            MOVE(XPBNAME, 'SAMPLE')
            CONVERT
            SETUP('MIN', 'RANGE', 'TOL', 'BOUNDS', 'B1')
            MOVE(XOBJ, 'COST')
            MVADR(ALL, FIRST)
            GOTO(LOOP)
MORE        RESTORE
            TALLY(COUNT, LOOP)
            GOTO(OUT)
LOOP        MVIND(XRHS, ALL, 3)
            PRIMAL
            SAVE
            SOLUTION
            ALL=ALL+3
            GOTO(MORE)
OUT         EXIT
COUNT       DC(2)
ALL         DC(0)
FIRST       DC('ZZ1', 'ZZ2', 'ZZ3')
            PEND
```

The above SAVE and RESTORE procedures are extremely efficient when solving problems with multiple objective functions or right-hand vectors. Let us return to Example A.2 and assume we have three resource vectors. ZZ1, ZZ2, and ZZ3. Table A.5a lists a control program handling such a series of problems. Additional procedures are added only to shorten the program. The names used in column 1 are reference names. For example, GOTO(LOOP) instructs MPSX to branch to LOOP which solves the problem with a new RHS vector. References COUNT, ALL, and FIRST are declared DC to be initialized to 2, 0, and the names of the resource vectors, respectively.

Let us consider each new procedure that is defined. MVADR(ALL, FIRST) puts in location ALL the address of the first member of FIRST. Then MVIND(XRHS,ALL,3) moves the next three bytes, or characters, into cell XRHS. This gives us the first resource vector. After this problem is solved and the optimal basis is saved, then ALL is increased by 3. This in effect gives MPSX the address of ZZ2 or the fourth character in FIRST. GOTO-(MORE) branches us back to restore the basis. Now since we only want to solve three similar problems, a procedure TALLY is used which tests COUNT with zero and then subtracts one from it. If the result is nonzero, MPSX branches to LOOP; otherwise MPSX branches to OUT and terminates.

TABLE A.5b. **Multiple Problems**

```
              PROGRAM
              INITIALZ
              MVADR(ALL2, FIRST)
              MVADR(ALL1, DNAME)
              MVADR(ALL3, COST)
              GOTO(LOOP)
MORE          TALLY(COUNT, LOOP)
              GOTO(OUT)
LOOP          MVIND(XDATA, ALL1, 6)
              MVIND(XRHS, ALL2, 3)
              MVIND(XOBJ, ALL3, 2)
              MOVE(XPBNAME, 'SAMPLE')
              CONVERT
              SETUP('MIN')
              PRIMAL
              SOLUTION
              ALL1=ALL1+6
              ALL2=ALL2+3
              ALL3=ALL3+2
              GOTO(MORE)
OUT           EXIT
COUNT         DC(2)
ALL1          DC(0)
ALL2          DC(0)
ALL3          DC(0)
FIRST         DC('ZZ1', 'ZZ2', 'ZZ3')
COST          DC('C1', 'C2', 'C3')
DNAME         DC('INPUT1', 'INPUT2', 'INPUT3')
              PEND
```

The same procedures are used to solve n linear programming models. A control program is given in Table A.5b which solves three linear programming problems with XDATA names of INPUT1, INPUT2, and INPUT3; associated objective functions C1, C2, and C3; and right-hand sides of ZZ1, ZZ2, and ZZ3.

A.6 Computer Output

In this section we discuss the output important (or of interest) to most users, in particular, the iteration log and the solution output.

Table A.6 gives an iteration log or a summarized account at each step of the simplex algorithm (tableau) for Example A.1. The heading tells us we are using the procedure PRIMAL with objective function COST and resource vector ZZ1. In the iteration log the vectors are given serial numbers. The first eight vectors correspond to the slack and surplus vari-

TABLE A.6. Iteration Log

MPSX–PTF9 EXECUTOR. MPSX RELEASE 1 MOD LEVEL 3

PRIMAL OBJ = COST RHS = ZZ1

TIME = 0.01 MINS. PRICING 7

SCALE =

	ITER NUMBER	NUMBER INFEAS	VECTOR OUT	VECTOR IN	REDUCED COST	SUM INFEAS
M	1	4	7	10	3.58222–	2244.44
	2		3	11	2.26993–	101.582
	3		8	12	2.00000–	93.6509
	4		11	9	.08811–	2.93115
M	5	0	2	3	.76365–	

FEASIBLE SOLUTION

PRIMAL OBJ = COST RHS = ZZ1

TIME = 0.01 MINS. PRICING 7

SCALE = 1.00000

	ITER NUMBER	NUMBER NONOPT	VECTOR OUT	VECTOR IN	REDUCED COST	FUNCTION VALUE
M	6	2	5	11	.08773–	51.3400

OPTIMAL SOLUTION

ables associated with the rows defined in the row section, i.e., the objective function row and seven constraints. The serial numbers for the structural variables then are defined to be $m + 2$ to $m + n + 1$ where m is the number of constraints in the model and n is the number of structural variables. Hence, in this example, x_1 = BARLY = $9, \ldots, x_4$ = SALT = 12. In the first iteration, we see that variable 10 enters the solution and replaces variable 7 (slack variable for row R6). Also, in the initial tableau there exist four artificial variables in the pseudo-objective function. The sum of infeasibilities is 2,244.44 at iteration 1 and 0.0 at iteration 5, implying that at this point the problem has a feasible solution. At the sixth iteration the problem is optimal with a functional value of 51.340.

Table A.7 gives the output, decomposed into the row and column sections. The same serial number index is used as described above. The row sections give the value of each row evaluated at $\mathbf{x}^0$ and the corresponding slack for each restriction. These are listed under ACTIVITY and SLACK ACTIVITY, respectively. The next two columns give the lower and upper limits (if any) imposed on each restriction. The last column in section 1 is the dual activity describing the rate of change in the optimum per unit change in the resource availability (for less-than or equal-to constraints it describes the rate of change if the resource is decreased

TABLE A.7. Output

Section 1. Rows

NUMBER	. . . ROW . .	AT	. . . ACTIVITY . . .	SLACK ACTIVITY	. . LOWER LIMIT .	. . UPPER LIMIT .	. DUAL ACTIVITY
1	COST	BS	51.34000	51.34000–	NONE	NONE	1.00000
2	R1	EQ	2000.00000	.	2000.00000	2000.00000	.04300–
3	R2	BS	208.37500	8.37500–	200.00000	NONE	.
4	R3	BS	63.20000	9.20000–	54.00000	NONE	.
5	R4	UL	90.00000	.	NONE	90.00000	.30000
6	R5	BS	1000.00000	600.00000–	400.00000	NONE	.
7	R6	UL	1000.00000	.	NONE	1000.00000	.00750
8	R7	EQ	5.00000	.	5.00000	5.00000	.03200

Section 2. Columns

NUMBER	. COLUMN .	AT	. . . ACTIVITY . . .	. . INPUT COST . .	. . LOWER LIMIT .	. . UPPER LIMIT .	. REDUCED COST .
9	X1	BS	465.00000	.02500	.	NONE	.
10	X2	BS	1000.00000	.02800	.	NONE	.
11	X3	BS	530.00000	.02200	.	NONE	.
12	X4	BS	5.00000	.01100	.	NONE	.

one unit; for greater-than or equal-to constraints it describes the rate of change of the optimum if the resource level is increased one unit).

The last column of section 2 gives the change in the optimum that would result if one unit of a new activity is forced in the solution. Under the column labeled AT we have a two-character code denoting the status of a row (or column) activity, namely,

BS in the basis and feasible
** in the basis and infeasible
EQ nonbasic, artificial or fixed
UL nonbasic, activity at upper limit
LL nonbasic, activity at lower limit

The column section is almost identical to the row section with a column of cost coefficients defined for each variable.

A.7 Parametric Programming

The MPSX parametric programming procedure is now described. Assume for illustration purposes that we are interested in what happens if the price of barley and corn is allowed to increase at the same rate. Hence we want to consider the objective function

$$C(\theta) = (.025 + \theta)x_1 + (.028 + \theta)x_2 + .022x_3 + 0.11x_4 \qquad \text{for all } \theta > 0$$

In the input stream, we have to define an additional row, CHANGE. Adding the following cards to our input stream

```
X1    CHANGE    1.0
X2    CHANGE    1.0
```

and defining the new row as N in the row section are the only modifications needed in the input.

The only thing remaining to be altered is the control program; we need only to define after PRIMAL

(1) the change row
(2) the lower value of θ, initially zero
(3) the upper value of θ
(4) the step sizes by which θ will be increased

Suppose we want to consider $\theta \in [0, .05]$ in steps of .01. Then the control program will have the following cards after PRIMAL:

```
MOVE(XCHROW,'CHANGE')
XPARAM = 0.0
XPARMAX = .05
XPARDELT = .01
PARAOBJ
SOLUTION
.
.
.
```

A.8 Range Output

After an optimal solution has been reached the procedure RANGE is used to give a postoptimal or sensitivity analysis of the objective function coefficients and resource levels. The computer output of the range analysis consists of four sections. Sections 1 and 3 contain information for rows at their lower or upper limits and intermediate levels, respectively. Similarly, section 2 contains information of activities at their lower or upper levels and section 4 of activities at their intermediate levels.

Let us consider Table A.8 which consists of rows at their lower or upper limits. These are obtained by rows R1, R4, R6, and R7; hence the slack for these rows is zero. Columns 1 to 6 are self-explanatory. Columns 1 and 2 give the numeric and alphabetic names of the rows at their bounds; columns 3 and 4 give the status of these rows evaluated at $\mathbf{x}^0$; column 6 specifies the lower and upper limits given for these constraints. In column 7, LOWER ACTIVITY gives the lower limit to which the resource can be decreased at a reduction in profit per unit given by the corresponding UNIT COST of column 8. Below this limit, the change in profit per unit of resource (shadow price) will change. The UPPER ACTIVITY in column 7 shows the upper limit to which the resource level can be increased at the implicit cost given in column 8. Column 9 is not used in this section. If the values specified in the lower and upper activities are exceeded (as shown in column 7), the column labeled LIMITING PROCESS designates which activity enters the solution (basis).

Sections 3 and 4 are given in Tables A.9 and A.10. Section 2 for this example does not exist since no bound section was specified in the input data.

Section 3 has the same interpretation as section 1, but the unit costs are the reduced costs associated with the structural variables. Columns 7 and 8 therefore imply, for instance, that if b_2 (protein) cost 20 cents, 200.4249 units would be used; or the resource level of protein can be moved to 200.4249 given a reduced cost of 20 cents for every unit of this

TABLE A.8. Section 1. Rows at Limit Level

..ROW..		AT	...ACTIVITY...	SLACK ACTIVITY	.LOWER LIMIT. .UPPER LIMIT.	LOWER ACTIVITY UPPER ACTIVITY	.UNIT COST.. .UNIT COST..	UPPER COST. LOWER COST	LIMITING PROCESS.	AT AT
					1999.99915	1933.57061	.04300–		X1	LL
2	R1	EQ	1999.99915	.	1999.99915	2088.33246	.04300		X3	LL
					NONE	84.69999	.30000		X3	LL
5	R4	UL	89.99999	.	89.99999	94.64999	.30000–		X1	LL
					NONE	848.57074	.00750		X3	LL
7	R6	UL	999.99920	.	999.99920	1103.33245	.00750–		X1	LL
					5.00000		.03200		X4	LL
8	R7	EQ	5.00000	.	5.00000	71.42854	.03200–		X1	LL

TABLE A.9. Section 3. Rows at Intermediate Level

...ROW		AT	.ACTIVITY..	SLACK ACTIVITY	.LOWER LIMIT. .UPPER LIMIT.	LOWER ACTIVITY UPPER ACTIVITY	.UNIT COST.. .UNIT COST..	.UPPER COST. .LOWER COST.	LIMITING PROCESS.	AT AT
3	R2	BS	208.37490	8.37500–	199.99990	200.42490	.20000		R4	UL
					NONE	208.37490	INFINITY		NONE	
4	R3	BS	63.19999	9.20000–	53.99999	55.17428	.14151		R6	UL
					NONE	63.19999	INFINITY		NONE	
6	R5	BS	999.99895	599.9994–	399.99958	848.57049	.00750		R6	UL
					NONE	999.99895	INFINITY		NONE	

TABLE A.10. Section 4. Columns at Intermediate Level

. COLUMN .		AT	. ACTIVITY .	. . INPUT COST . .	. LOWER LIMIT . . UPPER LIMIT .	LOWER ACTIVITY UPPER ACTIVITY	. UNIT COST . . . UNIT COST . .	. UPPER COST . . . LOWER COST . .	LIMITING PROCESS .	AT AT
9	X1	BS	464.99987	.02500	.	464.99987	INFINITY	INFINITY	NONE	
					NONE	1146.42811	.00167	.02333	R6	UL
10	X2	BS	999.99920	.02800	.	848.57076	.00750	.03550	R6	UL
					NONE	999.99920	INFINITY	INFINITY–	NONE	
11	X3	BS	529.99999	.02200	.	77.54695–	.00214	.02414	R6	UL
					NONE	529.99999	INFINITY	INFINITY–	NONE	
12	X4	BS	5.00000	.01100	.	5.00000	INFINITY	INFINITY	NONE	
					NONE	5.00000	INFINITY	INFINITY–	NONE	

resource used. The second entry in column 7 for this example is the activity (level) of R2 at $\mathbf{x}^0$.

The computer output of sections 2 and 4 contains information for columns at their lower and upper limits and intermediate levels. The first five columns in these sections are self-explanatory. Columns 6 through 11 have two rows for each activity. Column 6 contains the lower and upper bound specified by the user on an activity in the bound section. Column 9 contains the range on the cost coefficient associated with each activity. In this example, since all structural variables are at some intermediate level, in section 4 we see that the range on c_1 is $[.02333, \infty)$, for c_2 it is $(-\infty, .03500]$, etc. Therefore, if c_1 assumes any value from .02333 to ∞ our present basic solution is still optimal. Columns 7 and 8 give the levels an activity assumes if the cost coefficient is moved by some α (column 8); in particular, if the cost coefficient of x_1 is decreased by .00167, then x_1 will come in the solution at 1146.42811 and constraint R6 will be at its upper level.

APPENDIX B

CONVEX CONES

DEFINITION B.1. *A set L in E^m is a cone if $k\mathbf{y} \in L$ for any $\mathbf{y} \in L$, scalar $k \geq 0$.*

DEFINITION B.2. *A cone L in E^m is convex if $\alpha\mathbf{x} + \beta\mathbf{y} \in L$ $(\alpha, \beta \geq 0$, $\alpha + \beta = 1)$ for any two vectors $\mathbf{x}$ and $\mathbf{y}$ in L.*

EXAMPLE B.1

Consider the set $L = \{\mathbf{y} \in E^2 \mid y_1 + y_2 = 0\}$. If $(\hat{y}_1, \hat{y}_2) \in L$, then $\hat{y}_1 + \hat{y}_2 = 0$; and moreover, for any scalar $k \geq 0$, $k\hat{y}_1 + k\hat{y}_2 = 0$. Therefore L is a cone. Also if two arbitrary vectors $\hat{\mathbf{y}}$ and $\tilde{\mathbf{y}}$ are in L, then $\alpha\hat{\mathbf{y}} + (1 - \alpha)\tilde{\mathbf{y}} \in L$ where $\alpha \in [0, 1]$. Thus L is a closed convex cone.

EXAMPLE B.2

The following example illustrates a cone which is not convex. Let $L = \{\mathbf{y} \in E^2 \mid \text{if } y_1 \geq 0 \text{ then } y_2 = 0, \text{ or if } y_2 \geq 0 \text{ then } y_1 = 0\}$.
For any $\mathbf{y} \in L$ and scalar $k \geq 0$, $k\mathbf{y} \in L$; therefore L is a cone. However, L is not convex since $(1, 0)$ and $(0, 1)$ are in L and for $\alpha = 1/2$, $(1/2)(1, 0) + (1/2)(0, 1)$ is not in L. (See Fig. B.1.)

DEFINITION B.3. *For any set L in E^m,*

$$L^* = \{\mathbf{y}^* \in E^m \mid \mathbf{y}^{*\prime}\mathbf{y} \geq 0 \quad \textit{for all } \mathbf{y} \in L\}$$

is called the polar cone of L.

EXAMPLE B.3

Consider again Example B.1, then

$$\begin{aligned} L^* &= \{\mathbf{y}^* \in E^2 \mid y_1^* y_1 + y_2^* y_2 \geq 0 \quad \text{for all } \mathbf{y} \in L\} \\ &= \{\mathbf{y}^* \in E^2 \mid y_1^* y_1 - y_2^* y_1 \geq 0 \quad \text{for any } y_1\} \\ &= \{\mathbf{y}^* \in E^2 \mid y_1(y_1^* - y_2^*) \geq 0 \quad \text{for any } y_1\} \\ &= \{\mathbf{y}^* \in E^2 \mid y_1^* = y_2^*\} \end{aligned}$$

since y_1 is unrestricted. (See Fig. B.2.)

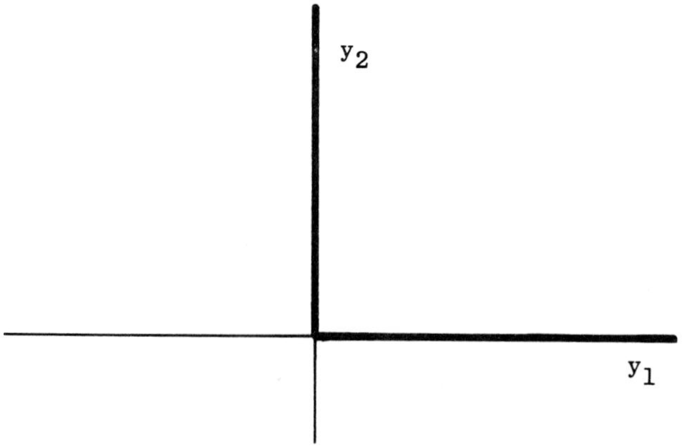

Figure B.1. Example B.2.

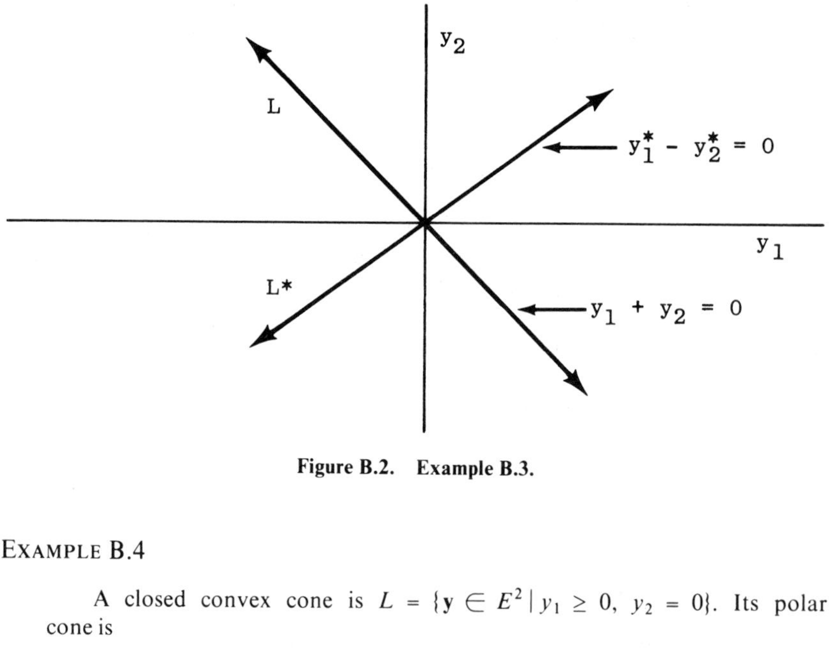

Figure B.2. Example B.3.

EXAMPLE B.4

A closed convex cone is $L = \{\mathbf{y} \in E^2 \mid y_1 \geq 0,\ y_2 = 0\}$. Its polar cone is

$$\begin{aligned} L^* &= \{\mathbf{y}^* \in E^2 \mid y_1^* y_1 \geq 0 \quad \text{for all } y_1 \geq 0,\ y_2^* \text{ unrestricted}\} \\ &= \{\mathbf{y}^* \in E^2 \mid y_1^* \geq 0,\ y_2^* \text{ unrestricted}\} \end{aligned}$$

Geometrically, the polar cone L^* consists of all vectors from the origin which form nonobtuse angles with all vectors of L. (See Fig. B.3.)

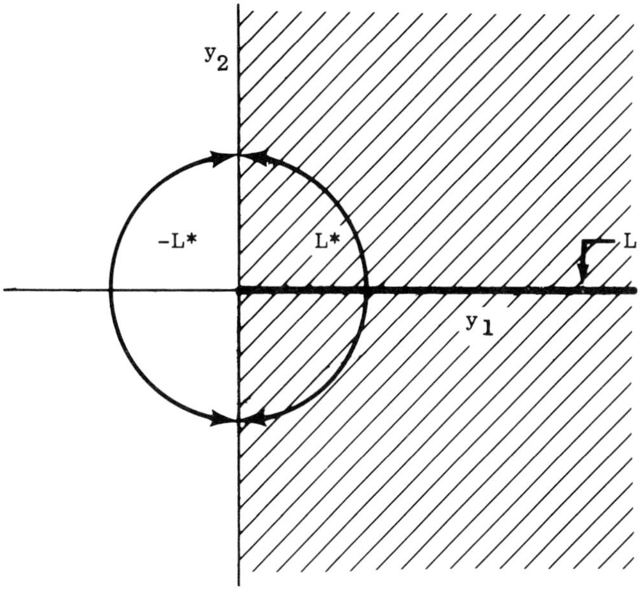

Figure B.3. Example B.4.

From the above definitions, it follows that L^* is also a closed convex cone and $(L^*)^* = L$ if L is a closed, convex cone.

DEFINITION B.4. *For L, a closed convex cone in E^m*

$$-L = \{\mathbf{y} \in E^m \mid -\mathbf{y} \in L\}$$

Clearly, $-L$ is also a closed convex cone.

LEMMA B.1. $-(L^*) = (-L)^*$.

Proof.

$$\begin{aligned}(-L)^* &= \{\mathbf{y}^* \in E^m \mid \mathbf{y}^{*\prime}\mathbf{y} \geq 0 \quad \text{for all } \mathbf{y} \in -L\} \\ &= \{\mathbf{y}^* \in E^m \mid \mathbf{y}^{*\prime}\mathbf{y} \geq 0 \quad \text{for all } \mathbf{y} \text{ such that } -\mathbf{y} \in L\} \\ &= \{\mathbf{y}^* \in E^m \mid -\mathbf{y}^{*\prime}\mathbf{z} \geq 0 \quad \text{for all } \mathbf{z} \in L\} \\ &= \{\mathbf{y}^* \in E^m \mid -\mathbf{y}^* \in L^*\} \\ &= -(L^*)\end{aligned}$$

Note that in view of Lemma B.1, brackets are not required; and $-L^*$ denotes both the left and right sides of the statement of Lemma B.1. Thus, for example, we have $(-L)^* = -L^*$.

DEFINITION B.5. *The dual cone L^0 of a set L in E^m is defined as*

$$L^0 = \{\mathbf{a} \in E^m \mid \mathbf{a}'\mathbf{y} \leq 0 \quad \textit{for all } \mathbf{y} \in L\}$$

Thus it follows that $L^0 = -L^*$.

EXAMPLE B.5

For a moment consider the set $Q = \{\mathbf{x} \in E^n \mid \mathbf{x} \geq 0\}$. Then for any vector $\mathbf{x} \in Q$ and any $k \geq 0$ we have that $k\mathbf{x} \in Q$. Also

$$\alpha\mathbf{x}_1 + (1 - \alpha)\mathbf{x}_2 \in Q$$

for any $\mathbf{x}_1, \mathbf{x}_2 \in Q$ and any $\alpha \in [0, 1]$. Thus Q is a closed convex cone in E^n, and moreover Q is its own polar cone.

Considering the set $-Q$, then clearly, $-Q$ is the dual cone of Q.

We denote the set Q in E^n by Q_n^+ and define it to be the positive orthant. Likewise, Q_n^- denotes the negative orthant.

LEMMA B.2. *For any two sets X and Y in E^m, if $X \subset Y$ then $Y^0 \subset X^0$.*

Proof. For any vector $\mathbf{a} \in Y^0$ we have

$$\mathbf{a}'\mathbf{y} \leq 0 \quad \text{for all } \mathbf{y} \in Y$$

but $X \subset Y$, therefore

$$\mathbf{a}'\mathbf{x} \leq 0 \quad \text{for all } \mathbf{x} \in X$$

Thus from Definition B.5, $\mathbf{a} \in X^0$ and $Y^0 \subset X^0$.

LEMMA B.3. *Let X be any set in E^m, then $X \subset X^{00}$.*

Proof. For all $\mathbf{x} \in X$ and $\mathbf{a} \in X^0$, we have that $\mathbf{a}'\mathbf{x} \leq 0$, which implies that $\mathbf{x} \in (X^0)^0$ since

$$X^{00} = \{\mathbf{x} \mid \mathbf{x}'\mathbf{a} \leq 0 \quad \text{for all } \mathbf{a} \in X^0\}.$$

DEFINITION B.6. *A set C is called a polyhedral set if C is defined by the intersection of a finite number of supporting half-spaces of the form*

$$\{\mathbf{x} \mid \mathbf{a}'\mathbf{x} \geq (\textit{resp.}, \leq)\, \beta\}$$

The smallest convex cone that contains X is called the convex cone spanned by X and is denoted by $\mathcal{P}(X)$

$$\mathcal{P}(X) = \left\{ \sum_{k=1}^{r} \alpha_k x^k \,\middle|\, \alpha_k \geq 0, \mathbf{x}^k \in X, k = 1, 2, \ldots, r, r \text{ arbitrary} \right\}$$

Note that a convex polyhedral set is also closed. Therefore, if X is a convex polyhedral set, $\mathcal{P}(X) = X^{00}$.

INDEX